Design for Product Success

Design for Product Success

Devdas Shetty

Society of Manufacturing Engineers
Dearborn, Michigan

987654321

No liability is assumed by the publisher with respect to use of information contained herein. While every precaution has been taken in the preparation of this book, the publisher assumes no responsibility for errors or omissions. Publication of any data in this book does not constitute a recommendation or endorsement of any patent, proprietary right, or product that may be involved.

Library of Congress Catalog Card Number: 2002102724
International Standard Book Number: 0-87263-527-9

Additional copies may be obtained by contacting:
Society of Manufacturing Engineers
Customer Service
One SME Drive, P.O. Box 930
Dearborn, Michigan 48121
1-800-733-4763
www.sme.org

SME staff who participated in producing this book:
Cheryl Zupan, Staff Editor
Walter Kelly, Consulting Editor
Rosemary Csizmadia, Production Supervisor
Kathye Quirk, Graphic Designer/Cover Design
Jon Newberg, Production Editor
Frances Kania, Production Assistant

Printed in the United States of America

Dedicated to the loving memory of my parents

Table of Contents

Preface

This book, *Design for Product Success*, delves into the techniques used by world-class companies to guide the design and development of high-quality products in a step-by-step manner, using analytical tools and case studies. In a global economy that is becoming more and more integrated, a shift is taking place in many companies. The price of a product marketed globally is dictated by world economy, and not by one's own economy or by a company's marketing edge. Successful companies are the ones that focus on a product and process delivery system and know how to transform process innovations into technical success.

In today's competitive environment of manufacturing, product cycles are short and markets are fragmented, thus making quality and speed critical. The automation of machines and processes has become pervasive, impacting every dimension of manufacturing. Micro-electromechanical systems, nanotechnology, and embedded systems have contributed to the development of intelligent products. Also noteworthy is the technology of rapid prototyping that assists in the creation of a product model before it is manufactured.

Driven by global competition and the demand for high quality, lower cost, and quick delivery times, the product realization process has seen several changes. To survive, companies must produce products that are globally competitive in a changing environment. Many companies have formed partnerships with suppliers, customers, and international partners.

Modern manufacturing industries are adopting the concept of lean thinking. The new generation of lean companies is expected to meet customer requirements with products and services that meet function, time, and cost requirements, while being flexible to changing customer needs.

Industries adopting lean practices will be expected to make changes in their policies and programs. As a result, product designers and product managers of the future must be more skilled at developing and implementing strategies in totality, rather than as a series of isolated steps. The measure of the abilities of product managers to make profits while reacting to sudden and unpredictable changes in customer service demands is very important. The challenge facing manufacturers today is the improvement of product development systems. As companies feel more competitive pressure from world-class manufacturers, it is essential that they critically examine their product design strategies. The design of a product determines its method of assembly, component tolerance, the number of adjustments, and the type of manufacturing processes used.

The objective of this book is to familiarize readers with the concepts, techniques, and tools that encourage creativity and innovation. It offers a strategic approach for organizing product design. Although its main emphasis is on the design of products with an engineering content, most of the approaches included are general in nature and can be applied to different types of products.

Design for Product Success includes design-oriented discussions that support successful product creation for designers, engineers, and technology managers. Some of the topics covered include: how to build product teams; characteristics of self-directed product teams; new product creation strategies and processes; creative design techniques; organizational aspects and principles of design for assembly, disassembly and manufacturing; and environmentally conscious design.

A new engineer's career involves designing, prototyping, and fabricating a wide variety of products. Many companies are expecting graduate engineers to be familiar with the steps involved in successful product creation. There is a need to familiarize students not only with design concepts, but also with a global view of manufacturing, techniques, and management. This need has caused many engineering programs in the country to re-examine the design content in their curriculums. This book can be used by students in product design and design for manufacturing courses in the mechanical, industrial, manufacturing engineering, and engineering management areas. The end-of-the book problems can be very useful to instructors teaching these courses. The chapters are organized to take the reader through the various steps of creative product design, starting with the concepts and ending with the final stages of production and marketing.

Using case studies, various product redesign techniques are explored. The book can be even more useful when combined with projects that require step-by-step problem solving procedures and the application of analysis techniques.

The text includes useful tools that can easily have an impact on the new product development process. It provides an understanding of methods to minimize the impact of design changes on the production launch and offers ways to improve and structure the relationship between the design and manufacturing departments.

Product design is a complex process that requires a systematic approach. The book deals with foundations of thinking that stress engineering design fundamentals. There are practical real-world problems that are easy to follow and implement. Successful techniques for building product design teams and managing innovation are introduced.

Each chapter of the book focuses on a different aspect of product design and includes examples. The first five chapters are devoted to product design strategies and the next three are devoted to manufacturing strategies. The last chapter features case studies of successful product designs.

Chapter 1 deals with the need for better design methodology. Considering the fundamentals of the product realization process, it provides an in-depth discussion of some essential elements of product development.

Chapter 2 is entirely devoted to the conceptual phase of the product design process. Various techniques used for concept generation are described, including axiomatic design, quality function deployment, TRIZ fundamentals, and product design using function analysis.

In Chapter 3, design considerations necessary to create a better product assembly are explained. The Boothroyd-Dewhurst, Hitachi assembly evaluation, and Lucas design methods are introduced. It details a number of case studies and projects to illustrate each step.

Chapter 4 introduces readers to various manufacturing processes and guidelines for efficient manufacturing. Recent advances in design for disassembly and design for life cycle are presented. Chapter 5 describes tools and techniques used in product design, and the tools of optimum design and decision-making using learning curves.

Chapter 6 describes methods of streamlining a process, workplace design, and Value Stream Mapping^{SM}. Chapter 7 deals with product creation using optimum design and business results. Using the Internet for information gathering and management aspects of product creation is also discussed. Chapter 8 discusses the importance of product design groups.

Industrial case studies of successful product design are described in Chapter 9. They illustrate the phenomenal success of integrated product and process developments and the value of using analytical tools. The critical aspects and constraints of time-driven product development are examined at companies such as Boeing and Pratt & Whitney.

Acknowledgments

The material presented in this book is a collection of many years of research and teaching at the University of Hartford, Cooper Union. It is also a result of the insight gained from working closely with industry affiliates such as Pratt and Whitney, Hamilton Sundstrand, Carrier Corporation, Otis Elevator Company (all of United Technologies), Wiremold Company, Jacob Vehicle Equipment Company, and many others.

I am grateful to a number of professors whose comments and suggestions at various stages of this project were helpful in revising the manuscript. I would like to acknowledge Prof. Richard Kolk of Carrier Corporation, Prof. Jean Le'Mee of Cooper Union for the Advancement of Science and Art, Prof. M.S. Fofana of Worcester Polytechnic Institute, Prof. R. Dukkipati of Fairfield University, Prof. Lee Tuttle of Kettering University and Prof. Zbigniew Bzymek of the University of Connecticut.

I warmly acknowledge the professional support given by Jim Rivera of Otis Elevator Co., Steve Maynard of Wiremold Co., Peter Carter of Carrier Corp., Joe Wagner of United Tool and Die, and Adish Jain of Jacob Vehicle Equipment Co. I thank Claudio Campana and Suresh Ramasamy of the University of Hartford for the detailed assistance that helped me to refine the material. I also want to thank Jun Kondo, Rafat Yamani, and S. Krishnamurthy, all of the University of Hartford, for helping with the data collection on many design methods.

I am indebted to many of my past graduate students who are now successful professionals. They include Ken Rawolle and Kiran Kolluri of Pratt & Whitney, Nilesh Dave of International Fuel Cells, Beth Cudney of Jacobs Vehicle Equipment Company, Troy Chicoine and Andrea Sidur of Hamilton Sundstrand, Rob Choquette and Zlatko Strbuncelj of Otis Elevator, John Breault of Loctite Corp., Brian Blair of EMC Corp., Walter Mori of Gems Corp., and Saat Embong and Noreffendy Tamaldin of Motorola Corp.

I want to acknowledge Prof. Don Leone and Prof. Ron Adrezin of the University of Hartford, and Prof. Prakash Persad of the University of the West Indies for stimulating discussions on product design. I also acknowledge the hundreds of students from the classes where I have tested the teaching material.

Funding from the Society of Manufacturing Engineers Education Foundation and the National Science Foundation is gratefully acknowledged.

The tremendous support and encouragement I have received from my colleagues has been invaluable. I am indebted to the faculty and administration of the University of Hartford for their valuable support.

Thanks are due to the staff at the Society of Manufacturing Engineers (SME). I appreciate the assistance of editors Cheryl Zupan and Phil Mitchell for their superb contributions.

My sons Jagat and Nandan have contributed greatly in reviewing the manuscript. I also want to express my gratitude to my wife Sandya for her love and support.

Chapter 1

Building Blocks of the Product Design Process

PROCESS OF PRODUCT REALIZATION

Designing a product involves a constant decision-making process that includes problem solving in a sequential fashion and analysis of constraints at each step.

The philosophy underlying the author's method of design is unique. Human beings are a special kind of designer and their design philosophy influences their own life and environment. In general, design represents an answer to a problem, an answer that has visible form, shape, and function. Various professions define design differently. Business professionals, physicians, architects, and engineers all have their own unique views on design and their own personal experience with its use. However, this book limits its discussion to professional design as practiced by engineers. Providing a set of rules for reorganizing the elements of creation toward some greater purpose is known as *design intent*. It is the design intent that has ethical and moral dimensions.

In broader context, design is, in fact, any purposeful, thought-out activity. It is a way of doing things characterized by decision-making. A typical product may be made up of many technical and non-technical components that factor into product design. The product may not succeed, however, if there is no balance between design and other important factors such as safety, aesthetics, ergonomics, and cost.

Product creation influences the process of design as well as overall product cost. Cost of a product grows from conception, through stages of technical research, design, development, market testing, use, maturity, and, finally, disposal. An organization has greatest control over a product at the early stages of its creation, when the market, its factory cost, operational cost, and life cycle are determined. At this stage, status of a product can be unstable as the organization tries to optimize product distinctiveness for greater market acceptance. Naturally, a product's features determine its performance and cost. Speed of product development, market testing, and manufacturing are all important aspects of the product life cycle.

Figure 1-1 shows the different stages of a typical product life cycle. The cycle is composed of many individual processes. First, it is initiated by the market need. It starts with the task of planning a product based on strategic goals of an organization and goes through stages of feasibility analysis, research and development, design and prototyping, market testing, commercial manufacturing, marketing, product use, maintenance, and disposal. Products that become well defined in the early design

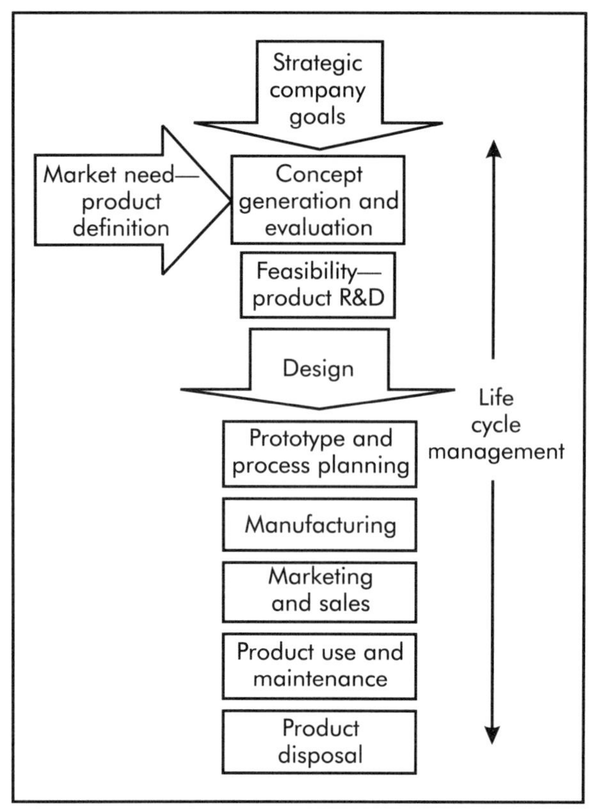

Figure 1-1. Product life cycle.

stages have higher chances of success than products that lack this preparation. Many studies have shown that conducting early design work, preparing sketches and prototypes, simulation studies, cost estimation, and talking to potential customers reduces uncertainties in a new product launch.

Design Approaches

Producing a marketable product from its initial requirements takes many steps. The probability of a product's success significantly increases if the design process is planned and executed carefully. It is imperative that processes integrate many different aspects of design into distinct logical steps.

Important contributions to the development of design methodology have come from various countries. Roth has developed a design theory characterized by a set of integrative steps, including problem formulation and functional and embodiment phases (Roth 1994). He also developed a catalog of design entities in graphical form. Nigel Cross developed a design methodology aimed at making the design process more algorithmic by dividing the process into descriptive, prescriptive, and systematic models (Cross 1994). Finger and Dixon developed a review of design methods and computer-based models for design (Finger and Dixon 1989). Pahl, Beitz, Hubka, and Schregenberger developed guidelines for design decision-making (Pahl and Beitz 1996; Hubka and Schregenberger 1987). John Dixon proposed a structure representing the relationship between engineering design and the cultural world (Dixon 1996). Stuart Pugh, Karl Ulrich, and Steven Eppinger introduced integrative methods for product design and development (Pugh 1991; Ulrich and Eppinger 1995).

Linguistic Approach

Hubka and Schregenberger introduced the idea of grammar being used to assist design. They found that the linguistic habits of designers affect their ways of conceiving design and approaching it (Hubka and Schregenberger 1987). Development of the science of engineering design lies in formulating it as a system of statements in a methodically consistent classification, such as in grammar. A comparison of the regions of Germany and Switzerland with those of the U.S. shows characteristic differences in the formulation of design-specific statements. Jean Le'Mee proposed that design processes could be expressed through a grammatical approach. He viewed design as a process involving a language right at the start. Le'Mee outlined a distinct relationship between design and grammar of the ancient language Sanskrit, which has the characteristics of being descriptive, illustrative, and creative (Le'Mee 1987). The language is grammar-oriented and verbs, in their semantic function, carry the meaning of *activity process* and *result of action*. Determination of the action factor deriving from observation of a given situation depends on the viewpoint of the observer. Depending on the context, additional aspects may influence the situation; but both action factors and results of action are implicitly present. As shown in Table 1-1, at each step of the iterative design process, six questions are asked to identify the activity process (Le'Mee 1987).

The Designer as an Agent for Action

The fundamental questions shown in Table 1-1 represent six activity and action factors of the design process. A design event begins by identification of the design situation, with the designer being a witness in the process of designing. The views and impressions about a particular situation produce a *design viewpoint*. This viewpoint changes with the designer's professional level or sphere of perception, which may be influenced by the designer's personality, immediate group, environment, and approach. Not only does the designer perceive, feel, and think, but he or she can also act. Therefore, the designer is an *agent for action*. Le'Mee notes that through oral or written communication or direct actions, design can influence surroundings. It therefore follows that the designer has a *standpoint for action*.

Action is a process leading to a result. This obvious statement is necessary to clearly distinguish the dual nature of action. It is, of course, also essential in linking an action to its consequences, and in exploring these consequences at the design stage. In the realization that the designer is a witness, it should be evident that actions of design should reflect an external reality. Humans perceive the world by a limited sense of understanding, influenced by sight, touch, taste, smell, and thereby mental processes. Therefore, it is justifiable to consider the domain of the mind and explore the way people think and feel, since their perceptions for so many years have been built into the whole design process. Roozenburg and Eekels propose that designers do not make a decision based on mental logic alone, but are influenced by the reality of some purpose and interaction (Roozenberg and Eekels 1995).

It is the idea of purpose that best characterizes design as shown in Figure 1-2. If design is thought of as an interaction between the domains of mind and material represented by matter, it is influenced by a

Table 1-1. Six factors of activity and action (Le'Mee 1987)

1. Agent—Who is the independent agent to carry it out?
2. Recipient—Who does it benefit?
3. Method/means—By what means can it best be carried out?
4. Conditions/environment—Under what conditions of time and place is it occurring?
5. Purpose—What is the purpose of the action?
6. Where—From where does the action spring? What is the reference frame?

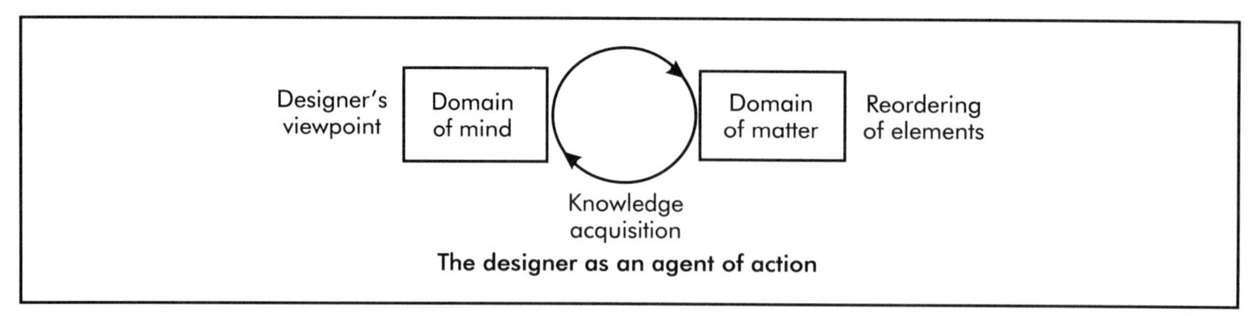

Figure 1-2. Influencing parameters in a design situation.

designer's viewpoint (reasoning) and knowledge acquisition from the external world. The designer also plays the role of agent of action.

Kansei

The Japanese use the word *kansei* when things engineered are fully accomplished. In Japanese, kansei expresses the sensation when all five elements (earth, water, fire, air, and space) come together in harmony and it feels just right. Our five senses link us to these five subtle, yet fundamental elements, but we also apprehend and deal with the world according to intellectual, emotional, and active modes.

The action of design engineering is to process and give form to materials, energy, and information—the substance of economic life. Through research in natural sciences, it is possible to refine applied aspects of the elements. These five elements also possess a human dimension which, when properly understood, may illuminate work by engineers, architects, and artists.

This is a new generation where, with help of automation in design and manufacturing, the designer may have much influence on the finished product. The computer has become smart and powerful and assists a great deal in coordinating interactive cycles of materials, energy, and information; but it still must defer judgment to the designer. In the final analysis, it is the design and the designer who are most important.

Design as a Problem-solving Exercise

Key factors that distinguish a good from a not-so-good designer are the attitude with which he or she approaches a design problem, his or her aggressiveness in looking for a solution, and the depths to which design methodology is used. Smart designers believe in following the steps of the design process, and the effective use of tools and techniques such as heuristics, feedback, model development, and analysis. They take enormous effort to understand various relationships that exist, re-describe the problem situation, create a mental picture, ask themselves questions, and break the design problem into several subproblems. The building blocks of the product design road map consist of the stages shown in Figure 1-3.

Design is Information

Design relies on information (User et al. 1998). This information is transformed into material through the medium of energy. Good design can be characterized as that which achieves its intended goal with a minimum of means. Good design optimizes the use of resources:

- at the information level, by minimizing information generated;
- at the energy level, by minimizing energy degraded; and
- at the material level, by minimizing the amount of materials used, processed,

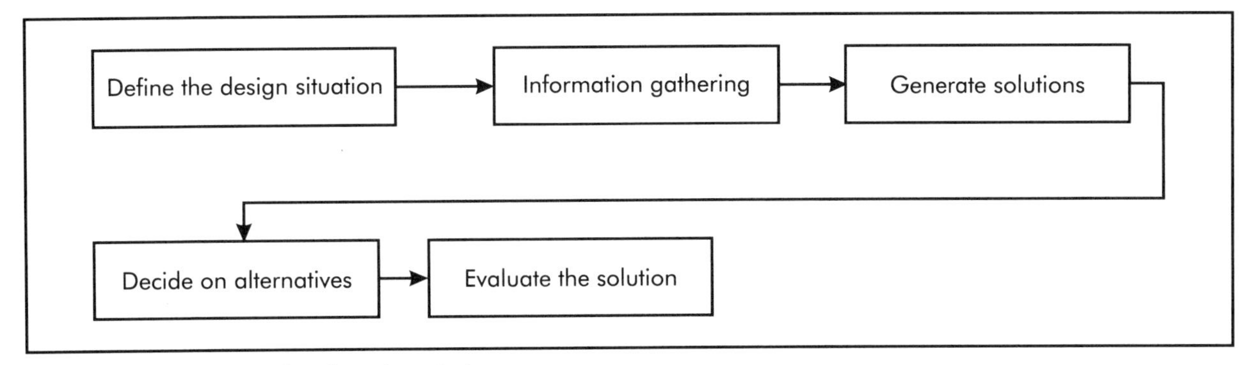

Figure 1-3. Building blocks of product design.

and recycled in the whole life cycle of the product—from conception to disposal and eventual reuse.

Engineering designers conceive design as a process. Industrial designers conceive design as a product. The solution of a design problem may take several forms. It may be a particular device or a product, such as an automobile, submarine, or space shuttle. A solution for a design problem could be a process such as a technique for preserving food, or a new procedure for cutting metals. It could also be a procedure or set of action plans. Although various professions define it differently, in general, design is some answer to a problem; an answer that has visible form, shape, or function.

In product design and manufacturing, emphasis on the *volume* of production received a great deal of attention from 1950-1970. The emphasis shifted to *quality* during the 1980s and 1990s. Now, emphasis on *time* is coming to the forefront. The trend in product design is toward shorter development time through integration of the design process, computerization of as many functions as possible, increased flow of communication, concurrent engineering, nanotechnology, the team approach rather than a departmentalized approach, and increased flexibility at all levels, from the design process to the total enterprise. In keeping with its importance, database technology has steadily improved and evolved. With the fast emergence of Web-based solutions, database management has assumed an important role.

Foundation of New Product Development

Design and development of a product invariably involve considerable investment of time, effort, and money. It is essential that a new product is thoroughly examined and reviewed before it is presented to the public. A company's credibility, reputation, and finances rest on the launching of each of its products. Fundamental questions a product designer has to look into are the *Why? What? How? Who?* and *When?* pertaining to the product (see Table 1-2).

The question *why* regards the business-approach strategies of the company and looks into evaluating success and failure. Success does not only depend on design, technology, and marketing, but also on realistic planning. There may be several reasons for launching a product. Usually, profit is the main motivating factor. The financial setting of the company—including cash inflow and outflow and net payback time for investment—are among the issues that influence the decision. However, there are instances where a successful new product

Table 1-2. Step-by-step design methodology

Definition of design situation

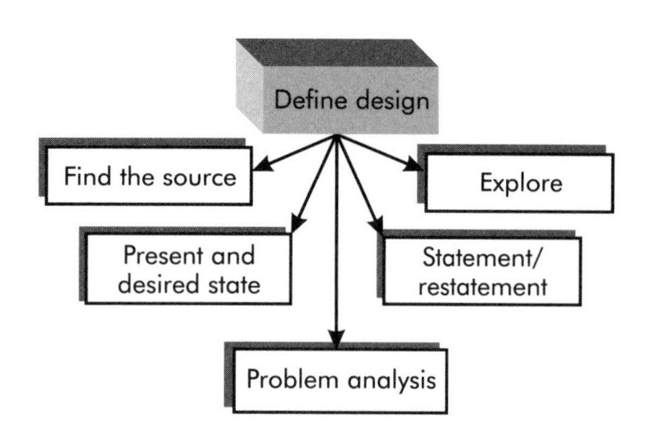

The identification and proper definition of the problem is very important. The definition is an expression of a desire to achieve a transformation from one situation to another. This stage gives a real understanding of the problem, its history, and availability of resources. Defining the problem correctly is important as the solution depends on how the problem is defined. The designer proceeds from recognizing the needs to be satisfied, to the completion of these goals. This process involves a significant amount of reporting and feedback, as well as consideration of value-related issues of the solution. Identifying the problem correctly at this stage saves time and money, and makes it easier to reach a satisfactory solution. It is important to find out where the design problem originates. Some of the steps used by experienced designers are:

- Gather and analyze the information.
- Talk with the people familiar with the problem—look past the obvious; question the basic premise.
- View the problem first hand, if possible.
- Reconfirm the findings.
- Find out where the problem came from.
- Are the reasoning and assumptions valid?

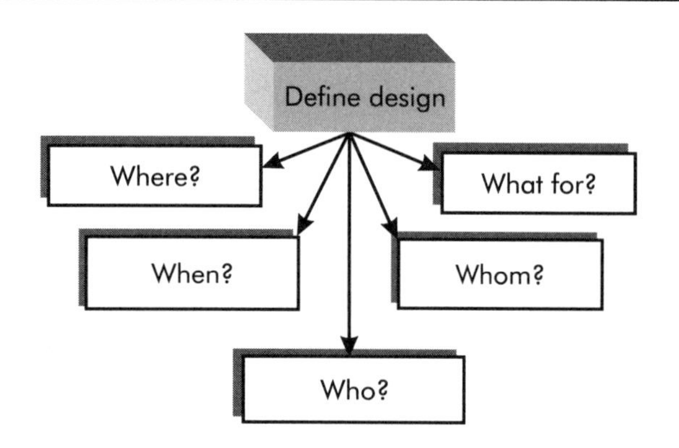

The following questions can function as a guide in defining the problem:

- From *where* does this action spring, that is, what is the frame of reference?
- *What* is the action for? (Why?)
- *Whom* does it benefit?
- By what means can it be best carried out? (How?)
- In what conditions of time and place is it occurring?
- *Who* is the independent agent to carry it out?

may translate into enhancing the prestige or even maintaining survival of the company.

The question *what* pertains to the product under consideration and the market segment. The market segment influences the function of a product and the technology segment determines the manner in which the product can carry out the function. Market segments are dynamic and constantly evolve. They have to be well specified before a product is planned and resources are allocated. The designer should consider adoption

Table 1-2. *(continued)*

Establishing design criteria

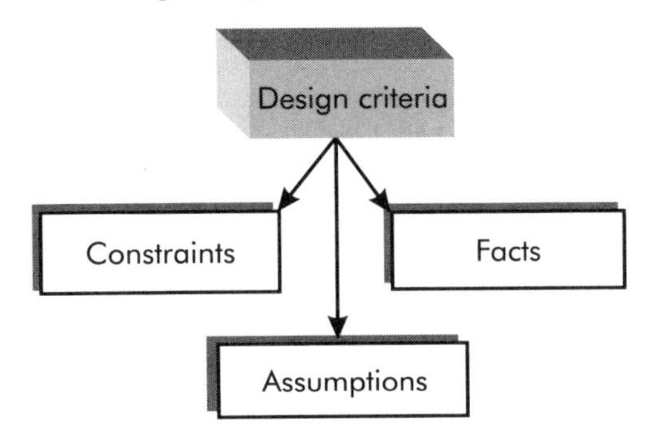

Constraints are factors that affect the outcome of the project and cannot be changed. They may show reflections of the values of the customer and the society.

Facts are listed to help clarify what is known, and what may need to be found out prior to proceeding with the project.

Assumptions are facts or statements that are accepted as true, without doubt. The first step is to clarify the assumptions in regard to the problem. Once clarified, assumptions can often be modified to simplify the problem and make it solvable.

Sources of information needed

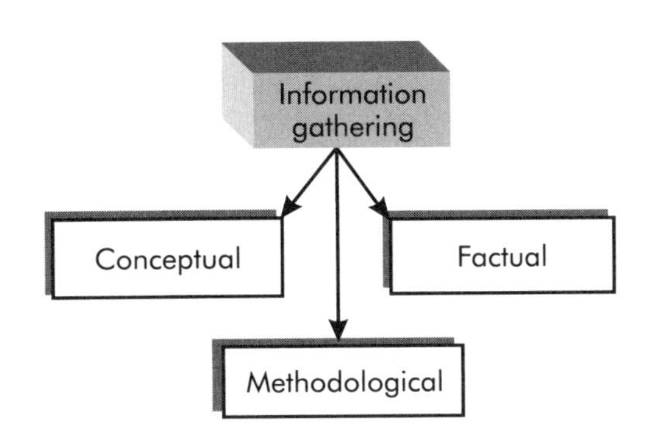

Conceptual information covers the fundamental principles and laws, such as the principles of conversion of mass, energy, momentum, etc. Being principles, they do not change unless a fundamental and radical philosophical change has taken place through the perspective of the world's scientific community.

Factual information, such as the properties of substances, can be found in handbooks. It is the sort of information that keeps growing and changing as new substances are invented, new products are developed, etc.

Methodological information represents a link between conceptual and factual information. It is knowledge of the methods and ways by which conceptual information can be applied to generate factual information or more conceptual information. It is a skill, a set of attitudes and procedures, which can only be acquired experientially.

Generating options and solutions

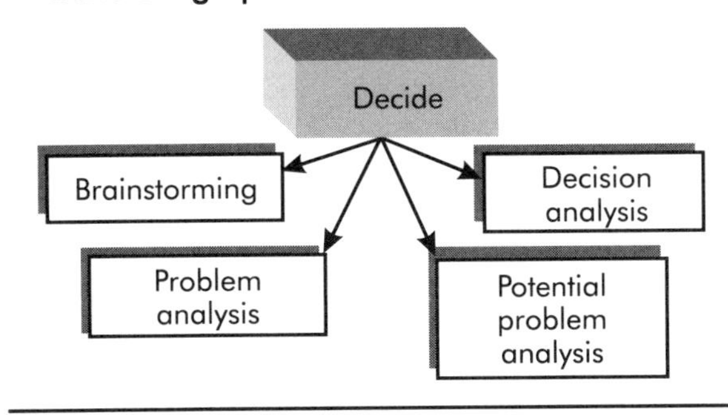

The goal in this step is to generate as many potentially useful solutions to the problem as possible. Solutions need to be innovative and the designer should maintain an open, receptive mind to new ideas. An appreciation for the unusual or extraordinary is also important. This ability comes with a great deal of practice as this skill is developed and continues to develop throughout the designer's career.

Table 1-2. (*continued*)

Evaluation of options

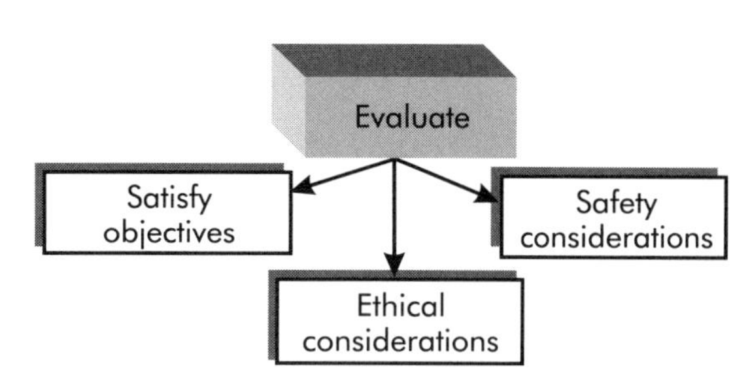

What is most desirable? The selection of a particular solution is based on understanding how it relates to certain ecological, social, political, and cultural aspects of life. Values play a strong role in this step.

- Does the solution satisfy the basic objective of the project?
- Is the solution theoretically feasible?
- Is the solution practical?
- Is the cost within the means?
- Is the proposed solution safe to operate?
- Is it the optimum solution?
- Does it satisfy the constraints?
- Does the solution satisfy all the human, social, and ecological factors involved?
- Is it aesthetically acceptable?
- Is it legal?
- Can the project be completed in the time allotted?

Feasibility study of preferred option(s)

At this stage, a feasibility study has to be prepared. It is a proposal outlining what is necessary to analyze. The proposal should indicate:

- the solution(s) being investigated;
- the reasons for these selections;
- the coordination with other groups on the project;
- the constraints, assumptions, theories, variables, and parameters being used;
- the goal of the analysis;
- the elements into which the problem has been broken down; and
- how the newly acquired information resulting from the analysis will be used by the group and disseminated to other groups.

Ranking of options

In this step, the first-, second-, and higher-order impacts of the solution on the society are identified. Further, based on an understanding of the value issues associated with these impacts, the alternatives are ranked based on the benefits, as well as costs and risks for each alternative.

After the alternative solutions have been evaluated and ranked, the alternative that maximizes benefits and minimizes costs and risks is chosen. This is a complicated process as any quantification is based on the values of individuals, groups, etc.

Detailed analysis of the chosen solution

This step involves separating a potential solution into its meaningful elements. These parts will be scrutinized through comparison with other known facts, theories, and views.

or modification of an existing product. It is also possible to create a product through collaboration, licensing, or joint venture from external agencies.

Scientific study of marketing research can provide quantitative data, allowing the analysis of past demand and predictions about demand in the future. It can provide information on the trend of technology, the influence of federal, state, and local governmental restrictions, and the impact of industrial restructuring in a particular field. A major factor at this early stage is the importance of choosing among various product ideas. Ideas have to be compared and ranked, depending on business history, technological infrastructure, and marketing thrust.

The question *how* deals with aspects of technical and commercial implementation. Product implementation to a large extent is based on company strengths, weaknesses, and basic capabilities. It is based on human and financial resources, and could be related to the physical infrastructure of the company. Basic exploratory activity is very important before heading forward with technical developments. Also important is creating a comprehensive survey translating customer requirements, concept definition, feasibility documentation, and prototype demonstration. Side by side with market research, these are essential elements of technical and commercial implementation. Commercial development is the heart of the whole process, involving establishing a list of strategies, analyzing market competition, benchmarking with respect to other companies, pricing, and market distribution. Obtaining and enforcing intellectual property rights and trademarks has to be systematically planned, along with technical and business-related development of the product.

The question *who* defines accountability and responsibility for the new product development procedure. Whatever may be the organizational structure, people involved in product development have to understand the institutional setup and layout. Many product failures are attributed to breakdowns in communication and inadequate follow-up at critical junctures. It is necessary to effectively communicate—throughout the company—the business objectives, goals, and various work functions of participants, as well as financial and legal issues. Responsible information retrieval and delivery are crucial in new product development.

The product developer has to be sensitive to timing of the product launch. Time is a critical factor because it is closely related to money and investment. Time also is linked to a window of market opportunity. A product that enters the market before its time or after it is needed will certainly not be as profitable to the company.

Table 1-3 provides a summary of five key questions involved in the new product design and product launching process.

Why do Promising Products Fail?

Why do products with high expectations fail?

One reason may be that any innovative product, especially if it has advanced technological content, meets customer resistance and sells slowly until consumers perceive it as safe.

Neil Rackham suggests that the main problem is the way highly innovative products are launched to the sales force, which, in turn, influences the manner in which the product is sold (Rackham 1998). Figure 1-4 identifies the steps for introducing a new product to the sales group. Focusing on new features that make the product highly innovative draws the attention of the sales force away from the most important issue in the process: the needs of customers. The

Table 1-3. Summary of five key questions involved in a new product launch

Question	Strategic Objective	Financial Consideration	Market Consideration
1. Why	Survival of the company Business opportunity	Financial reward	Synergistic impact
2. What	Product definition Adopt or modify existing design	Joint venture Collaboration Licensing	Projected product demand
3. How	Strengths Weaknesses Physical infrastructure	Financial benchmark	Market response
4. Who	Structured decision-making Human consideration	Accountability and responsibility	Information retrieval Information dissemination
5. When	Timing is critical	Money and investment	Market window

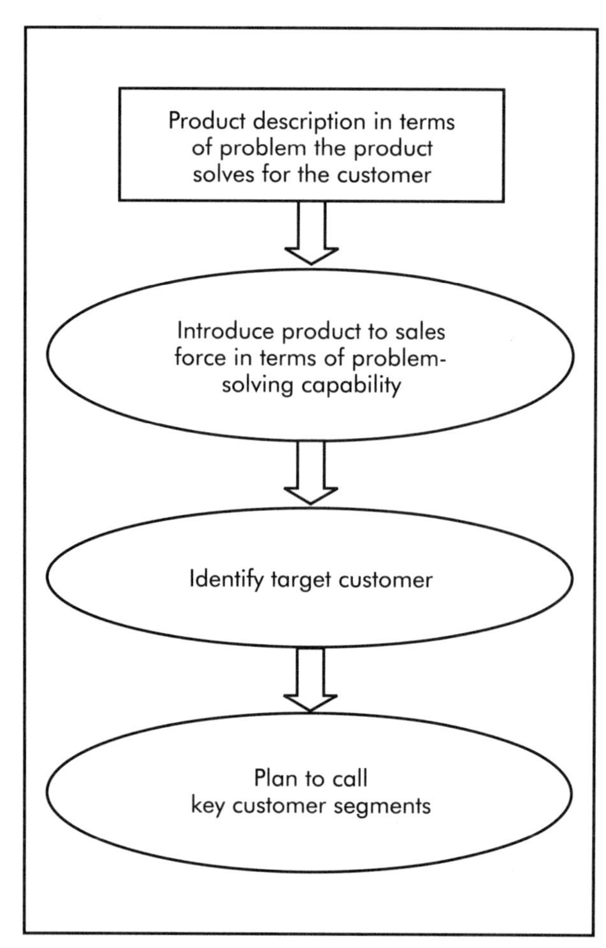

Figure 1-4. Steps for introducing a product to the sales force.

sales force learns product-centered information about its capabilities and communicates this information to customers in the way it was communicated to them. Launching the product becomes product-centered instead of being customer-centered, which decreases customer interest. A more customer-oriented approach—by dwelling on the product in terms of problems it solves for the customer—is more successful.

Building Blocks of New Product Design

Companies need a systematic design process that must be communicated to their designers. The actual process by which product designers implement their tasks and responsibilities is typically a function of the individuals involved. Their approaches, degree of documentation, and habits are unique and randomly acquired. Hence, it is sometimes difficult for one person to follow up another's work, not being familiar with design philosophy and approach of the former.

Streamline Product Development

A well-developed product development process enables companies to select, propose, design, develop, and market new products effectively. A systematic product design and development process is a key element to the infrastructure of an organization. Such a process can provide a stable structure for strategic planning, decision-making, operation, effective communication, implementation, and control. Figure 1-5 shows the components of product development.

Companies around the world have been influenced by studies showing that improved production structure can take care of fluctuations in workloads, reduce variation in products, and eliminate bottlenecks. Paul Adler has studied a dozen companies that have started applying process management techniques to product development, including: Raychem, Motorola, Harley-Davidson, Hewlett Packard, General Electric, AT&T, Ford, General Motors, and NEC. The results of this study have shown three significant observations (Adler et al. 1996):

1. Product development projects are accomplished quickly if the organization takes on fewer projects at a time.
2. Investment made by the company to relieve project bottlenecks results in early market launch of the product.
3. Eliminating unnecessary variation in workloads and work processes removes distractions and delays, thereby freeing up the organization to focus on key areas of the project.

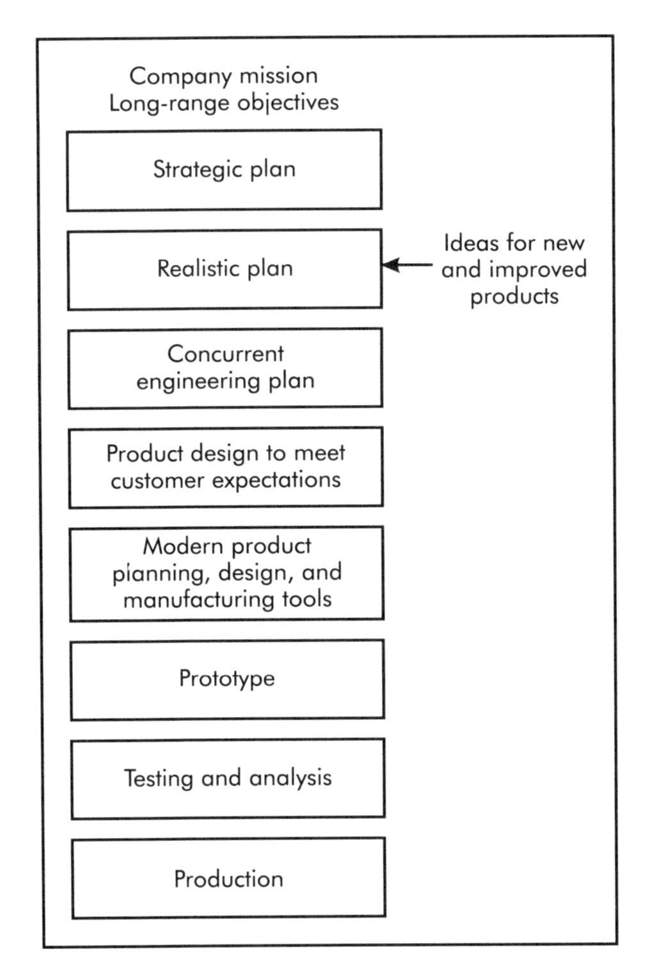

Figure 1-5. Components of product development.

Comprehensive Strategic Plan of the Organization

The strategic planning process is a major vehicle for policy deployment and execution in an organization. It consists of a mission statement, situation analysis, identification of objectives and strategies, preparation of action plans, and implementation of policy issues.

A strategic plan:

- identifies the company's mission;
- identifies the company's strengths and weaknesses, and establishes long-range business objectives;
- selects market segments to be pursued;
- formalizes the process for selecting products for development;
- selects products and identifies the strategic and tactical issues that the company must resolve to facilitate success; and
- projects financial returns expected from selected markets and products.

Figure 1-6 lists the components of a strategic planning process.

The strategic plan starts with a mission statement that should provide a clear prod-

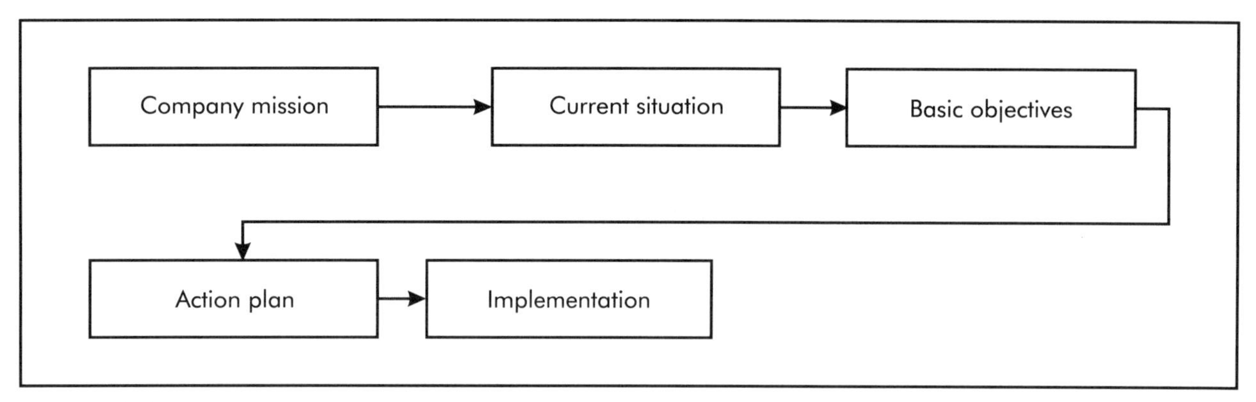

Figure 1-6. Strategic planning process.

uct and market focus, identifying sources of opportunities. It should be brief and be presented clearly. The strategic plan also includes a situation analysis that addresses changes in the company, changes in marketplace and competition, technical readiness of the company, standards, and regulations. While market analysis focuses on market share, competition, and customer perception, internal analysis of the company looks at the strengths, weaknesses, threats, and growth opportunities for different products. Objectives are outlined based on these considerations. They should be presented in a clear timetable of the various undertakings involved.

The action plan details strategies for achieving the objectives. The plan could be product specific and is very much based on customer perception. At this stage, it is necessary to examine issues that impact the strategic plan, which include market regulations, new technology, and ways to measure improvement. At the implementation stage, a few targeted projects are identified for deployment. Annual objectives for implementation are recognized along with a plan to monitor and assess the projects step by step. The strategic plan has to be distributed throughout the company to allow employees to develop personal objectives under the scope of the strategic plan.

Realistic Plan

Plans for successful development of any product have to be well thought out. The team involved in the work must comprehensively develop these plans. It is important that managers not force plans upon the group; rather the product teams should agree on a team plan and have the organization support it.

Design in a Global Environment

Economic growth of countries has been marked by a series of important technological landmarks in materials, methods, consumer products, transportation, agriculture, and pharmaceuticals. Computerization and new materials have made production of just about anything cheaper and more efficient, and also have made quality easier to maintain. Countries have roared into the twenty-first century wired to the Web, facing and consuming new technology as soon as it is introduced. New products have been exerting a more serious impact on our life and environment than ever before. These technological innovations have caused higher rates of productivity, and therefore higher return rates on investment. Unfortunately, these technological changes have not always taken environmental impact into account, and have become major causes of environmental crises.

There is no doubt that globalization, new economies, and information technology will have an effect, not only on business, but also on the types and forms of products that new markets require. The impact these events will have on the cycles of materials, energy, and information worldwide is tremendous. Global product design and use has enhanced the relationship between producer and user. It is logical to conclude that the design stage has to be enlarged, first in terms of environmental impact of products, and second, in terms of the dualistic mental framework in which design activity and activity of the organization itself are viewed.

There is a powerful trend in product design toward material substitution—using new, synthetically designed materials that have highly desirable characteristics of toughness, lightness, durability, and flexibility. These technological advances are spurred by higher rates of productivity and, therefore, higher return rates on investment. In all these cases, new technology exerts a more serious impact on the environment than the older one if it is not used with care. Thus, the major cause of environmental crises has been technological innovations that have originated from a faulty design—failing to take the environment into account in the quest for narrowly defined efficiencies. There should be concern at the corporate level for the larger consequences of design. By neglecting such safety factors in design, companies end up borrowing on the future and running an environmental repair and clean up.

Environmental effects in product design are an important area of concern. Just as manufacturers cannot put up a plant without preparing an environmental impact statement, a product needs to be studied for its impact before it is put on the market. This aspect is becoming critical as the life cycle of the product gets shorter and shorter. George Seielstad points out that, "Each member of an advanced industrial society requires excavation of 20 tons of minerals a year" (Seielstad 1989). For a North American population of roughly 300 million, this means six billion tons of minerals shuffled every year. On top of this, he says fossil fuel used to process these minerals adds an additional 1% a year to carbon dioxide in the atmosphere. Industry uses large amounts of energy and as a consequence contributes substantially to the energy-related environmental problems. According to Graedel and Allenby, in the Unites States, manufacturing activities account for some 30% of all energy consumed and much of that energy is very inefficiently employed (Graedel and Allenby 1996). This has consequences on pollution of soil, air, and water.

Automotive industry approach. There is a growing concern regarding disposal of used products. For example, today in Germany, two million cars are dumped each year. As a result, there are about 130,000 tons of discarded plastics. In the U.S. over the next four years, an estimated 230,000 tons of plastics will be generated by the automotive industry. In the case of automobiles, over the past decade, the use of carbon steel, iron, and zinc castings has dropped significantly and high-strength steel, aluminum, and plastics use has risen substantially. Virtually all material in today's automobiles can be recycled. The challenge facing engineers is to make this recycling process economical—especially for materials in components such as seats and instrument panels. Recycling these components requires that different materials be separated so that each can be recycled individually. This separation can be accomplished either manually, where workers disassemble and sort the vehicle by hand, or mechanically, where the vehicle is shredded and the materials are sorted by properties such as conductivity and density.

Today, when an automobile is recycled, it is broken down as follows:

1. All reusable and remanufacturable components are manually removed from the automobile and resold. Most of these parts consist of power train components.
2. Materials of high value are manually removed (aluminum, magnesium, and any other large pieces of pure metal).
3. The vehicle is sent to the shredder. The pieces are mechanically separated depending on their material properties.
4. Ferrous metals are separated from the nonferrous.
5. The last 25% of the car is referred to as automotive shredder residue (ASR). ASR consists of rubber, plastics, glass, dirt, fluids, etc.

In general, most metals from automobiles are recycled today. This constitutes the largest portion of the vehicles (approximately 75%). Automotive manufacturers are attempting to further reduce environmental impacts from vehicles they produce by looking at the remaining portion of the vehicle that is not recycled. Recycling efforts exist because a profit can be made. Without this incentive, vehicles would simply be sent to landfills. Cost often plays the biggest roll in the level of effort put toward recyclability. The driving forces that cause companies to recycle include:

- Profits can be made from recycling.
- Government regulations are forcing more and more industries to think about recycling.

Increased emphasis on recyclability in the automotive industry has caused major automotive manufacturers to create a partnership dedicated to this cause. This partnership, called the Vehicle Recycling and Development Center, serves as the headquarters for a cooperative recycling effort among U.S. automakers.

Modularity in design. Another area of concern is the appliance and computer industry. Several companies take back their equipment (for example, copiers) to recycle them. The devices are disassembled, components are sorted and cleaned, and parts that are reusable are sent back to the factory floor for remanufacture. Just as many people today must recycle their bottles, cans, plastics, and newspapers, more and more industries will be forced to do a better job in the future.

Modularity is an important consideration in product maintenance. Modularity allows the product to be repaired easily since parts are removed and replaced in modular form. As a result, recyclability of these modules must be considered during the design phase. This also brings the life cycle of the modules into question. In general, the life cycle of any product can be broken down into four areas:

1. design and development;
2. production;
3. operational use and maintenance support; and
4. retirement and material disposal.

It is becoming increasingly obvious that, during design phase, engineers must think about modularity. With modularity, some devices from existing designs can be used for newer designs. Electromechanical products, such as computers, telecommunication devices, and peripherals, are most affected by this fast-changing technology. The best example is the personal computer (PC). It is estimated that processor speed increases 1.5 times every 18 months. This essentially means that, while the processor becomes obsolete very quickly, other devices within the CPU may not become obsolete. The hard-drive, video card, modem, monitor, and RAM may be appropriate to use again with the newer CPU. Since all these devices are modular, many customers upgrade their PC by replacing the processor only. Product modularity is an important design-for-maintenance issue for electromechanical products such as computers, telecommunication devices, and peripherals. The short technology life cycle of many functions in these products, combined with customer demand for a wide variety of features, requires product designers to optimize the modularity of components for manufacturability and serviceability.

Design for recycling. Designing products for recycling is an evolving process. Planning for product retirement will become a critical issue in the near future. The best time to make these plans is during the initial product design. Until manufacturing companies have a financial incentive to incorporate design for recycling into their product line, progress in this area will be very slow. In many cases, design for disassembly may be able to take design for recycling into consideration. As a result, these two areas are tightly related. Design for disassembly and recyclability makes it possible to reuse, remanufacture, and recycle materials in an efficient manner. Reuse and remanufacture will save many resources by prolonging the useful life of products.

CONCURRENT ENGINEERING IN DESIGN

Key concepts of new product development include:

- customer orientation;
- major decisions upfront;
- concurrent development of product design and production processes;
- using cross-functional teams; and
- use of efficient design and manufacturing techniques.

It is important to convert customer inputs into specific product functions, features, and specifications. The voice of the customer must be translated into product requirements that meet the needs and expectations of customers. This is an important step and must be done in the beginning. Customer satisfaction is a primary factor in the success of a product. These expectations can be summarized as: better quality, reliability, free maintenance, and lower price. The product development process uses customer expectations as an input and concurrent engineering as a design approach.

In addition to the voice of the customer, there is increased pressure to get products of ever-higher quality to the market in ever-shorter times. Customers are scrutinizing product price/performance ratios more carefully. The traditional serial approach to product design and development reduces the ability to compete effectively in the global market because it has the following weaknesses:

- insufficient definition of the product;
- inadequate studies undertaken on the influence of design on manufacture and assembly;
- no clear guidelines before production on how designs will develop in detail;
- inadequate cost analysis; and
- changes occurring in the design process.

The appropriate response to these weaknesses is a concurrent and multidisciplinary approach in the early stage of the product development process. It requires a new approach and change in a company's culture. The new approach initiates design of the product and associated processes, which is not the case in the traditional approach. Studies show that more time spent early in the design process is more than compensated for by time saved when prototyping takes place. It offers the possibility to reduce costs and time to market, while at the same time increasing product quality and customer satisfaction.

Concurrent engineering is a design approach in which design and manufacture of a product are merged in a special way. It is the idea that people can do a better job if they cooperate to achieve a common goal. It has been influenced partly by recognition that many of the high costs in manufacturing are decided at the product design stage. The characteristics of concurrent engineering are:

- better definition of the product without late changes;
- design for manufacture and assembly undertaken in the early design stage;
- well-defined process of product development; and
- better cost estimates.

The cost of design engineering is generally less than 5% of the total budget of a project, but its influence on total costs is significant. This is shown in Table 1-4.

Figure 1-7 shows the relationship of cost incurred and committed during a product's life cycle. With design decisions so critical, design is a small piece of the development

Table 1-4. Influence on product cost

	Initial Project Cost	Cost Committed
Design	5%	70%
Material	50%	20%
Labor	15%	5%
Overhead	30%	5%

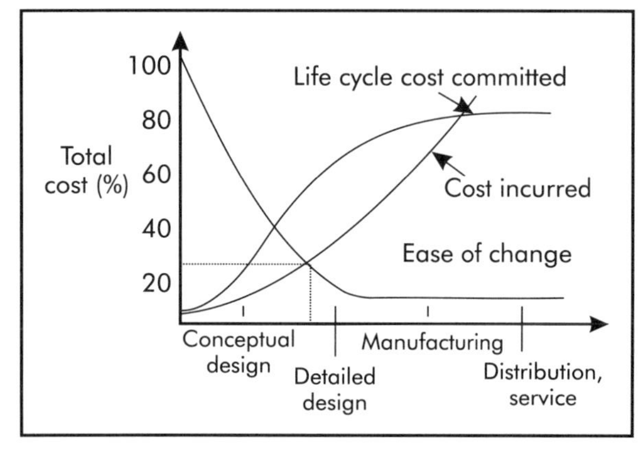

Figure 1-7. Cost incurred and committed during a product's life cycle.

chart, but it locks in the bulk of later spending. Decisions made in the design stage result in profound effects on product cost. Concurrent engineering involves more than just design and manufacturing. Even during the design stage, it is involved with customer perception, market analysis, optimized performance, life cycle performance, quality, reliability, and sales.

Concurrent engineering philosophy places emphasis on concurrent product design and process planning using a team approach. The team must develop a sound insight into the nature of the activity. Therefore, most decisions relating to design are made early in the process by a design team that consists of experts from the different stages of product life, starting from the marketing stage to maintenance and service stages. Cross-functional team members, while individually performing their respective functional design responsibilities, simultaneously work very closely with the rest of the team. For this reason, good communication is essential to effective operations in any company. Poor communication leads to faulty decision-making and is a major obstacle to a solid design process. Effective managers function as facilitators, while decision-making is delegated to people doing work commensurate with their responsibilities. Multifunctional design teams have to create a product that addresses the following requirements in a product:

- robustness;
- design for manufacture and assembly;.
- reliability; and
- environmentally friendly.

Using concurrent engineering principles as a guide, a designed product is likely to meet four basic requirements:

1. high quality;
2. low cost;
3. time to market; and
4. customer satisfaction.

Figure 1-8 shows the basic concurrent engineering model. Concurrent design improves the quality of early design-related decisions and has a large impact on the life cycle of the product. Creating a quality culture—introducing a team focus and product focus approach to project management—generates the total philosophy of concurrent engineering in the organization. It is well suited for team-oriented project management, with emphasis on collective decision-making. Among activities of different members of the team, coordination is an important element. Successful implementation of concurrent engineering is possible by coordinating adequate exchange of information and dealing with organizational barriers to cross-functional cooperation.

A major step in the product design schedule is the development of design for manufacturing (DFM) intent. *Design for manufacturing* is a technique for developing a product that meets the desired performance specifications while optimizing the design through the production system. It uses concurrent engineering based on a fundamentally different way of looking at how products are conceived, produced, and supported. DFM follows a procedure intended to help designers consider all elements of the product system life cycle from conception through disposal—including quality, cost, and user requirements. Use of DFM methods tackles the problems of improper definition of the product and addresses the issue of building standard and reliable products for industry.

MATERIALS AND MANAGEMENT IN DESIGN

Rapid developments currently taking place in certain research and development fields will have tremendous effects on product design. Changes witnessed in the electronics industry in the last 30 years, from the vacuum tube to the present-day computer chip, are just a glimpse of what is in store for the future. Wide spectrums of information technologies and biotechnologies, along with new materials, are influencing the market significantly.

In the field of materials, interest in such subjects as superconductivity is high. The engineering community is developing new materials with special characteristics, such as high-performance plastics, super glues, new alloys, and high-technology composites that are light, never wear out, and can withstand high temperature. These products are on their way to replacing more familiar materials. The new materials are known for

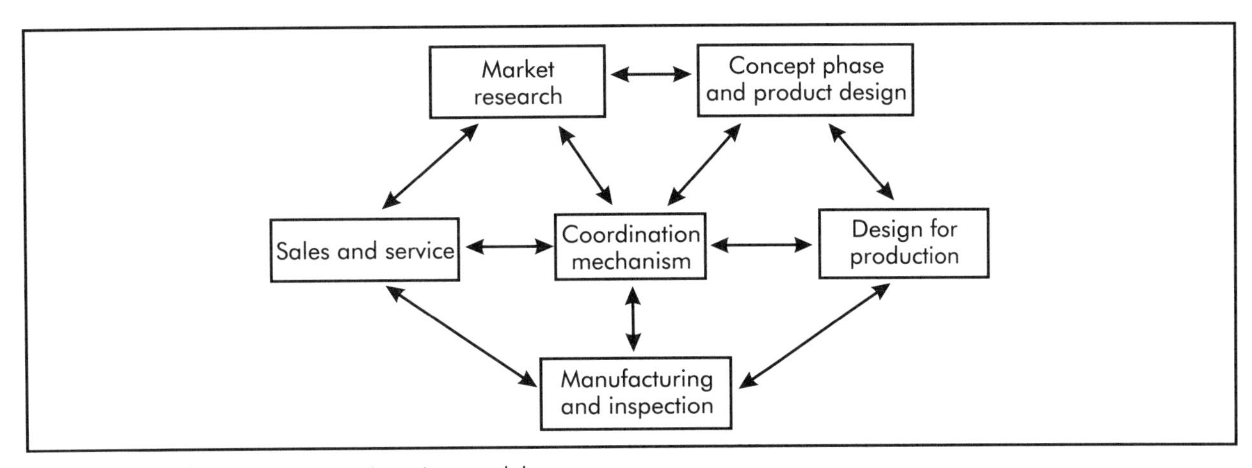

Figure 1-8. Basic concurrent engineering model.

their efficiency and power-to-weight ratios (many are used in the automotive industry). They work at much higher temperatures and at the same time do not require lubrication or cooling. Therefore, design changes and substitution of one material for another are what can be expected on a larger and larger scale.

Micro miniaturization has been going on in the electronics industry side-by-side with miniaturization of mechanical products. Micro-electromechanical systems are now being made in laboratories and are finding application in manufacturing, bioengineering, and medicine. From microns for typical parts of machinery, manufacturers are now down to parts measured in fractions of microns. And already, in the distance, nanotechnology provides measure in billionths of a meter, literally the molecular level. These are products and processes that emulate the mechanisms of biological life.

The importance of managing the design process—of not letting things just happen—is that it systematically creates favorable conditions that promote innovation and new ideas. The importance of people, of cooperation within multidisciplinary teams, and of communication is that these skills and knowledge are decisive in ensuring the success of the design process and the product. Balance within the team and in its interaction with the customer is crucial. Emphasis on production, use, and consumption are important. It clearly demonstrates that designers are decision-makers; design is nothing but decision-making. Now these decisions have implications and consequences that often go far beyond the simple designer-customer relationship.

ADDITIONAL CONSIDERATIONS FOR NEW PRODUCT DEVELOPMENT

General considerations in product design consist of cost and function. Cost is an obvious consideration, although a small number of parts may not necessarily mean a low product cost. It is possible that a design, solely being considered for correcting a functional problem, may result in a costly new product. Market design considerations involve the customers' needs, breadth of product line, product customization, expansion, upgrading, time to market, and future designs. Factory considerations are: delivery, quality and reliability, ease of assembly, ability to test, ease of service and repair, and shipping. Social design considerations include human factors, appearance, style, and safety. In considering environmental factors, related issues are product pollution and ease of recycling a product.

Impact of Modern Manufacturing on Product Design

As humans transition into the 21st Century, the marketplace has become truly global. Most companies have much wider product ranges. These companies are introducing new products more quickly with a sharp focus on the market.

In the early 1980s, most industries were involved in getting manufacturing operations under control through the use of formal production and materials planning, shop-floor scheduling, and enterprise resource planning. This happened with varying degrees of success. In the 1990s, many industries attempted to achieve world-class status by implementing total quality management methods. Some introduced just-in-time manufacturing techniques like cellular manufacturing, quick changeover procedure, one-piece part flow, kanban, and other techniques resulting in inventory reduction. Spurred by success stories, industries moved to team-based continuous improvement and experimented with self-directed work teams. Studying best practices used by others and benchmarking them has become a standard procedure.

As designers move into the new century, they need to incorporate these improvements to develop a truly agile product development process. Agility is the ability to succeed in an environment of constant and unpredictable change. There is a trend toward a multiplicity of finished products with short development and production lead times. The impact of this is seen in many companies in responsiveness, inventory, and organizational structure. Mass production does not apply to products where customers require small quantities of highly customized, design-to-order products, and where additional services and value-added benefits like product upgrades and future reconfigurations are as important as the product itself. An agile approach to manufacturing faces the reality that companies must serve customers with small quantities of custom-designed parts with perfect quality, 100% on-time delivery, and very low cost. Companies are forced to organize themselves in such a way that high-quality products can be developed very quickly in response to customer requirements.

Focus on the Customer

Customer requirements and competitive pressures have resulted in a need for companies to decrease product development costs and overall product costs, reduce product development cycle time, and improve quality. World-class manufacturers have placed great emphasis on being close to the customer. Having customers fully participate in the design of the product can significantly enhance the design process. Customers bring their design skills to bear on the project and the manufacturing company adds its production skills to the equation. In some cases, suppliers and outside process vendors also can be integrated into the design process so that the product is designed to meet customer needs effectively. This close cooperation allows for development of service-rich products that can evolve over time, as the customer and the company work closely together. Products may be designed to not only meet current needs but to be reconfigurable to meet a customer's future needs. Attention is paid to configurability, modularity, and design for the longer-term satisfaction of customer requirements. The advantage of a close relationship with the customer is that it helps ensure that the product being developed really meets the customer's requirements.

Focus on Product Teams

Use of cross-functional product development teams has a major effect on both cycle time and quality. With people from different functions working together, development gets done faster because activities can be done in parallel rather than in series. Quality improves because people from different functions work together to understand and solve development problems. The process is quicker and quality is better; the net result is that it is less expensive. In a cross-functional product development team, product developers from different functions work together and in parallel. Team members come from functions such as marketing, design, service, quality, manufacturing, engineering, testing, and purchasing. Often, key suppliers are included in the team. Sometimes, customer representatives also are included, allowing the voice of the customer to be heard throughout the development process. Team members work together, sharing information and knowledge, and producing better results faster than they would have done if operating in a traditional product development mode. The end result is that products get to market faster, costs are reduced, and quality is improved.

Focus on Information

Skills and knowledge of people within the company become a paramount consideration as a company develops results-based marketing. This knowledge includes product knowledge and experience, but it also includes a rich depth of knowledge of customer needs, anxieties, and service requirements. Increasingly, the best way to create close customer awareness is to provide people within the company, and the customers themselves, with a great deal of information. This may be product information, company information, education and training, product upgrades, manuals, instructions, and specifications. Orders can be placed automatically from the customer and scheduled within the plant, yielding the customer accurate delivery promises. Design requirements can be automatically picked up in the customer's information systems without drawings or specification being printed and passed. This enables the company to address customer needs with great speed. Design, delivery of information, history, accounts receivable, and customer service contact can all be integrated.

Some technologies required to achieve this level of information sharing and availability have only now become available. Wide access to the Internet and the World Wide Web has opened up a standard and direct method of access to information. The Internet, and other networks, allow the customer to have a simple and standard link to place orders, make inquiries, send messages, and specify needs.

Focus on Cycle Time

Cycle time has become a key parameter. Reduced lead time opens up new market opportunities and improves profits. It decreases market risk by reducing the time between product specification and product delivery. The sooner customers use a product, the sooner their feedback can be incorporated in a new, improved version. In fast-evolving technological environments, products become obsolete sooner. The decrease in time between product launch and product retirement erodes sales revenues. Since this phenomenon depends on factors beyond a company's control, the only way it can lengthen a product's life is to get it to market earlier.

Bringing products to market quickly means that product offerings will be fresher and the latest technology can be included because less time passes between definition of the product and its arrival on the market. Less time in development means less labor and less cost. The company responds more quickly to customers, gets more sales, and sets the pace of innovation. A company good at developing new products can use this advantage to gain market share. While competitors are busy developing the same abilities, the leading company introduces new products and features faster, and also develops new abilities.

Focus on the Development Process

The cycle time of any development operation depends closely on the development process. Any attempt to improve cycle times involves an investigation of how to improve the process. A clearly defined and well-organized product development process lies at the heart of an effective engineering environment, yet only a few companies have realized the potential advantages it offers. To make improvements, the process has to be analyzed and understood in detail. A new, fast, waste-free, low-cost process has to be defined and then implemented. Probably many existing tasks have to be removed, and some new ones added. Overall organization of the process changes significantly.

Without a well-defined development methodology, it is unlikely that members of the product development team work to maximum efficiency. Rules for working together during the development of a product have to be delineated. A clearly defined approach to development appropriate for the product family and understood by all team members provides best results.

Product development is a complex process involving many poorly understood variables, relationships, and abstractions. It addresses a wide range of problems, and is carried out by a wide variety of people, using a wide range of practices, methods, and systems, working in a wide variety of environments. Converting a concept into a complex multitechnology product under these conditions is not easy. It requires a lot of effort, definition, analysis, investigation of physical processes, verification, trade-offs, and other decisions. Companies without a well-defined product development process do not get the benefits they expect from initiatives to improve engineering performance. Without a clearly defined methodology, it is not known which systems and practices are most appropriate—so the necessary integration of an initiative is difficult to carry out. Any gains that come from the use of an initiative in one place are likely to be lost in another place because a coherent solution has not been prepared. Companies that understand this and put in place a clear product development process supported by a well-defined development methodology have every chance of becoming market leaders (see Figure 1-9). They use a methodology as the basis for involving people at all levels and in all functions to define, design, and produce the best product and get it to market faster.

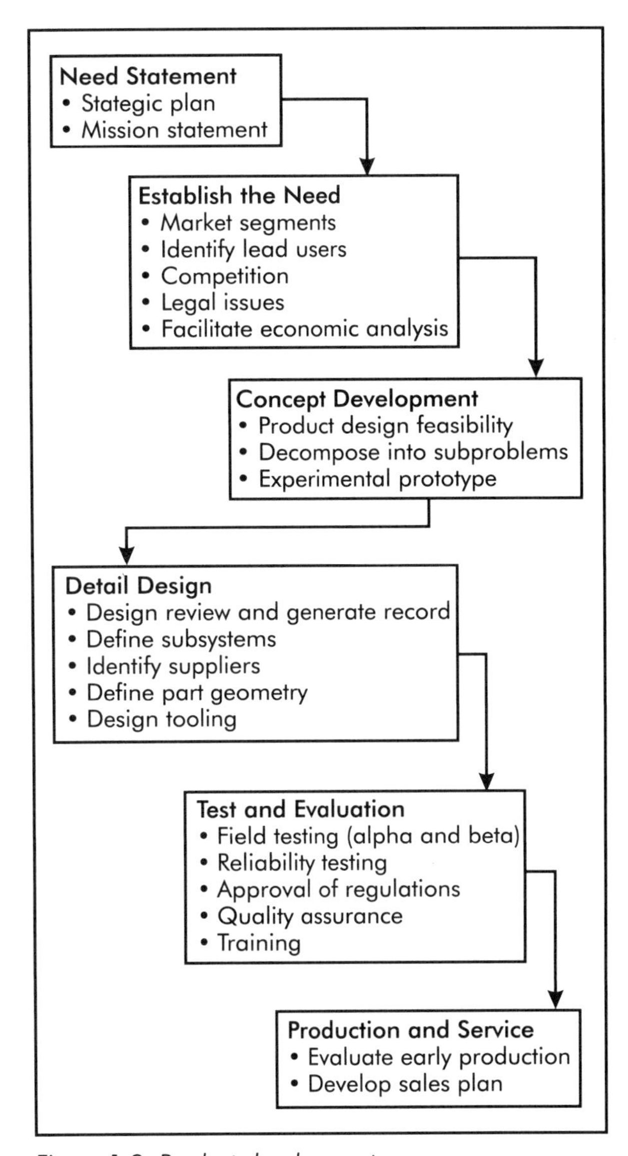

Figure 1-9. Product development process.

Focus on the Supplier

A successful product development process requires that a product get to market faster; offer customer satisfaction, and have reduced development costs. There are many possible approaches to reorganization. Many of them increase reliance on suppliers. Companies that focus on upstream product specification and design will want to outsource downstream activities. In the latter, they are not cost-effective or are less competent than specialized organizations. Consequently, suppliers have a greater role to play in these areas. For many companies, the cost of purchased materials accounts for more than half of expenses, so it is a good place to try to reduce costs. In addition to increased use of suppliers, the company involves them earlier in product development.

Focus on Prototyping

Computer-based simulation and rapid prototyping provide fast, low-cost proof of design concepts. In recent years, there has been a lot of pressure on companies to improve product development performance. They must develop products faster, at a lower cost, and with better quality. Simulation and rapid prototyping techniques have helped them meet these objectives.

Simulation is carried out to study the performance of a system, product, or process before it has been physically built or implemented. It involves development and testing of a computer-based model of a part or product. *Rapid prototyping* is production of a physical prototype directly from a computer-based model of a part or product.

The benefits of simulation come from the use of computer-based models. There are savings in reduced material costs. There also are savings because defining the process for making the prototype—and then building it and testing it—may no longer be needed. Quality is improved because it is possible to define and test many more potential designs using a computer-based model of the part than when using physical prototypes.

Companies that do not use simulation and rapid prototyping find their product development cycles longer, and their development costs higher, than companies that do use them. Both practices offer the advantages of reduced development costs and cycles, and improved quality.

Benchmarking is the continuous process of measuring products, services, and practices against a product development organization's toughest competitors or those renowned as industry leaders. If other organizations are found to have more effective operations, the product development organization can work out why they are better, then start to improve its own operations.

Activity-based costing (ABC) is a costing technique used to overcome the deficiencies of traditional product costing systems, which may calculate inaccurate product costs. The reason for these errors is often that the attributes chosen to characterize costs related to a particular product are attributes of unit products (such as direct labor hours per product) whereas many costs (such as setup time) are related to batches of products. ABC is based on the principle that it is not products that generate costs, but the activities performed in planning, procuring, and producing products. Resources necessary to support these activities result in costs being incurred. ABC calculates product costs by determining the extent to which a product makes use of the activities.

Tools and Techniques

The goal is to develop the product that is the best for function, manufacturing, reliability, and servicing. This is the ideal design, but one that is difficult to achieve. Thus, the process has to be managed in steps, which ultimately have to be successfully integrated. Table 1-5 shows the various tools and methodologies and links them to the appropriate product development phase.

REFERENCES

Adler, Paul, Mandelbaum, Avi, Nguyen, Vien and Schnerer, Elizabeth. 1996. "Getting the Most of Your Product Development Process." *Harvard Business Review*, March-April: p. 134-151.

Table 1-5. Tools and techniques for product development

Phase	Tool
Concept development	Market studies Voice of the customer House of quality
Design and development	Function analysis Design for manufacturing Design for disassembly Product modeling using CAD/CAM Simulation Optimization Design for six-sigma analysis Rapid prototyping Design for environment and service
Analysis and testing	Failure mode and effects analysis Robust design Statistical reliability analysis Design for life cycle
Product creation	Workplace design Flexible automation tools Value Stream Mapping℠

Cross, Nigel. 1994. *Engineering Design Methods—Strategies for Product Design*, second edition. New York: John Wiley and Sons.

Dixon, John R. 1996. *Design Engineering: Inventiveness, Analysis, and Decision Making*. New York: McGraw Hill.

Finger, Susan and Dixon, John R. 1989. "A Review of Research in Mechanical Engineering Design" (two parts). *Research in Engineering Design*. New York: Springer International. Vol. 1.

Graedel, T.E. and Allenby, B.R. 1996. *Design for Environment*. Upper Saddle River, NJ: Prentice Hall.

Hubka, V. and Schregenberger, J.W. 1987. "Path Toward Design Science." Proceedings of the International Conference on Engineering Design, Boston. Zurich, Switzerland: International Society for the Science of Engineering Design, WDK.

Le'Mee, Jean. 1987. "Grammatical Approach to Design." Proceedings of the International Conference on Engineering Design, Boston.

Pahl, Gerhard, and Beitz, Wolfgang. 1996. *Engineering Design—a Systematic Approach*. New York: Springer Verlag.

Pugh, Stuart. 1991. *Total Design—Integrated Methods for Successful Product Engineering*. Wesley, MA: Addison.

Rackham, Neil. 1998. "From Experience: Why Bad Things Happen to Good New Products." *Journal of Production and Innovative Management*: 15: p. 201-207.

Roozenberg, N.F.M. and Eekels, J. 1995. *Product Design Fundamentals and Methods*. Chichester, UK: John Wiley and Sons.

Roth, K. 1994. *Designing with Design Catalogs*. Berlin and New York: Springer Verlag.

Seielstad, George. 1989. *At the Heart of the Web*. New York: Harcourt Brace Jovanovich.

Ulrich, Karl T. and Eppinger, Steven D. 1995. *Product Design and Development*. New York: McGraw Hill.

User, John, Roy, Utpal, and Parsaei, Hamid. 1998. *Integrated Product and Process Development*. New York: John Wiley and Sons, p. 125-147.

Chapter **2**

Creative Concept Generation

Creativity involves the human mind manipulating past experiences by combining concepts to produce new ideas. Although there is no single definition for creativity, most definitions identify creativity as a combination of experience, intelligence, and motivation. These definitions are as shown in Table 2-1.

TECHNIQUES OF CREATIVE CONCEPT GENERATION

Creativity is important at various stages of product development. It has been said that innovative ideas do not result from straightforward analytical procedures or complicated algorithms. Ideas come from the creativity of designers. Vertical thinking, lateral thinking, and brainstorming are three procedures that apply to understanding creativity.

Vertical Thinking

Analytical thinking or deductive reasoning is also called *vertical thinking*. In vertical thinking, the individual always moves forward in sequential steps only after a positive decision has been made, based on available information. Vertical thinking is analytical, judgmental, critical, and selective. If no positive decision can be made, the vertical-thinking pattern ends abruptly. Vertical thinking is then used to focus ideas into real working solutions to problems. It moves in a straight line until it is stopped by a positive or negative conclusion.

Table 2-1. Definitions of creativity

Creativity has been called the combination of seemingly disparate parts into a functioning and useful whole (Adams 1976).

Creativity is the operating skill that intelligence needs to act upon experience for a purpose (DeBono 1970).

A creative act is a combination of previously unrelated structures in such a way that one gets more out of the emergent whole than one puts in (Koestler 1969).

A creative person is an individual who is motivated, curious, self-assertive, aggressive, self-sufficient, less conventional, persistent, hard-working, self-disciplined, independent, autonomous, constructively critical, informed, open to feelings, aesthetic in judgment, adapts values congruent with the environment, and an achiever (Stein and Heinze 1960).

Lateral Thinking

In *lateral thinking*, the individual moves in many different directions, combining different bits of information into new patterns until several solution directions are exposed. Then, all of the solution directions are developed further until several possible alternatives are completely exposed. Lateral thinking is random, sporadic, non-judgmental, and generative. The basic function of lateral thinking is to take experiences and reform them a few times to generate new ideas. Lateral thinking moves randomly in fits and starts, without a need to reach a conclusion.

Brainstorming

Brainstorming is the oldest and best-known creative-thinking technique. *Brainstorming* has also come to mean a serious effort to think out a problem. The objective of a brainstorming session is to use the disconnected ideas of individuals to trigger new ideas in each participant. The technique relies heavily on group interaction for exchanging ideas and provides an excellent means of building upon other participants' ideas. Triggering ideas in others is key to successful group brainstorming.

Environment

A creative idea should be nurtured and protected. Thus, a creative environment has to be set up so that an idea will be developed further and interact with other ideas to develop a pattern. Continual exposure to new experiences in the fine arts, sports, industry, sciences, music, and literature provides a format from which new patterns are made.

Guidelines for successful brainstorming sessions are:

- Get as many ideas as possible. The more ideas generated, the greater the probability of hitting a great idea.
- No negative comments or judgments are made at this stage. An individual can come up with as many ideas as possible and the merits of each are not evaluated. Statements like, "that is too expensive," "not enough time," "this will not work," or "it is against the company policy" should be avoided.
- In group discussions, people tend to criticize views as soon as they are expressed. A good facilitator can create a sense of security within a group and ensure that ideas are not criticized at this stage.
- An intense creative-thinking exercise might generate a number of ideas, which can be followed by an incubation period during which ideas are sorted out.
- Idea improvement is sought. In addition to contributing ideas of their own, participants should suggest how the ideas of others can be turned into better ideas, or how two or more ideas can be joined into still another idea.

The brainstorming group should consist of a small group of individuals with different perspectives on the product and with different degrees of knowledge about the problem. The group leader explains the rules of brainstorming and gives the group several practice exercises in their application. The group leader states the specific problem as accurately as possible and the session begins. Each participant shouts out ideas as rapidly as they come to mind. There should be no sense of formality or order to exchange ideas. To prevent boredom, the group leader should end the idea exchange when it becomes obvious that there is a mental block. Alternately, the leader can distribute the unedited list of ideas and call together another group session in a week or two. The most important role of the group leader is to stimulate the flow of ideas by maintaining an environment that is free from criticism.

Overcoming Mental Blocks

Scott Fogler and Steven LeBlanc explained the causes of mental blocks (Fogler and LeBlanc 1995). Common causes of conceptual mental block include:

- a very narrow definition of the problem;
- an assumption that there is only one answer;
- frustration caused by not having immediate success;
- attacking symptoms, rather than the problem;
- getting attached to the first answer that comes to mind;
- a mental wall preventing the creative designer from correctly perceiving the problem.

The most common mental blocks are characterized as perceptual block, emotional block, intellectual block, environmental block, and expressive block (see Table 2-2).

A fair number of structured techniques are available to overcome mental block. Some techniques address attitude adjustment by focusing on a positive aspect of the problem and trying out bold new design alternatives. Some accepted techniques used for idea generation are:

- random simulation;
- Osborn's checklist;
- attribute listing;
- morphological analysis;
- futuring;
- others' views; and
- synectics.

Random Simulation

Random simulation is a way of generating ideas that are totally different than ones previously considered, so that users can get out of a mental block mode. The mind looks for similarities in patterns and then groups these experiences. It immediately rejects ideas that are totally unrelated to each other. Some suggested procedures are as follows:

- Use a dictionary to produce a random word; this random word is then used as a trigger to generate other words that can stimulate the flow of ideas.
- Select some object in the room.

The dictionary technique is simple and offers the greatest number of possible combinations of objects. Procedures such as rolling dice or looking at a table of random numbers can be used to locate a page number. One of the important attributes of this

Table 2-2. Common types of mental block

Perceptual block	A perceptual block prevents the designer from clearly seeing the problem itself and the information needed to solve it. A major cause of perceptual block is the combination of information overload and saturation.
Emotional block	An emotional block interferes with the designer's ability to conceptualize. Fear of risk taking, approaching the problem with a negative attitude, and a lack of challenge are listed as examples of an emotional block.
Intellectual block	An intellectual block can happen if the designer does not have the necessary background, training, or knowledge to solve the problem.
Environmental block	An environmental block results from a lack of physical and organizational support to translate creative ideas into practice.
Expressive block	An expressive block is the type of mental block that results in an inability of the designer to communicate in written form.

technique is the ability to leapfrog from one idea to another. It is also a common practice to use a journal or newspaper to locate the random word in a similar manner to the dictionary method. Another idea that can be used is picking a random picture in a magazine or catalog. This kind of activity also can be fun for participants.

Osborn's Checklist

Osborn's checklist can help a group build on one another's ideas (Osborn 1957). It is a thinking technique based on the following set of questions used to stimulate the mind to change its perspective of the problem:

- Adapt? How can the product be used? What are the other adapted uses of it?
- Modify? Can the shape, color, material, or focus of the product be changed?
- Magnify? Can a new feature be added? Can it be made longer, thicker, and higher?
- Substitute? Who else can use it? What else can be used? Where else can it be developed?
- Rearrange? Can parts be interchanged? Can the positive be changed to negative? What about a different pattern?
- Combine? Can different components or ideas be combined? Is there a compromise? Can it be blended?

The nature of questions is not important. Questions are merely a mechanism to change an individual's viewpoint of a problem. The leader often introduces a checklist or idea-stimulating questions during a group problem-solving session.

Attribute Listing

The first step in this technique is to write down all design attributes of the problem, similar to the process of writing down product specifications. The second step is to apply a list of modifiers to each attribute one at a time to generate new alternatives.

- What shape?
- How deep?
- Are they adjustable?
- Are they removable?

The difficulty with attribute listing is that the designer must be familiar with the product and its features. Attribute listing looks at each parameter of a product in isolation, but ignores interactions between two attributes that might lead to a different solution.

Morphological Analysis

Morphological analysis is an organized method that enables designers to make comparisons among various attributes of a problem. It is a method for creating new forms of a design. The objective of the morphological chart is to encourage the designer to identify novel combinations of elements and recombine them to derive a solution. It helps the designer generate the complete range of alternative design solutions for a product, and hence to widen the search for potential new solutions. The steps involved in the morphological chart method are as follows:

1. List the product's essential features.
2. For each feature, list the means by which it can be achieved.
3. Prepare a chart that contains sub-solutions.
4. Identify possible combinations of the sub-solutions to make a product.

Nigel Cross discusses the generation of a morphological chart for a forklift truck. The first step is to identify essential and common features (see Table 2-3). Table 2-4 shows that by combining different combinations of the sub-solutions, unique design possibilities can be created for differing environments.

Table 2-3. Morphological chart for the creation of an alternate forklift truck (Cross 1994)

Feature	Means				
Support	Wheels	Track	Air cushion	Slides	Pedipulators
Propulsion	Driven wheels	Air thrust	Moving cable	Linear induction	
Power	Electric	Petrol	Diesel	Bottled gas	Steam
Transmission	Gears and shafts	Belts	Chains	Hydraulic	Flexible cable
Steering	Turning wheels	Air thrust	Rails		
Stopping	Brakes	Reverse thrust	Ratchet		
Lifting	Hydraulic ram	Rack and pinion	Screw	Chain or rope hoist	
Operator	Seated at front	Seated at rear	Standing	Walking	Remote control

Table 2-4. One selected combination of sub-solutions from the morphological chart (Cross 1994)

Feature	Means				
Support	**Wheels**	Track	Air cushion	Slides	Pedipulators
Propulsion	**Driven wheels**	Air thrust	Moving cable	Linear induction	
Power	**Electric**	Petrol	Diesel	Bottled gas	Steam
Transmission	Gears and shafts	**Belts**	Chains	Hydraulic	Flexible cable
Steering	Turning wheels	Air thrust	**Rails**		
Stopping	**Brakes**	Reverse thrust	Ratchet		
Lifting	Hydraulic ram	Rack and pinion	**Screw**	Chain or rope hoist	
Operator	Seated at front	Seated at rear	Standing	Walking	**Remote control**

Futuring

Futuring is another technique used to overcome a mental block by imagining a solution currently not feasible, but that could be in the future. Questions are asked about an ideal solution, benefits of the ideal solution, and how to devise ways to achieve it. As an example, participants should consider designing a special product. During the production process a product gets fabricated, but it also results in manufactured waste. Since treating waste is expensive, the product is not only expensive, but it also contributes to environmental problems. As a futuring exercise, the group tries to imagine an ideal solution where the product is not only profitable, but also does not produce any scrap. While generating solutions, participants should imagine processes that have no waste.

Others' Views

At times, it becomes easier to solve a problem when it is examined from different viewpoints. The problem definition then becomes different, depending on whose viewpoint is selected. Scott Fogler and Steven LeBlanc give an example of differing viewpoints of a problem with a space capsule (Fogler and LeBlanc 1995):

- Problem: Space capsule burns upon entering the atmosphere.
- Project manager: The project gets completed on time.
- NASA accountant: Solve problem but keep cost low.
- Engineer: New material should not interfere with capsule performance.
- Material scientist: Find a material that can handle the high temperature on reentry.
- Astronaut: Does not care about the capsule, wants to return alive.
- Solution: Allow surface of the capsule to be destroyed protecting astronauts.

Synectics

Synectics means joining together different and apparently irrelevant elements. Synectics deals with problem solving and the way that creative people think. It is a process that places the most emphasis on the emotional component and understanding of the irrational element in decision-making. Synectics research has shown that creative efficiency in people can be increased if they understand the psychological process by which they operate (Nierenberg 1986).

The first step is to define the problem in a way that can be understood by the people solving it. The problem solver initially analyzes a strange situation to look for bits of the problem that are familiar. Using small and familiar pieces, the mind rearranges the problem into a situation that it can visualize. This step is known as problem-statement formulation.

Synectics research recommends stopping once the problem statement is clearly formulated, instead of continuing to analyze it. Once the problem statement is understood, it must be twisted into a totally different form. The main focus of synectics is its emphasis on distorting the problem so that the perspective changes dramatically. It is like making a familiar situation into a strange situation. By forcing a change in a viewpoint, this technique generates very unusual solutions.

Synectics research has generated four mechanisms for idea generation. Each of these mechanisms is intended to distort the problem:

1. personal analogy,
2. direct analogy,
3. symbolic analogy, and
4. fantasy analogy.

Personal analogy. This is where an individual is placed in a position to study the product. If the product being studied is an elevator, the designer places himself inside the elevator driving mechanism to experience the movement of the carriage going up and down. The person is now able to get an inside view of the problem while looking out at external forces.

Direct analogy. This mechanism involves the ability to make comparisons between similar but different technologies. Biological systems provide a lot of comparisons between human biology and everyday engineering mechanics.

Symbolic analogy. Using a *symbolic analogy*, the problem is described with poetic inspiration and metaphors.

Fantasy analogy. While using a *fantasy analogy*, the individual tries to imagine an ideal solution, but at the same time suspends judgment on whether the product is feasible or not.

Design groups using synectics require proper training in analogy mechanisms. For example, a particular bicycle manufacturer may be facing a loss of sales to competitors who have a new product. This may motivate the company to design a new bicycle that could be appealing to the younger generation and to adventurous bicyclists.

Bicycle Example

Using the techniques of brainstorming, random thinking, Osborn's checklist, attribute listing, and a morphological chart (see Table 2-5) enable a company to produce a standard-size bicycle with a new body shape available in three colors.

Brainstorming. The brainstorming session generates some basic requirements. For instance, the bicycle is unique because it will have a bright color, a sporty shape, durability, extra features to hold water bottles, special grip handles, and because it will be easy to repair and maintain.

Random thinking. The random dictionary word selected for this product is "release." Based on the random word, ideas generated for the bicycle include:

- easy removal of flat tire;
- easy removal of handlebar;
- ejection of seat;
- self-contained bicycle lock;
- body shaped like an "R";
- quick mounting and dismounting;
- slogan for marketing such as, "release the animal in you"; and
- quick release to market.

Osborn's checklist.

- Adapt—use for road racing, mountain biking, dirt bike riding, or normal

Table 2-5. Morphological chart for a new bicycle

Feature	Means				
Body material	Steel	Plastic	Carbon fiber	Aluminum	Titanium
Transmission	Belt	Gears	Chain and sprockets	Rope and reel	
Capacity	1 person	2 persons	3 persons		
Power	Feet	Legs	Hands	Arms	
Peddles	Clip	Clipless	Straps	Attached shoes	No peddles
Drive	Front wheel	Rear wheel	All-wheel drive		
Steering	Front wheel	Rear wheel	All-wheel drive		
Stopping	Disk brakes	Drum brakes	Rim brakes	Parachute brakes	
Seat	Padded	Unpadded	Contoured	Flat	

transportation. It can be used by everyone for every purpose and by every age.
- Modify—modify design to a more appealing shape and color. Add more features like water bottle, clock, timer, and storage compartment.
- Substitute—use more durable tires and lightweight material.
- Combine—combine attractiveness with usability.

Color. It will provide a choice of three colors, each appealing to men, women, and youth.

Tires. Standard tires are available with option of street, race, or dirt track tires.

Seat. There is an adjustable and replaceable seat for different sizes of bicyclists.

Shape. There is a standard shape and size for all riders. Necessary modification can be made if needed.

DEFINING THE PROBLEM

A satisfactory definition of the problem is crucial to its success. Time spent defining the problem properly, and then writing a complete problem statement, results in efficient problem solving. It has been said that a well-defined problem is critical to finding a workable solution. The goals of a product-design project should be expressed in very broad terms at the beginning. It is a mistake to plunge headlong into a problem solution before setting appropriate goals and finding the real needs.

Product goals generally are derived from needs analysis. The needs analysis process consists of listing the needs of the customers, contractors, marketing agents, trade associations, government agencies, technicians, and servicing agencies. The product designer has to understand and weigh each of these basic needs. For example, the customer may want a product to be functional, aesthetically appealing, durable, and inexpensive. The marketing group may want it to have sales appeal, a minimal service factor, be easily transportable, and have a good profit margin. The manufacturing group may want it to be easily fabricated, have a low labor cost, be made with pre-approved materials, and come from an existing supplier. These are varying preferences and needs that may be difficult to satisfy.

All viewpoints should be considered in a needs analysis. In addition, there are technical needs, time needs, and cost needs to be considered. The study of these factors may take the form of a detailed benefit-cost analysis or a detailed cost estimate of manufacturing the design, including the profit and marketing costs. Most product-design problems have certain boundaries or constraints within which the solution must be found. Legal constraints on engineering design are becoming increasingly important. Federal and state regulations pertaining to environmental pollution, energy consumption, public health, and safety are examples of these limiting factors.

Problem definition is based on identifying the real needs of the customer and then formulating them as product goals. The problem statement expresses what is supposed to be accomplished to achieve the goals. Design specifications are a major component of the problem statement. A key role of the problem statement in the design process is shown in Figure 2-1. The essential elements of the problem statement are shown in Figure 2-2.

Find the Source

A few recommended tips for problem definition are:

- Collect and analyze the information and available data.
- Consult people familiar with similar products.
- Inspect the problem personally.
- Verify the collected information.

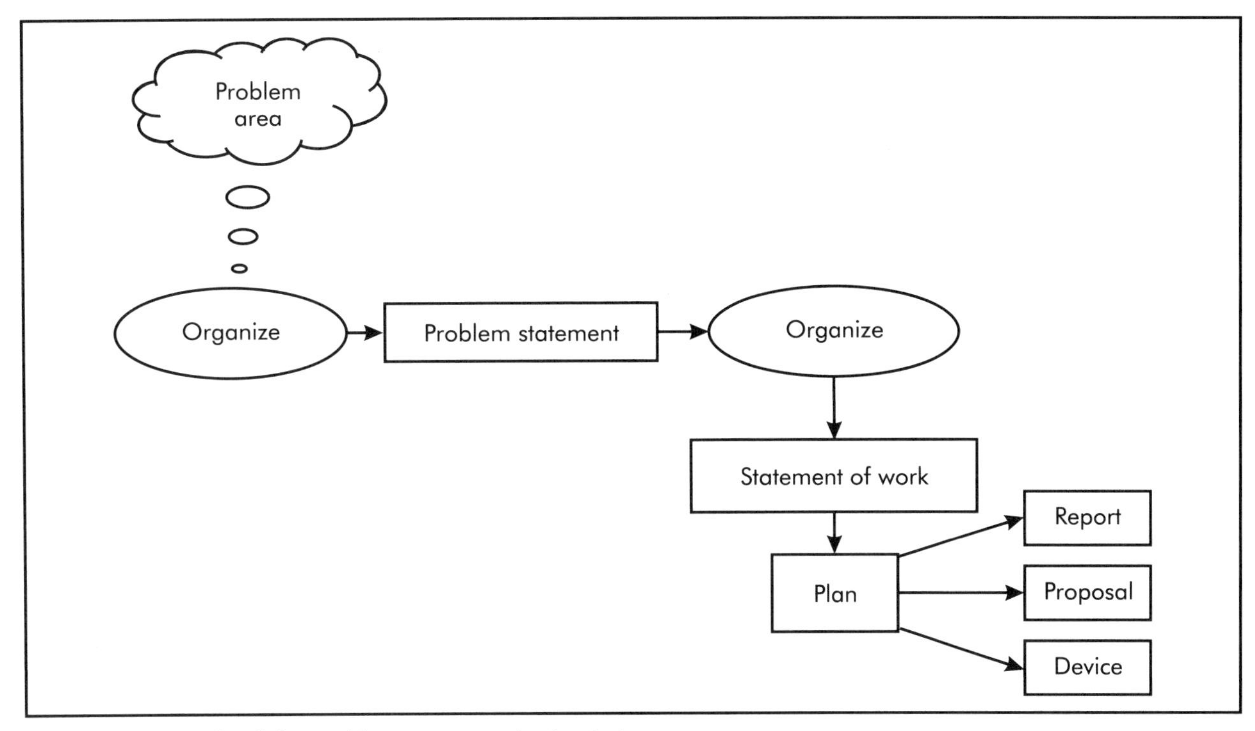

Figure 2-1. The role of the problem statement in the design process.

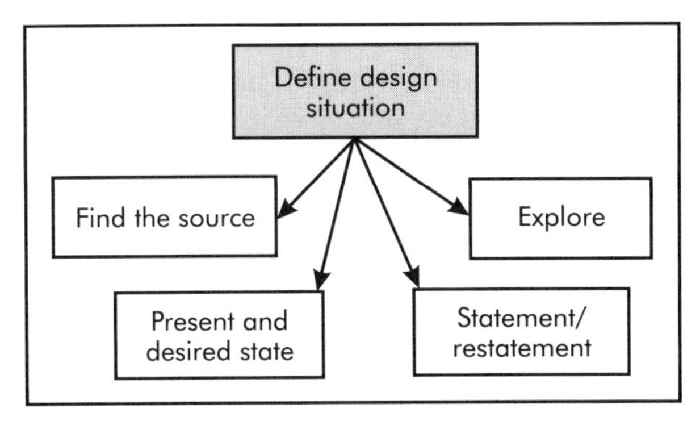

Figure 2-2. Elements of the problem statement.

It is very important to find out if the problem at hand reflects the real situation. To find the real source of the problem, the following questions should be asked. Where did the problem originate? Who needs the product? Who initiated it? Can the person responsible explain how they arrived at a definition of the need? Are the assumptions valid? Have all viewpoints been considered?

Explore

The five-point strategy (define, explore, plan, act, and reflect) is a technique that helps users to understand and define the real problem. Based on the strategy, the following steps can be used to explore a problem.

- Identify pertinent relationships among inputs, outputs, and other variables of importance.
- Recall past experiences.
- Discover real problems and constraints.
- Consider short-term and long-term implications.
- Collect missing information.
- Hypothesize, visualize, idealize, and generalize.
- If users cannot solve the proposed problem, they should first solve some related problems or part of the problem.
- Sketch out a pathway that will lead to a solution.

- After using some of the above activities, write a statement defining the real problem.

Define the Present State and Desired State

The present state (PS) and desired state (DS) technique helps an individual to visualize the starting point and where to proceed. Then, an appropriate path can be found to the desired state that represents solution goals. The designer should try to modify the statement of the present state or desired state until a satisfactory correlation is found between them. It is important that the present-state statement and the desired-state statement contain solutions that go to the heart of the problem. The following sections outline the use of the PS-DS technique.

The Situation

In schools with engineering programs, it is typical that the freshman year dropout rate from such programs is high. There are several ways this problem can be addressed. Users of the PS-DS technique first verbalize where they are and where they want to go. For example:

PS: The freshman engineering dropout rate is high.
DS: Increase the freshman retention rate in engineering programs.

Since there is no one-to-one match here and the PS does not have anything in common with the DS, the statements need to be modified. For example:

PS: The freshman dropout rate needs to be reduced.
DS: Make freshman courses more interesting to increase the freshman retention rate in engineering programs.

Because there is still a difference between the PS and the DS, it is necessary to revise the statements so that the gap between them is narrowed. For example:

PS: The freshman dropout rate is high because students are only exposed to non-engineering courses such as mathematics and physics. Students do not see a connection with the courses and the engineering profession.
DS: First-year students need to be exposed to engineering principles in combination with physics and mathematics so that they can learn why engineering is interesting.

There is a relationship between these two statements. Differences between the two statements should be made clear enough though. For example:

PS: The reason for the high freshman dropout rate is that students in the engineering program are unable to see a connection between the courses in engineering and other courses such as physics and mathematics.
DS: First-year students should be provided with integrated engineering courses, where engineering principles are taught in combination with physics and mathematics as integrative learning blocks.

PS and DS Using the Dunker Diagram

Dunker diagrams help the designer examine possible paths from the starting point to the desired state. By going through various paths, it can also lead to solutions. There are two types of solutions:

1. Examining the path to be followed and actions to be taken to achieve the desired state.

2. Solving the problem by making it acceptable not to reach the desired solution. This transforms the DS until it matches the PS. This procedure eliminates the need to achieve the DS.

Figure 2-3a shows the principles of the three-stage Dunker diagram. The first stage represents general solutions. The second stage represents functional solutions. Functional solutions are possible paths to a desired state, but they need not necessarily examine the feasibility of the solution. These solutions consider "what if" situations. The third stage represents specific solutions to implement functional solutions. Figure 2-3b presents a Dunker diagram that addresses the problem of attracting high-school students to engineering. In this case, the Dunker diagram analyzes two types of situations: one that involves attracting more students to the engineering path, and one that analyzes how to modify the desired state so it corresponds to the present state.

Statement-restatement Technique

The statement-restatement technique tries to achieve objectives by rephrasing a problem in a number of ways. Various problem-statement triggers can be applied such as varying the stress pattern on certain words, changing positive terms to negative terms, substituting explicit definitions of certain terms in the statement, etc.

Problem-statement Triggers

A.F. Osborn has developed a technique where an individual uses words and questions to trigger different thoughts. The triggers focus on possible changes in a problem statement by rewording the concept statement (Osborn 1957). Steps in the technique include:

1. Vary the stress pattern; try placing the emphasis on different phrases and words.
2. Choose a word that has an explicit definition and substitute the explicit definition in each place where the term appears.
3. Make an opposite statement, change positive to negative, and vice versa.
4. Replace persuasive and implied words in the problem statement (such as "obviously" and "clearly") with the use of such phrases as: "Is this reasoning valid?" "What is the evidence for such reasoning?" and "If the reasoning is invalid, how can we modify the statement of the problem?"
3. Express words in the form of an equation or a picture and vice versa.

An example from the aerospace industry. Jet engine failure is sometimes the cause of an airplane crash. It could happen because of many reasons. One is due to defects in the engine turbine blades. The designer perceives a need for a methodology to reduce the surface defects in the jet engine turbine blades. Based on this information, the initial problem statement is made as follows:

> "Surface irregularities on the turbine blades of a jet engine cause blade failure, which creates a danger to the aircraft."

Trigger 1: Vary the stress pattern and examine if the focus of the problem itself has changed. (Notice the different stress patterns in the following identical sentences.)

> The *turbine blade* failure is due to the surface irregularities *in jet engines* for commercial aircraft.
>
> The turbine blade failure is due to the *surface irregularities* in jet engines for commercial aircraft.
>
> The turbine blade failure is due to the surface irregularities in *jet engines* for commercial aircraft.

Trigger 2: Substitute a word with a more explicit definition in each place that a term appears.

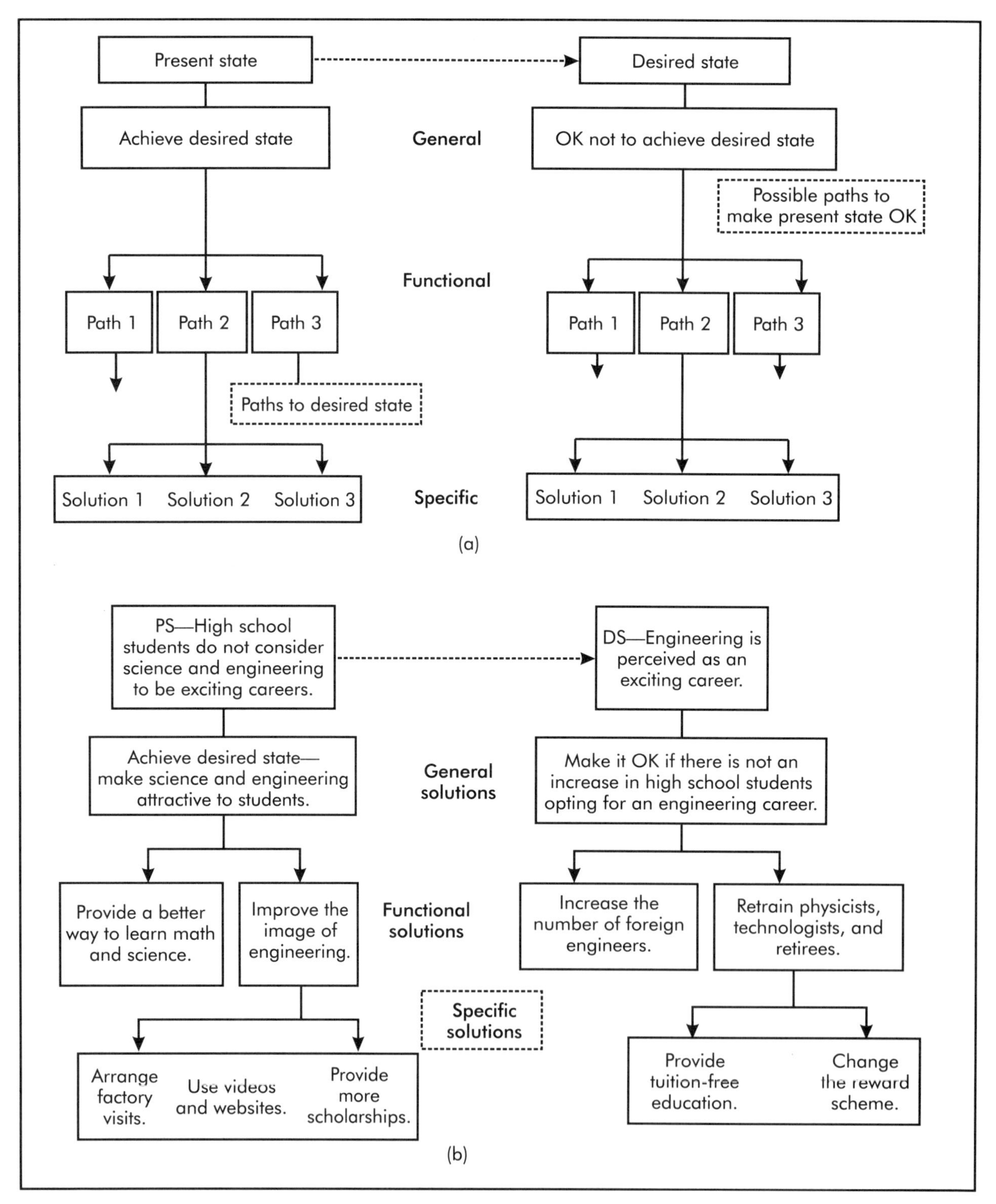

Figure 2-3. (a) Three-stage Dunker diagram; (b) Dunker diagram for an engineering career.

The *physical surface characteristics of a jet engine component* cause the engine of an aircraft to fail.

This makes Osborn users think about the physical characteristics of the component and how it can be designed not to fail.

Trigger 3: Change positive to negative and vice versa.

How can designers find a way to make worse surface characteristics so that the engine always fails?

This makes system users think about how to go about maintaining and measuring surface characteristics and how to control them.

Trigger 4: Change persuasive words—such as "every" to "some," "always" to "sometimes," and "sometimes" to "never." This trigger assists users to basically challenge the fundamental assumption in the problem definition.

The physical surface characteristic of a turbine blade is *always maintained* to prevent failure in the engine of an aircraft.

Why shouldn't the blade always be maintained by 100% inspection? This opens new areas of discussion.

Trigger 5: Express words in the form of an equation or a picture and vice versa.

Surface roughness = F (cutting tool geometry, vibration of the machine, machining process)

DESIGN CONCEPT DEVELOPMENT METHODOLOGY

The concept development phase needs coordination among many functions. This is the front end of the product development process. In the concept development phase, the needs of the customer are identified; customer needs are translated into technical terms; target specifications are established; alternative product concepts are generated and evaluated; and one or more concepts are selected. At this stage, the product team explores various possibilities of products to meet customer requirements. These include external as well as internal searches, brainstorming, and an exploration of various ideas. Such exploration requires the involvement of most functions of the company, customers, outside suppliers, and government agencies.

Figure 2-4 shows a four-step concept-generation methodology, starting with the first step of clarifying the problem by breaking it into subproblems and focusing on those that are critical. The second step involves a Web-based information search, literature and patent search, customer feedback reports, consultation, and benchmarking. The third step involves a systematic exploration of ideas, morphological classification, and combination charts to investigate the best possible scheme. The last stage examines solutions, feedback, and revision of the problem statement.

Concept Selection Using a Function Diagram

Consider a problem statement for the electrical receptacle diagrammed in Figure 2-5. In a typical installation, two standard 15/20-amp receptacles are installed in the raceway to allow the user to plug in an electrical device. A new receptacle design would need to be installed by electricians using common tools. It must meet all code requirements for this type of device. It should also be easy to install.

In Figure 2-5, the input is electrical energy and the output is the generation of power to switch the machine on/off. The diagram shows how the problem can be decomposed into subproblems. The function of the product is subdivided into two subfunctions:

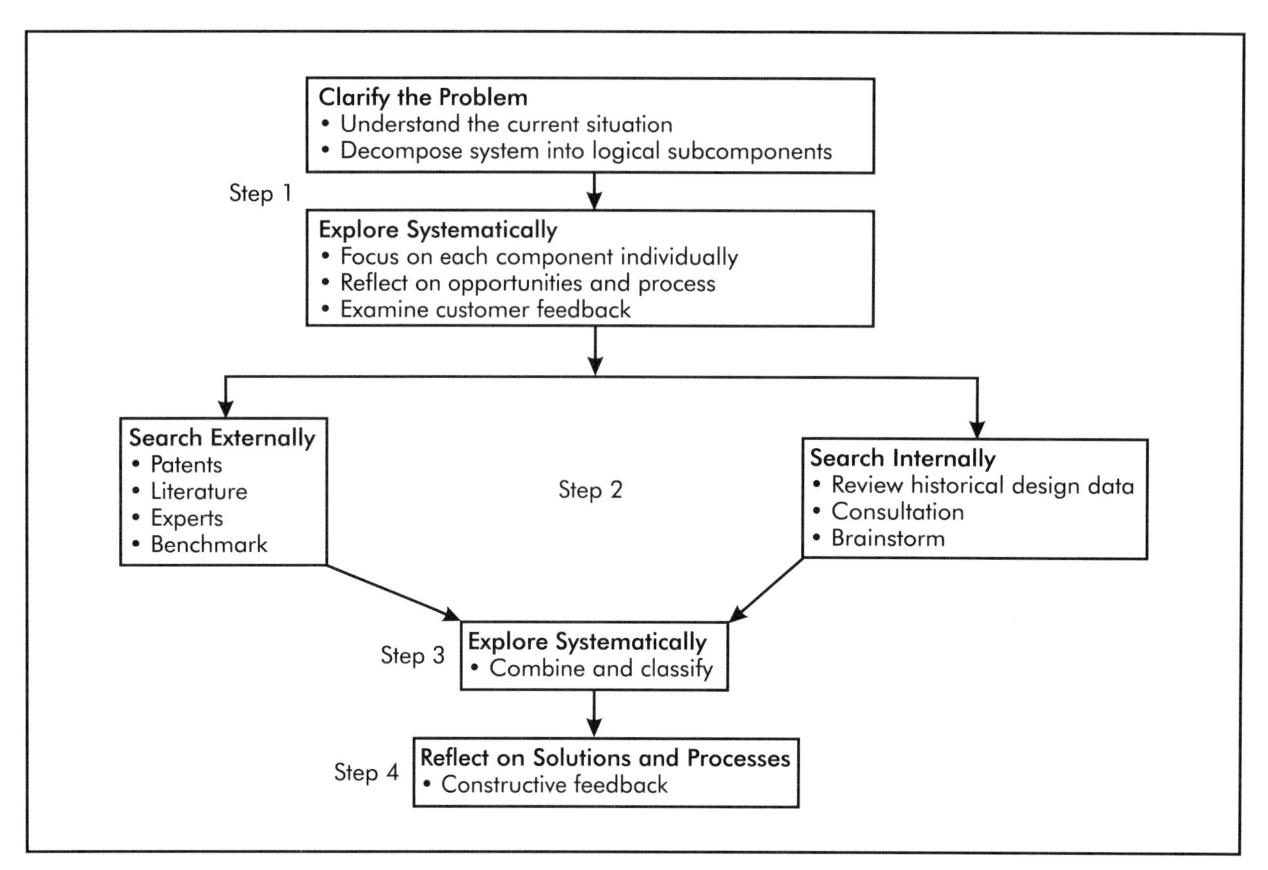

Figure 2-4. Concept generation methodology (Ulrich and Eppinger 1995).

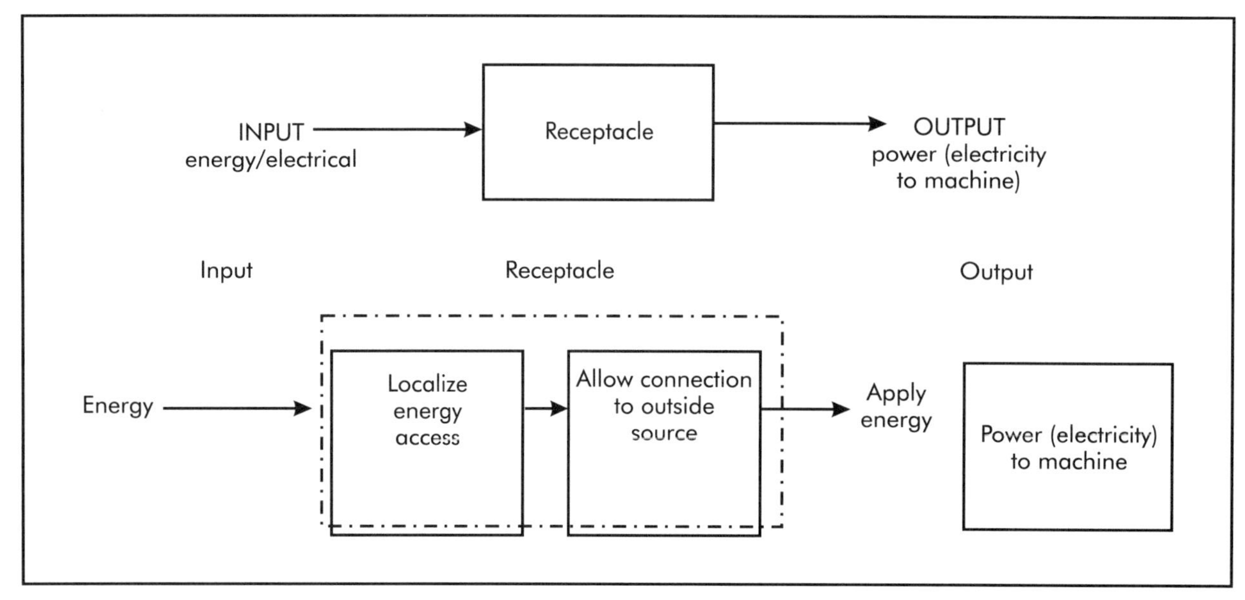

Figure 2-5. Function diagram of an electrical receptacle.

1. localize energy access, and
2. allow connection to outside source.

For example, the problem of a receptacle design can be stated and restated in a number of ways by varying the stress pattern. For example:

- A *receptacle* is too time consuming to install. (This will make the designer think about other products that are faster to install.)
- A receptacle is too *time consuming* to install. (Can it be made faster to install?)
- A receptacle is too time consuming to *install*. (This will focus attention on the installation process.)
- Develop a receptacle that is *easy* to wire and install. (Emphasis is on the design features.)
- Develop a receptacle that is easily *moved from place to place*. (Emphasis is on portability.)

Table 2-6 shows a morphological chart that arranges the subfunctions and their proposed solutions.

The means of achieving the subfunctions are shown in Table 2-7. By examining these solutions, the designer formulates the right combination of sub-features and components.

Hand-held Nailer

Karl Ulrich and Steven Eppinger have discussed the example of a hand-held nailer, a function diagram for which is shown in Figure 2-6 (Ulrich and Eppinger 1995). It operates using a solenoid, which compresses a spring and then releases it repeatedly to drive the nail with multiple impacts. The motor winds a spring that accumulates potential energy, which is then delivered to the nail in a single blow.

The motor repeatedly winds and releases the spring, storing and delivering energy over several blows. Multiple solutions arise from combining a motor with a transmission, a spring, and single impact as shown in Figure 2-7.

Cordless Drill/Driver

The cordless drill/driver unit shown in Figure 2-8 is being developed for the home repair market. After establishing a set of customer requirements and target requirements, product specifications are established. The cordless drill/driver unit will have a 12-volt supply and will be provided in two speed ranges of 0–600 rpm and 0–1,500 rpm. The unit is for heavy-duty use, and the maximum drill size is 3/8 in. (9.525 mm). The unit will have an adjustable clutch and electric brake for disabling it. It is also operable with batteries.

A function diagram (see Figure 2-9) represents the problem, with inputs and outputs shown around a gray box. The gray box is operated on input represented by an AC/DC input and a start/stop signal. Rotational energy, inserted drill bits, and adjustable torque control represent output. Figure 2-10 depicts the functional decomposition of the cordless drill/driver.

Table 2-6. Morphological chart for electrical receptacle

Sub-function	Solution		
Connect to building wiring	Terminal post	**Insulation displacement**	Soldering
Accept plug	**Contacts**	Stake on terminals	Clamps
Provide separation from electricity	**Plastic**	Stone	Wood

Table 2-7. Combination chart for electrical receptacle

Feature	Means		
Mounting	Screws	Snaps	Glue
Electrical contact	Terminals	Contacts	Welding
Wiring	Soldering	Terminals	Insulation displacement
Looks	Flat	Rounded	Obtrusive

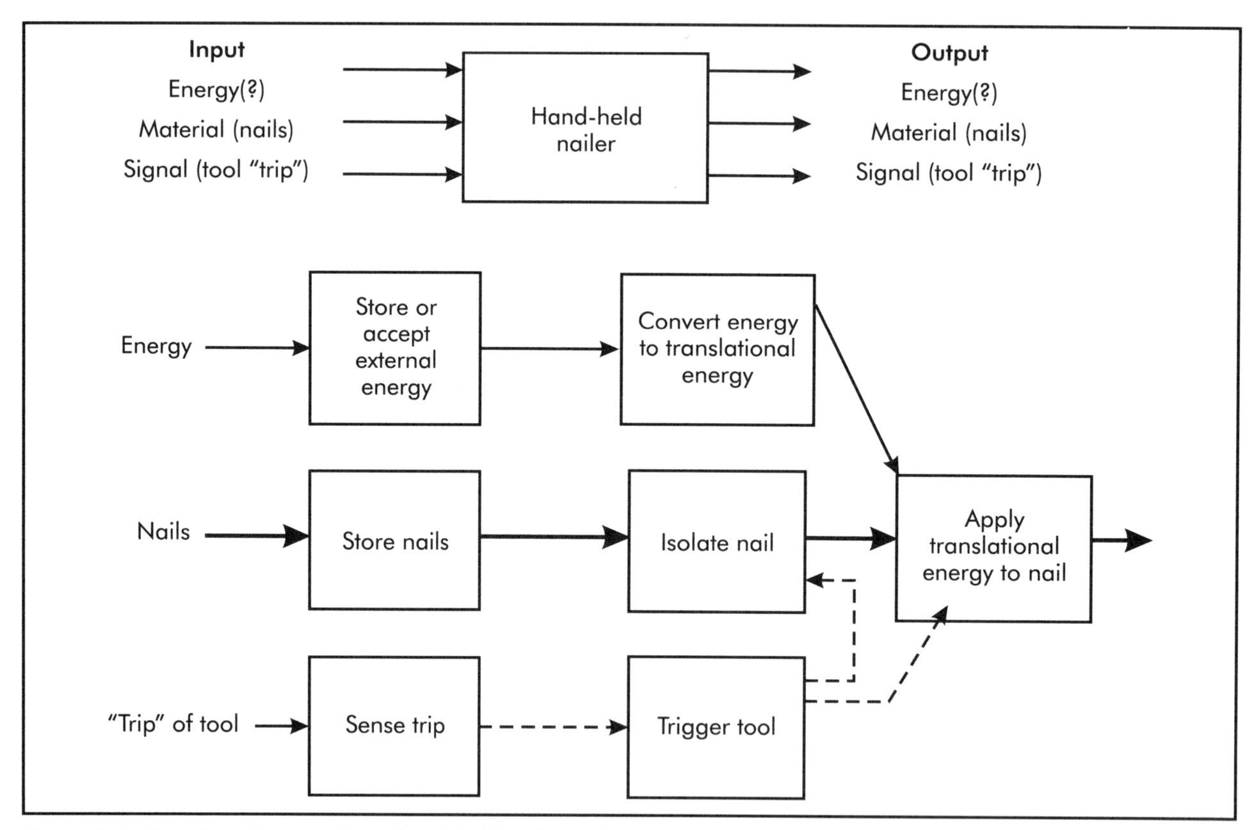

Figure 2-6. Function diagram for a hand-held nailer (Ulrich and Eppinger 1995).

Concepts are classified, as shown in Figure 2-11, and used to identify alternative solutions to the energy source, which could be electric, chemical, hydraulic, pneumatic, or nuclear. The choice of an energy source is narrowed down to three or four sub-concepts. Multiple solutions arising from combining the rechargeable batteries/power pack, fuel cells with torque sensors/piezo-resistive sensors, and variable speed drives are available. In this solution, rechargeable batteries are used along with a combination of torque sensors and a variable clutch.

DESIGN FOR FUNCTION

Design for function is of interest not only to product designers, but to the people involved

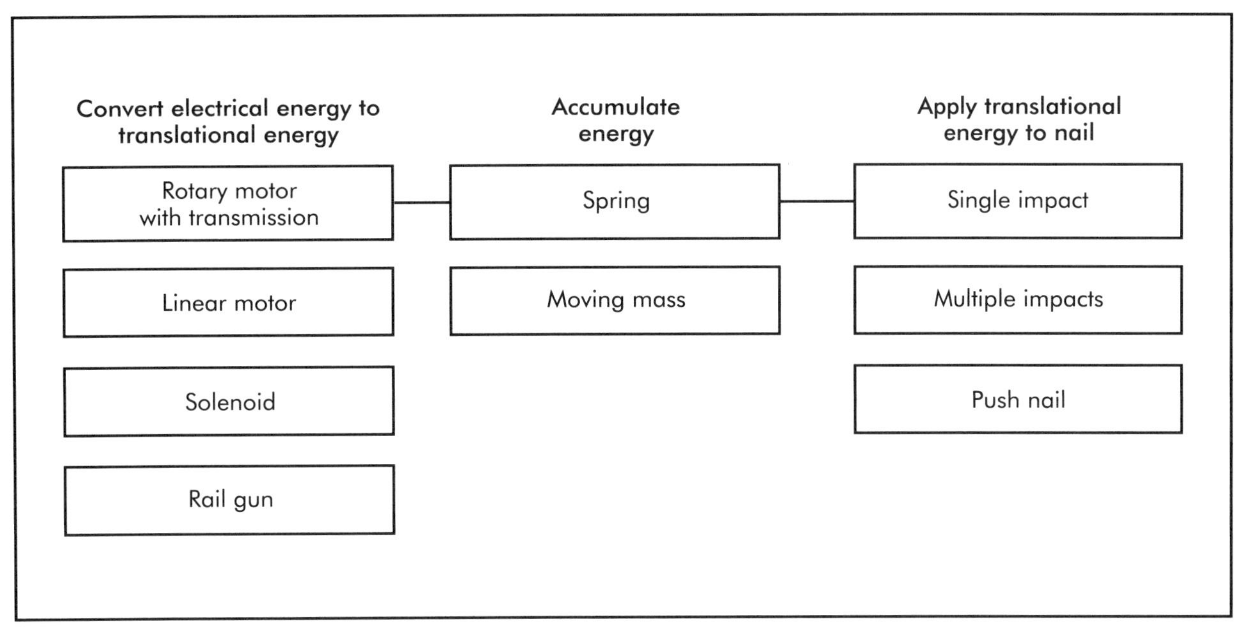

Figure 2-7. Multiple solutions for the hand-held nailer (Ulrich and Eppinger 1995).

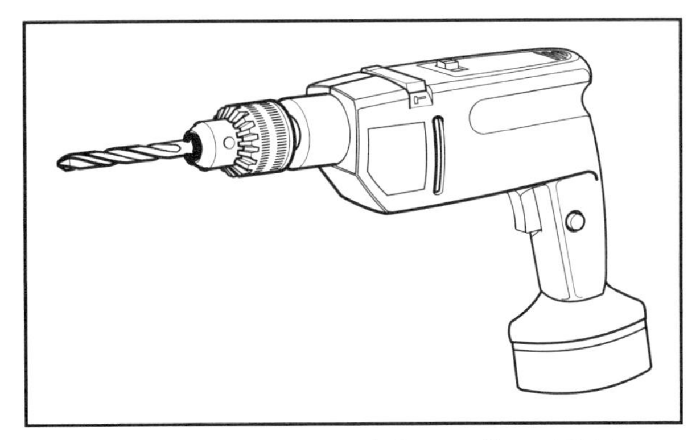

Figure 2-8. Schematic view of the cordless drill/driver.

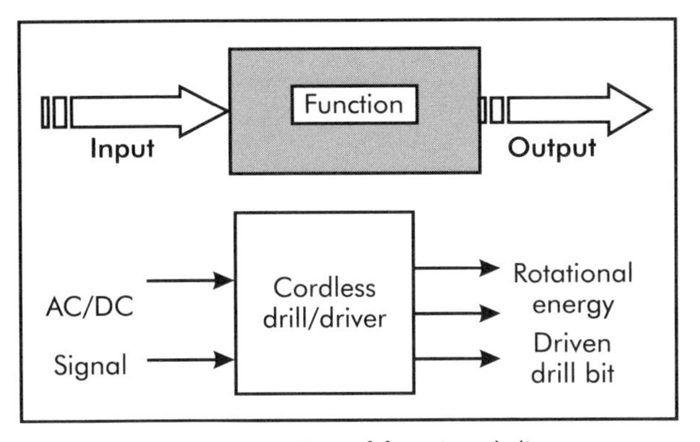

Figure 2-9. Representation of functional diagram.

in several business aspects—including product improvement, process improvement, and quality improvement. The designer should understand the function of a product and its parts before analyzing how to improve the assembly or manufacturing processes. The functional description of a product is its description at an abstract level. Function can be described in normal language, as a mathematical expression, or as a black box.

Function analysis methods such as value engineering, function analysis system technique (FAST), and verbal models are well represented in industry. The first two methods—value engineering and function analysis system technique—focus more on the design of individual parts than on redesigning the entire product.

Value engineering techniques have been in use for many years. Such techniques define the function, value, worth, and cost of the product, as well as individual parts along with their functions. Value engineering effort has most often focused on reducing the

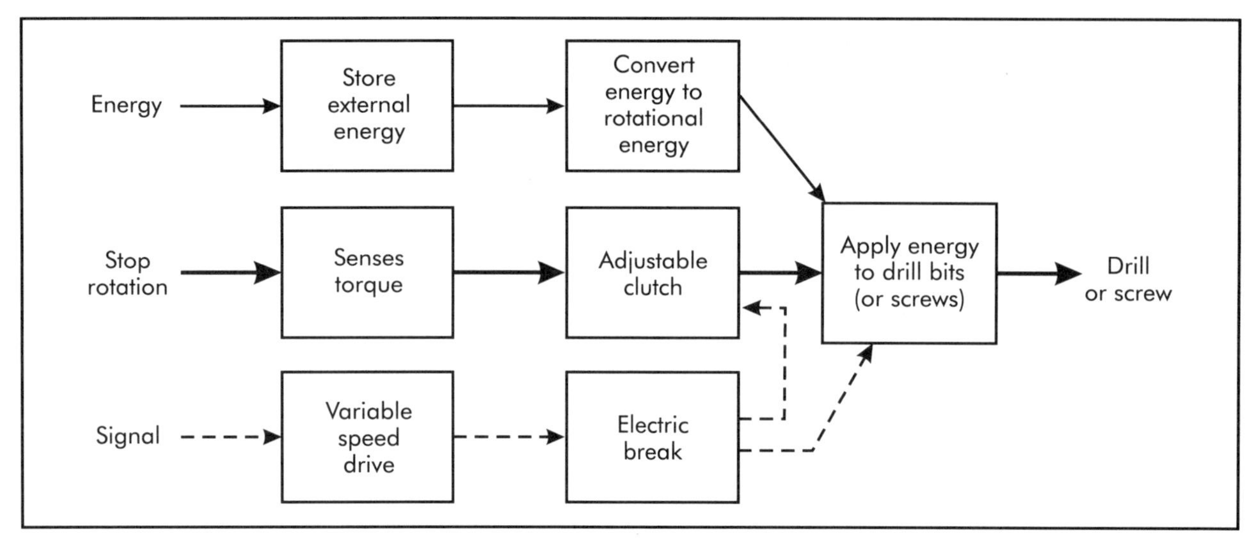

Figure 2-10. Functional decomposition of cordless drill/driver.

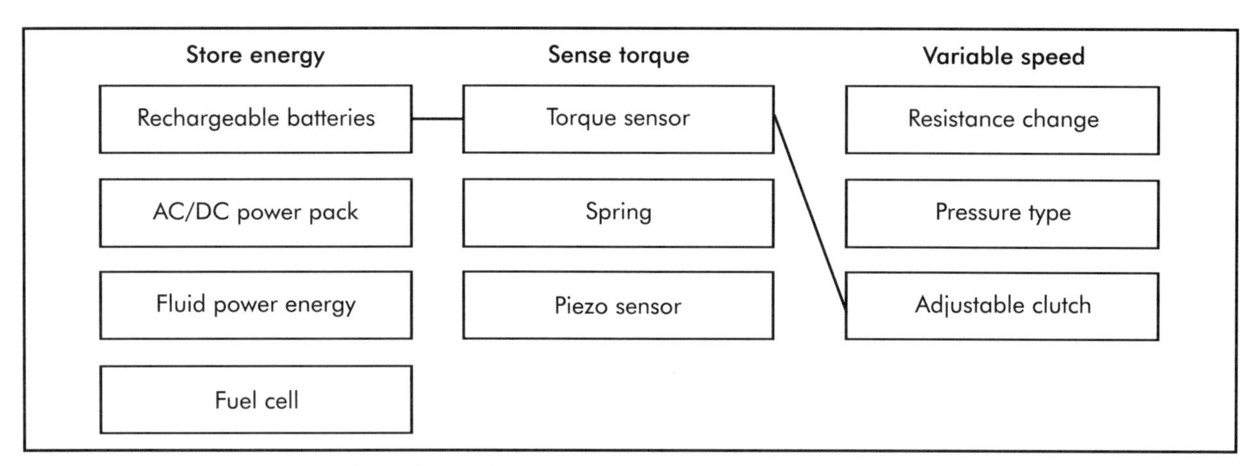

Figure 2-11. Multiple solutions from the combination of ideas.

material and manufacturing costs of individual parts (Miles 1972).

FAST identifies the functional relationship between parts at an individual part level. This technique is less effective as a design tool for manufacturability because it focuses on design at the individual part level. Function analysis using verbal models was proposed by Kaneo Akiyama. The objective of defining the function of a product is to identify, in generic terms, the interrelationship of that product with its environment (Akiyama 1991). Examples are shown in Table 2-8. The function is simply defined with two words: a verb and noun.

Verb: What does it do?
Noun: To what does it do this action?

In the concept design stage, two things are known: function and form.

Function: The designer specifies the function of the product by studying a customer's needs and desires.

Table 2-8. Verbal model of function analysis (Akiyama 1991)

Product	Function Verb	Function Noun
Clock	Decorate	Wrist
	Indicate	Time
Fountain pen	Release	Ink
	Store	Ink
Fuse	Cut off	Excess current
Binder	Hold together	Document
Travel report	Convey	Travel activities
Receipt	Confirm	Amount paid

Form: The designer generates several different concept designs (forms) that may satisfy the customer's needs to a greater or lesser degree.

Function Analysis for Product Design

Function analysis provides a clear picture of the objectives of design. It is a loosely structured methodology derived from previous ideas. It is based on customer-derived functions rather than engineering-conceived forms. Customer functions are translated into product functions, manufacturability is evaluated through analysis, and alternate products are created. Function-analysis methodology has seven major steps in the product-design process. The steps are:

1. Determine the needs, desires, and views of the customer through a customer/competitor analysis.
2. Establish a need for each product function from the perspective of the customer.
3. Translate the customer's needs, desires, and views into a functional product.
4. Develop a symbolic image of the product by constructing a function-family tree.
5. Perform design for assembly analysis to identify manufacturability difficulties.
6. Use creative thinking techniques to generate new concepts based on function.
7. Select a product design based on function and form, using the concept selection process.

The importance of using a function design procedure is that it frees team members from a vision of the old product and allows their minds to create a totally new product based solely on function.

Function Types

There are six types of functions: use, basic, secondary, aesthetic, necessary, and unnecessary functions (see Figure 2-12). Each function type is defined as follows.

- *Use functions* relate to functions involving the purposes or goals.
- *Basic functions* are the primary characteristics of a product or a part that fulfills a user need.
- *Secondary functions* are required to allow: a designer the choice of means for accomplishing the basic function; functions of a part that are not absolutely needed for basic product function; and functions that satisfy only the user's desires.
- *Aesthetic functions* provide only esteem appeal and occasional customer

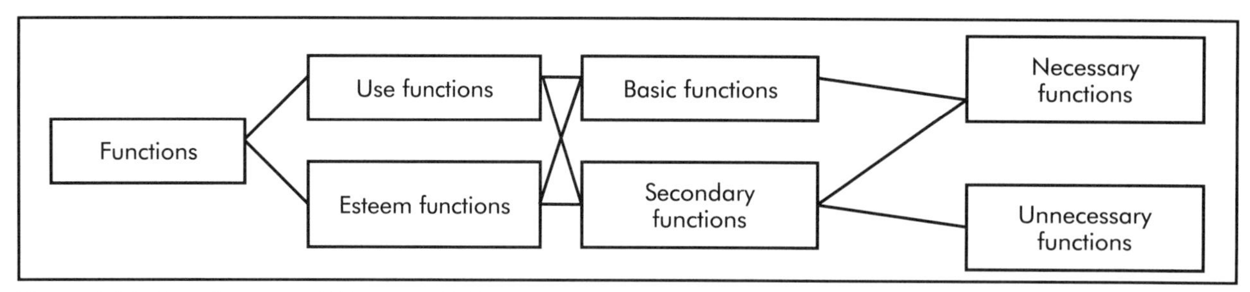

Figure 2-12. Function types (Akiyama 1991).

preferences, and are intended to give sensory satisfaction to the user.

- *Necessary functions* are those demanded by the customer.
- *Unnecessary functions* satisfy neither the needs nor desires of a customer in today's market. As designers consider the cus-tomer's perspective, it becomes apparent that a category is needed for product features that may have been useful at one time, but today have no value.

The function of an overall product is first defined and then it is followed by a definition of the subfunctional group of parts. This is followed by detailed functions of each part. A subfunctional group often utilizes a portion of parts in several subassemblies. Parts may work together to perform a function even though they are not assembled together into the same subassembly. Subassemblies indicate only the order in which parts are joined together on the assembly line, not the functional relationships between the parts.

Function Family Tree

The function family tree (FFT) is a block diagram indicating the functional relationship between a product and its environment, or between various product segments. The FFT is particularly important in the product concept phase because it allows for breaking major products into smaller, more manageable design problems. It also gives more freedom in design by removing any visual form from the function, allowing a view of function in a new perspective (see Figure 2-13).

The development of an FFT begins with a definition of the system boundaries at a level of interest. At the product level, system boundaries are easy to distinguish. Input of the user to the product represents the left boundary. The output of a product to its environment represents the right boundary. At the product level, system boundaries are the interface between the outside world and the product.

The function of the open/close circuit is to send current. The goal is to provide light. A means to the goal of providing light involves converting electricity to light and sending the current.

Concept Selection

In the Pugh concept-selection methodology, product-design criteria are written in the form of customer functions (Pugh 1981, 1991). Product functions should be listed from the top down, starting with the primary functions, then the secondary functions, and finally, the aesthetic functions. Concept sketches should be arranged simply, in the order that they were generated. The description of the function should be brief, open, and understandable by both the designer and the customer. The function of a product is defined clearly in a two-word, noun-verb descriptor. The methodology consists of the following:

1. Define the functions of the product based on the needs and desires of the customer.

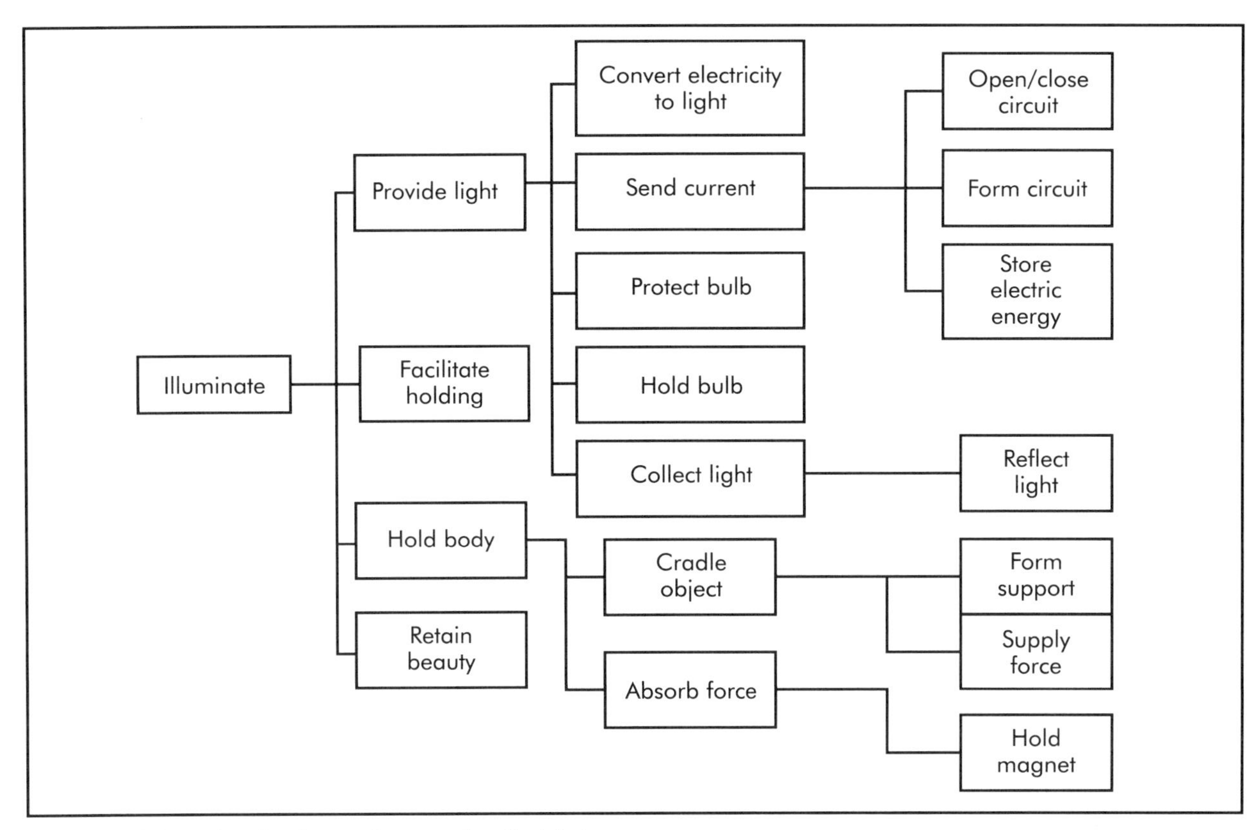

Figure 2-13. Function family tree for a pocket flashlight.

2. Using creative-thinking techniques, develop a dozen different design concepts (forms) that might satisfy the functions. All concept designs must be developed to the same degree of detail as their sketches. Concepts need not be in a final form.
3. Establish a concept comparison and evaluation matrix to compare each concept to a datum.

Table 2-9 shows the basic Pugh concept selection matrix. In the first step of the Pugh concept selection process, concepts are evaluated relative to the original product-design reference datum. If this were a conceptual design stage and a new product creation, one of the new designs would be chosen as a datum. Each concept is compared function-by-function to the datum design. The concept designs are not compared among themselves. They are only individually and separately compared with the datum. Evaluation can be made using the following criteria:

- +: This concept is clearly better than the datum.
- –: This concept is not as good as the datum.
- S: This concept is (about) the same as the datum.

If no clear concept selection can be made after the first evaluation, a second selection process is performed. Three or four most promising design concepts are now compared. One of the good redesign concepts is selected as the datum and each redesign concept is compared function by function with the new datum. The concept with the high-

Table 2-9. Selection matrix (Pugh 1981)

	Concept							
Functions	1	2	3	4	5	6	7	8
A								
B								
C								
D								
E								
F								
Score								

est score is the one that should be used for redesign (Pugh 1981, 1991).

The effects of a function analysis on product design are summarized in Table 2-10. Product design using function analysis is an integration of different concept development techniques. The importance of using this methodology is that it frees team members from a vision of the old product and allows their minds to create a totally new product based solely on function.

UNDERSTANDING THE CUSTOMER

Customer-driven Product Development

During the past two decades, the emergence of a competitive global economy, markets influenced by customer product preferences, and technological change has caused a major shift in quality. The quality management practices of Japanese and American companies provided an opportunity to influence the cost and lead times of

Table 2-10. Effects of function analysis on product design

Product and Design Process Activities	Effects of Using Function Analysis
Recognition of the object of design	Functional perspective provides a clear picture of the goals of the design.
Creation of design	Design develops smoothly based on functions.
Evaluation of design	Functional perspective provides clear criteria for design evaluation.
Transfer to drawings	It becomes easier to create drawings when demands are seen as functions.
Assignment of functions	Function analysis provides a clear identification of product function assignment.

new products. Early quality initiatives focused on reducing process variability in manufacturing. Later efforts focused on using concurrent-engineering philosophy in the company for product and process development. Advantages that come from cutting the time to market and continuously developing quality products are so great that the balance in some sectors is shifting in favor of companies that adopt new strategies. Companies that introduce new products and react quickly to external changes are racing ahead of their competitors.

One essential element affecting the entire development process is the proper determination of customer needs. Worldwide competitiveness has brought a greater focus on customers' views. The customer needs analysis projects future needs, not merely current needs from the marketing department. Encouraging the product design team to participate in customer needs analysis can enhance its creative contributions. It enables the team to see opportunities that it might not see by simply reading a market report. Gathering information from customers involves getting information from both internal and external customers. Internal customers are normally people that are connected with corporate management, manufacturing personnel, sales, and field service. External customers are normally the end users of the product.

Quality Function Deployment

The *quality function deployment* (QFD) methodology is ideally suited for supporting a total quality initiative in a company. It provides a framework for product or program design, which starts with customers and finds out what they want. It identifies a mechanism so that the organization can respond to customers' needs. QFD methods are now employed by some larger U.S. companies to obtain better quality products in a shorter cycle time.

One of the objectives of any company is to bring new products or improved products to the market with low cost and high quality sooner than its competitors. The procedure of QFD, based on the *house of quality*, provides a way to do this. QFD as an organized procedure is used for early product design. Basically, this technique involves deploying customer requirements and expectations into product design characteristics and then monitoring them through the stages of design, planning, and manufacturing. More precisely, QFD utilizes a conceptual map or group of matrices known as the *house of quality* to relate customer attributes to product specifications and design operations.

In QFD, the term *quality* takes on a much broader meaning than the more conventional definition, which is limited to a product's adherence to a manufacturer's specification. The basic definition of QFD, which is translated from the Japanese words—*hin shitsu ki nu ten kai*—is "a system for translating customer requirements at each stage from research and product development, to engineering and manufacturing, to marketing/sales and distribution." Ultimately, the customer receives a product that meets his or her demands with a minimal amount of design changes (Adams 1976).

What makes QFD a powerful tool in the manufacturing sector is direct input from the customer, along with interdepartmental communications between marketing, engineering, manufacturing, and management. In a typical company, the interdisciplinary group—usually referred to as the quality team—is charged with the responsibility of integrating customer requirements into the product design. This group usually meets on a regular basis and conducts brainstorming sessions. The end result is a system that delivers a product meeting specific quality standards.

QFD is totally driven by the concept of quality and results in the best possible product being brought to the market. It requires a paradigm shift from traditional manufacturing's quality control to product design quality control. The old paradigm involved quality control by inspecting physical products through observation and measurements; this is *inspected-in quality*. But the new QFD paradigm designs quality into products and manufacturing processes so that products are produced error-free; this is referred to as *designed-in quality* (see Figure 2-14.)

The advantages of using QFD are:

- reduction of product design time;
- cost reduction;
- exposing design tradeoffs early;
- providing written documentation of design decisions;
- reducing design errors and corrections;
- providing clarity for decisions; and
- incorporating the collective experience base of a multifunctional team capable of making sound decisions.

QFD is very essential in current industry, as customer demands tend to vary with changing times. This process needs to be incorporated in the design accordingly and accurately. It is the only comprehensive quality system aimed specifically at satisfying the customer. Further, QFD allows customers to prioritize their requirements and optimizes those features that bring the greatest competitive advantage.

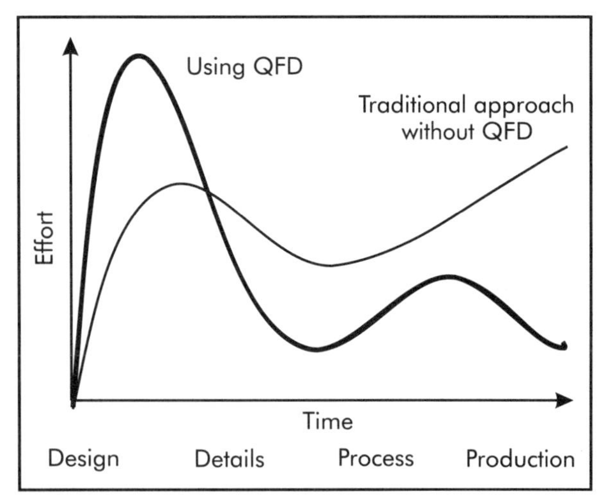

Figure 2-14. QFD versus traditional methods.

Phases

The four phases of QFD are:

1. product planning,
2. part deployment,
3. process deployment, and
4. product deployment (see Figure 2-15).

In the product-planning phase, the customer attributes are drawn based on surveys, interviews, observations, field contacts, focus groups, employee feedback, publications, and sales records. These attributes are converted into product characteristics. A relationship matrix between the customer requirements and the product characteristics is then drawn. This matrix has information about a competitor's products and market evaluation data—including the customer's expressed importance ratings. The matrix has information on current product strengths and weaknesses, measurable targets to be achieved, and selling points.

In the part-deployment phase, product characteristics are translated into component characteristics. At this stage, characteristics of the final product are converted into part details at the component level.

In the process-deployment phase, the process plan for manufacturing the component, subassembly, and assembly are identified, as well as the quality parameters. In the production-deployment phase, output from the process-deployment charts provides a measure of critical product and process parameters. At this stage, production operations for all critical components are identified. QFD identifies and prioritizes the voice of the customer; the theory of inventive problem solving (the Russian acronym TRIZ) helps to create new concepts (see Table 2-11).

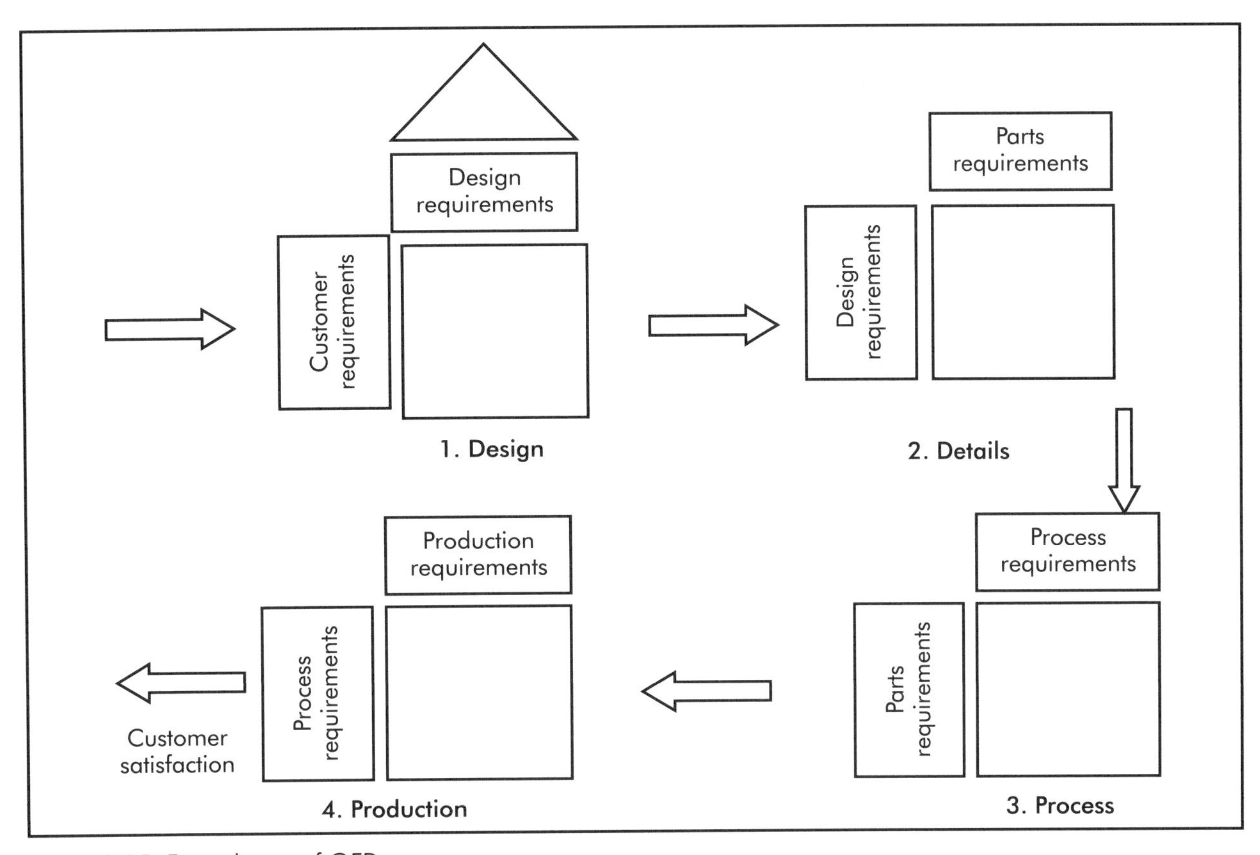

Figure 2-15. Four phases of QFD.

Table 2-11. Combining QFD and TRIZ

Development Phase	Benefit of Combining QFD and TRIZ
Market research	Use directed product evolution (DPE) with concept methods to show customers what new products will be like.
Preliminary research	To solve engineering bottlenecks and contradictions To eliminate contradictions discovered by roof of the house of quality To help determine target values in the quality planning table
Design	Use substance-field (Su-field) analysis and DPE to identify new functions to attract customers. Use anticipatory failure determination (AFD) to identify and prevent failure modes in new products. Use TRIZ to develop new concepts by DPE patterns. Use TRIZ to lower costs without resorting to tradeoffs.
Manufacturing	Remove design constraints due to limitations of equipment and manufacturability.
Production	Remove design constraints due to limitations of processes and people.
After service	Help in design for serviceability; remove bottlenecks.

House of Quality

The central body of the house of quality consists of the "whats" (what the customer needs), the "hows" (what the manufacturer controls), and the matrix of the relationships between the "whats" and the "hows" (see Figure 2-16). The customer's requirements, needs, and wants are known as the "whats." The counterparts of these technical characteristics are known as the "hows."

Attributes of the product. A product planner should develop a list of customer requirements for the product. The list can be obtained through surveys, interviews, observations, field contacts, focus groups, employee feedback, publications, sales records, and complaints. He or she should break down customer attributes into specific items or subdivisions, such as what is primarily important and what is secondary in nature. Each customer attribute should be weighed by its degree of relative importance. To get an objective weighing, a brainstorming session among product group members has to take place.

Product control characteristics. The planner should develop the engineering characteristics of the product that affect customer attributes. This requires that the product team conduct another brainstorming session. Engineering characteristics need to be broken into specific items as required

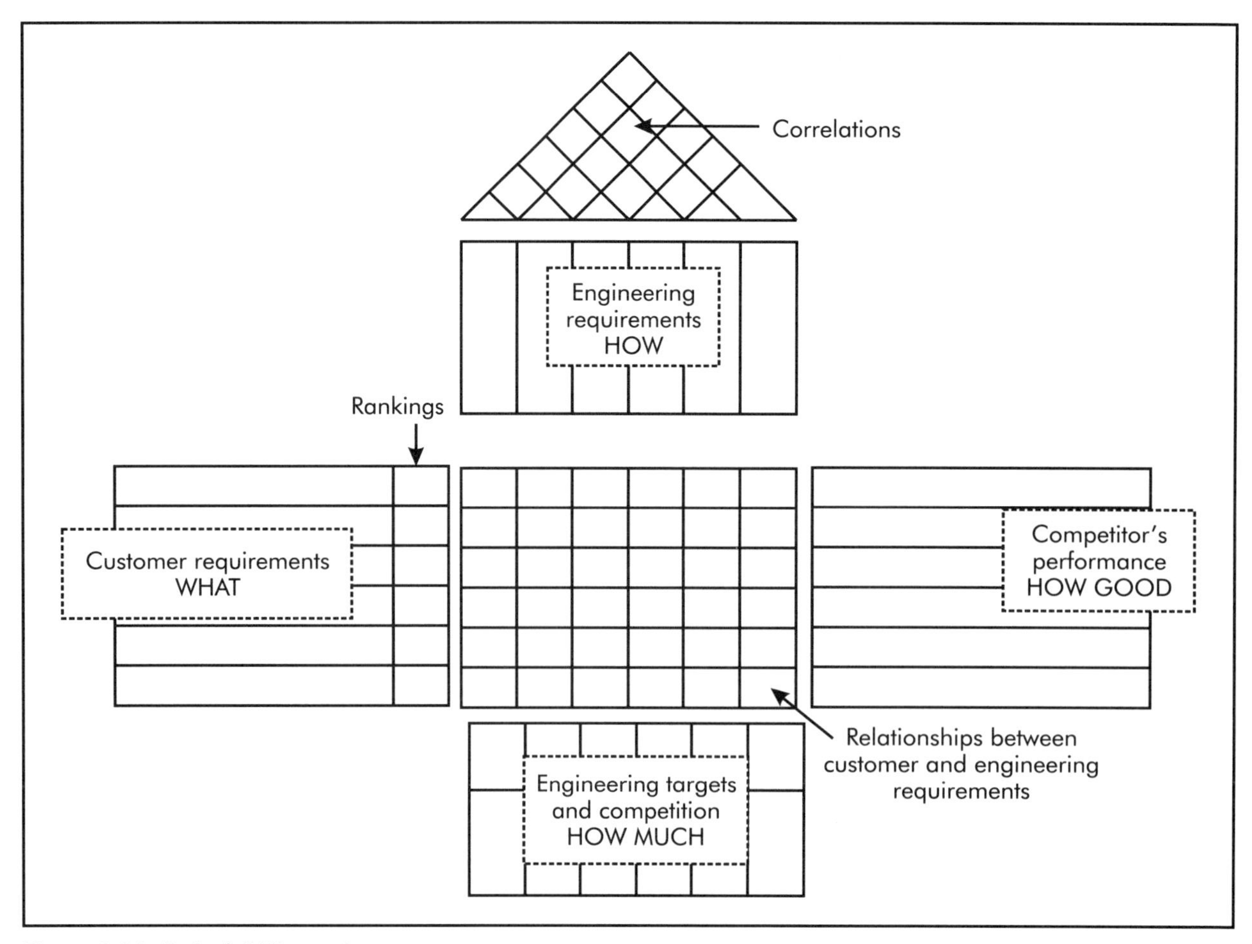

Figure 2-16. Typical QFD matrix.

and product characteristics need to be listed on the axis across the top of the house (see Figure 2-17).

Interaction matrix. The product group combines the two axes into a correlation matrix identifying the strong, medium, and weak correlations. A set of symbols is used to represent the relationship as shown in Figure 2-17. If the matrix shows a majority of weak relationship signs, it is an indication that certain customer requirements have not been met.

Interaction between parameters. The roof of the house of quality should be developed by building a diagonal matrix above the engineering characteristics. This allows the product team to rate weak-versus-strong relationships between different characteristics. Changing a parameter can influence other parameters. It is important to know the nature and strengths of these interactions.

Target values and technical analysis. The basement of the house of quality can be used when objective measurements need to be made of competitors' products and when it is necessary to compare the specifications of a company's product with a competitor's product. This kind of comparison provides an insight into the possibilities of product improvement and assists in setting up new target values to be followed. These components also allow difficult-to-meet and important requirements to be passed from one matrix to the next, thereby keeping a focused effort on design and manufacturing (see Figure 2-18).

QFD/House of Quality Steps

Step 1: voice of the customer. Identify the customer's needs, wants, and requirements. This ensures that product design decisions will be based on the customer and not just on the perceived customer needs. This step should involve all groups in a company that get any feedback from customers. The relative value that customers place on these items also should be identified.

Step 2: customer requirement refinement. The customer requirements that a product be dependable and economical can be expanded to more specific points (see Figure 2-19). This can be done with a "what-to-how" technique. A chart can be constructed. The list should be expanded until each point is a measurable quantity.

Step 3: begin laying out the planning matrix.

Step 4: fill out correlation matrix to determine how factors relate. Establish positive changes to any one of the control characteristics that affects other members. The product and process should be reconsidered if there are more negative than positive effects.

Step 5: complete the relationship matrix and importance rating values. A relationship between the "whats" and the "hows" is established through a matrix that assigns weights. For example, 0-9, where 0 is none, and 9 is very strong. The column values are added to give an importance rating. This should result in a few important features, and a few that are not important.

Step 6: customer importance rating and market evaluations. The opinions of the customer (as collected in Step 1) are quantified in terms of the importance of the requirement's "whats." Numbers are entered in the customer importance rating column. Overall ratings for products of the company and its competitor are ranked for each requirement from poor to good. These values are derived from information gathered in Step 1 and are entered in the market evaluation column. The sections clearly identify the strengths and weaknesses of the product.

Step 7: control characteristics' competitive evaluation. Competitor products and the internal product are compared technically here. Performance criteria are shown

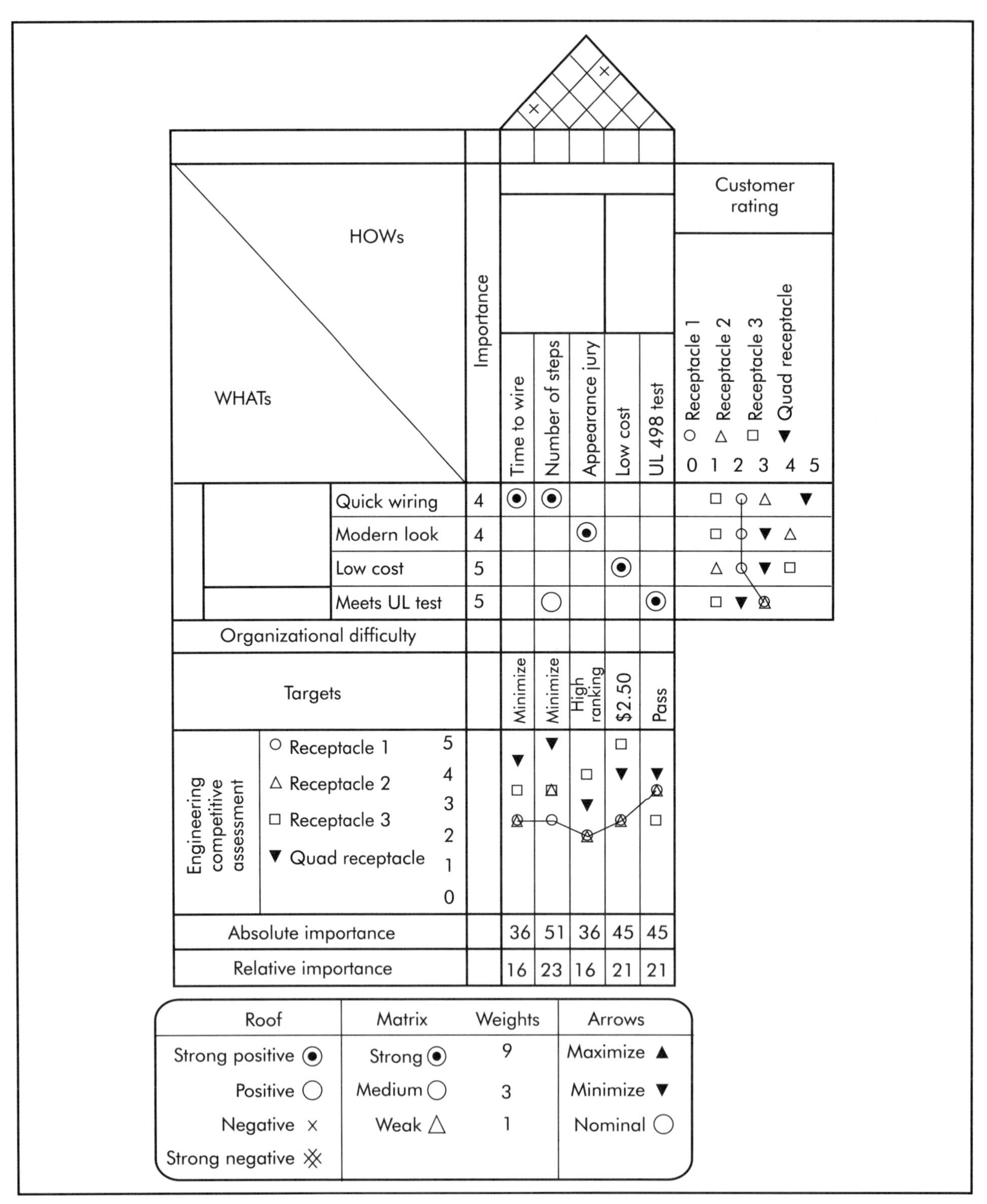

Figure 2-17. House of quality developed for a receptacle.

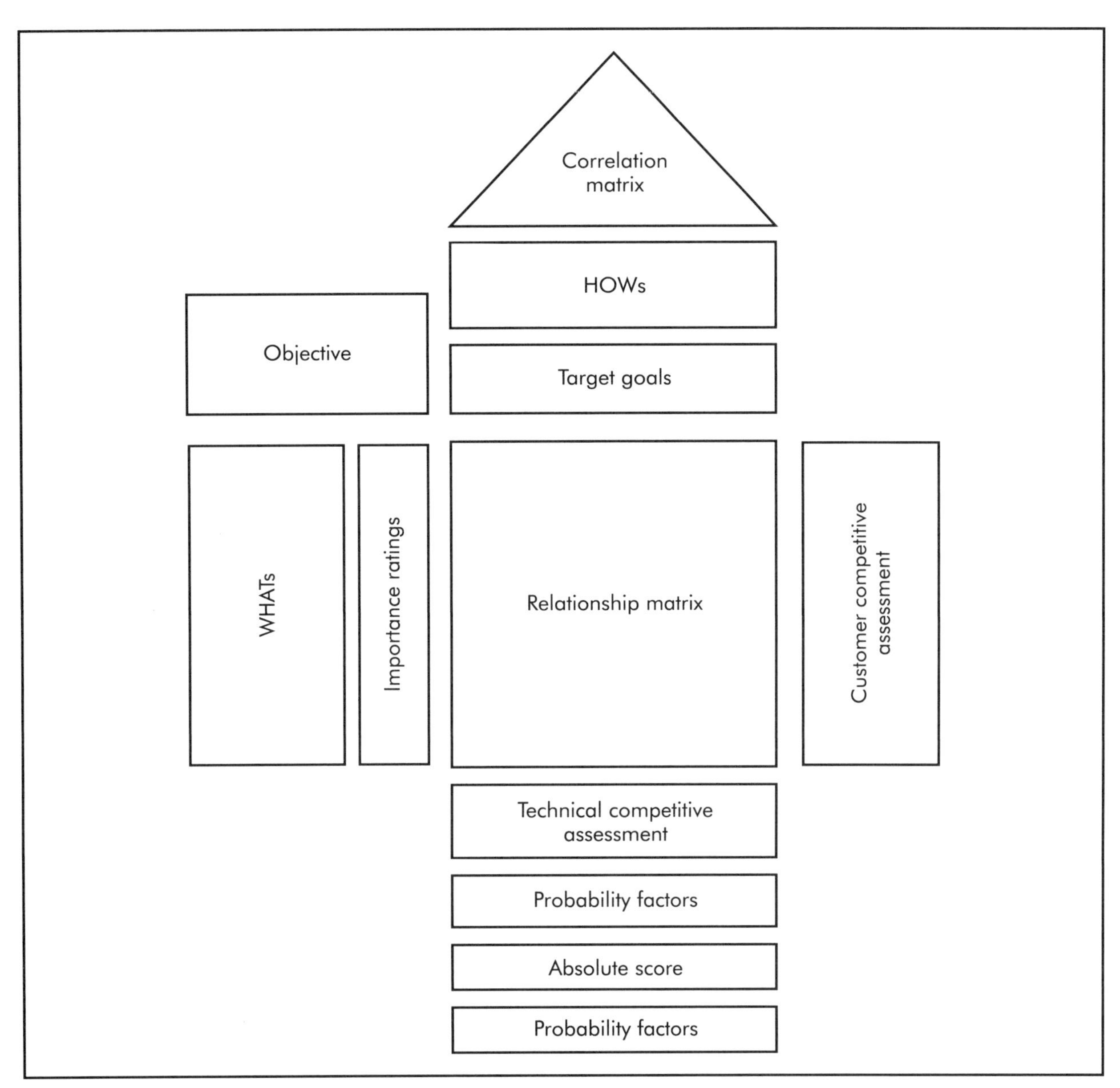

Figure 2-18. Components of a QFD model.

in terms of the final product's control characteristics. Values are entered in the control characteristics' competitive-evaluation section of the chart, and are then ranked from good to poor. When these numbers are compared to numbers in the importance rating row, the technical deficiencies of the product and its importance are clear.

Step 8: evaluate the chart. The chart at this point contains enough information to do some critical evaluations. The control characteristics' competitive evaluation and the importance rating indicate items that should be designated for a higher rating when the competitor product is rated better. The candidates selected should be checked

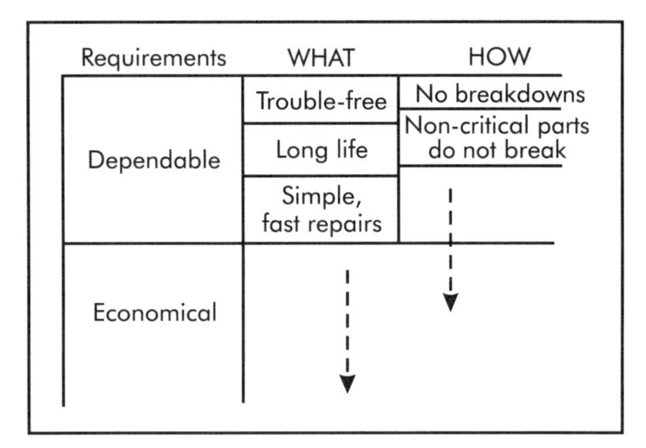

Figure 2-19. Primary, secondary, and tertiary requirements.

to see how they correlate to the customer importance rating and market evaluations. If the choice is not considered important to the customer, then it should be considered to be less important. If any of the customer requirements are unanswered, then the requirements' control characteristic list must be reconsidered.

Step 9: develop new target values. Using current design parameters, along with the relative importance exposed in the last step, new target values should be selected. The values determined for a competitor's product should be used, as well as in-house data for the product. A separate sheet, or document, may be used here because descriptions may become bulky.

Step 10: technical difficulty. Considering target values and previous production performance will help determine how the difficulty of achieving the target value should be estimated. A ranking for this is entered in the degrees-of-technical-difficulty row.

Step 11: deployment selection. Quality has a cost, and at this point the cost/benefit trade-off is made. One or more factors can be selected. If there are not a few clear choices, the process should be re-examined. The main objective is to select elements with the lowest technical difficulties, but with the greatest importance ratings. This decision will be slightly arbitrary, but it should not be far outside of what the chart suggests.

Step 12: deployment matrices. A deployment matrix is developed for each control characteristic selected in the last step. The top of the matrix is developed using factors discussed in the development of the planning matrix. The relationship between testable components and the "whats" of durability are inserted here. The control characteristics measured for all products and the target values are positioned below. The bottom matrix consists of system components that can be affected by the design. On the left are systems they affect and components in those systems. On the right are the measurable variables. In the center are the locations to track the relative quality of the components.

Step 13: design and test. The deployment matrix is used to do design work, test results, and compare the results to target values. The house of quality procedure is inherently a group work approach designed to ensure that everyone works together to give customers what they want. It has changed the way people think and brings quality into products and the manufacturing processes.

Case Example

The following is a case study of QFD techniques applied to an air cooler, resulting in its redesign and improvement (see Table 2-12).

The air cooler, unlike an air conditioning system, is used to control the humidity factor in areas having very dry climates. It is used in these regions, especially in homes, to introduce cool moist air. The design under consideration has a rectangular shape. A separate reservoir stores water (see Figure 2-20) and with the help of a one-way valve maintains a constant water level. An absorbent material (filter) then absorbs the water. A fan situated on one side of the filter

Table 2-12. Application of TRIZ contradiction matrix (partial representation)

	What Deteriorates (Undesired Result)			
What Should be Improved	Length of Movable Object	Weight of Movable Object	Length	Shape
Length of movable object	X	8, 15, 29, 34		1, 8, 10, 29
Weight of movable object		10, 1, 29, 34		
Length	8, 15, 29, 34		....	
Shape	1, 8, 10, 29		29	X

1 = Principle of segmentation
8 = Principle of anti-weight
10 = Principle of preliminary action
15 = Principle of dynamics
29 = Use of pneumatic and hydraulic structures
34 = Principle of rejection and regeneration of system parts

blows air through the moist filter. This generates moist, cool-air output from the air cooler. The customer survey indicates the following requirements:

- less noise during operation,
- smaller size,
- low overall cost,
- less maintenance cost,
- safe operation,
- better appearance,
- low power consumption,
- portability, and
- adequate cooling.

A planner could take this further by creating a house of quality and developing a chart showing the relationship between the voice of the customer and the design requirements. Based on the results of the QFD chart, he or she would formulate alternate ideas for the air cooler.

AXIOMATIC DESIGN METHOD

Many times, people identify a distinguishing piece of art or music, but find it difficult to explain why a particular combination of elements in a work causes it to be excellent. In other words, these results lack an absolute frame of reference, which often leads to differing opinions. Many results depend on intuition and experience when humans compose music or design a product or process. It is difficult to reduce these facts and observations to a consistent set of statements and descriptions. Nam Suh proposed the use of axioms to represent design based on the assumption that there is a fundamental set of principles that represents good design practice (Suh 1989, 1990). There are many similarities in the design methods of diverse fields—such as industrial design, architecture, mechanical design, software engineering, and development of management policies. In other words, it can be said that there are a set of common factors in a good design. These common factors can be applied to other design situations like natural laws in science problems.

Suh developed a set of axioms and corollaries to represent design. These were reduced to a set of two fundamental axioms, which, if followed, would result in a good design. This set is based on the following premises:

- *Axioms* are fundamental truths that are always expected to be true.
- *Corollaries* are propositions that follow from axioms.
- *Functional requirements* (FRs) are characterizations of the perceived needs for a product or process. In addition, they are the minimum set of independent

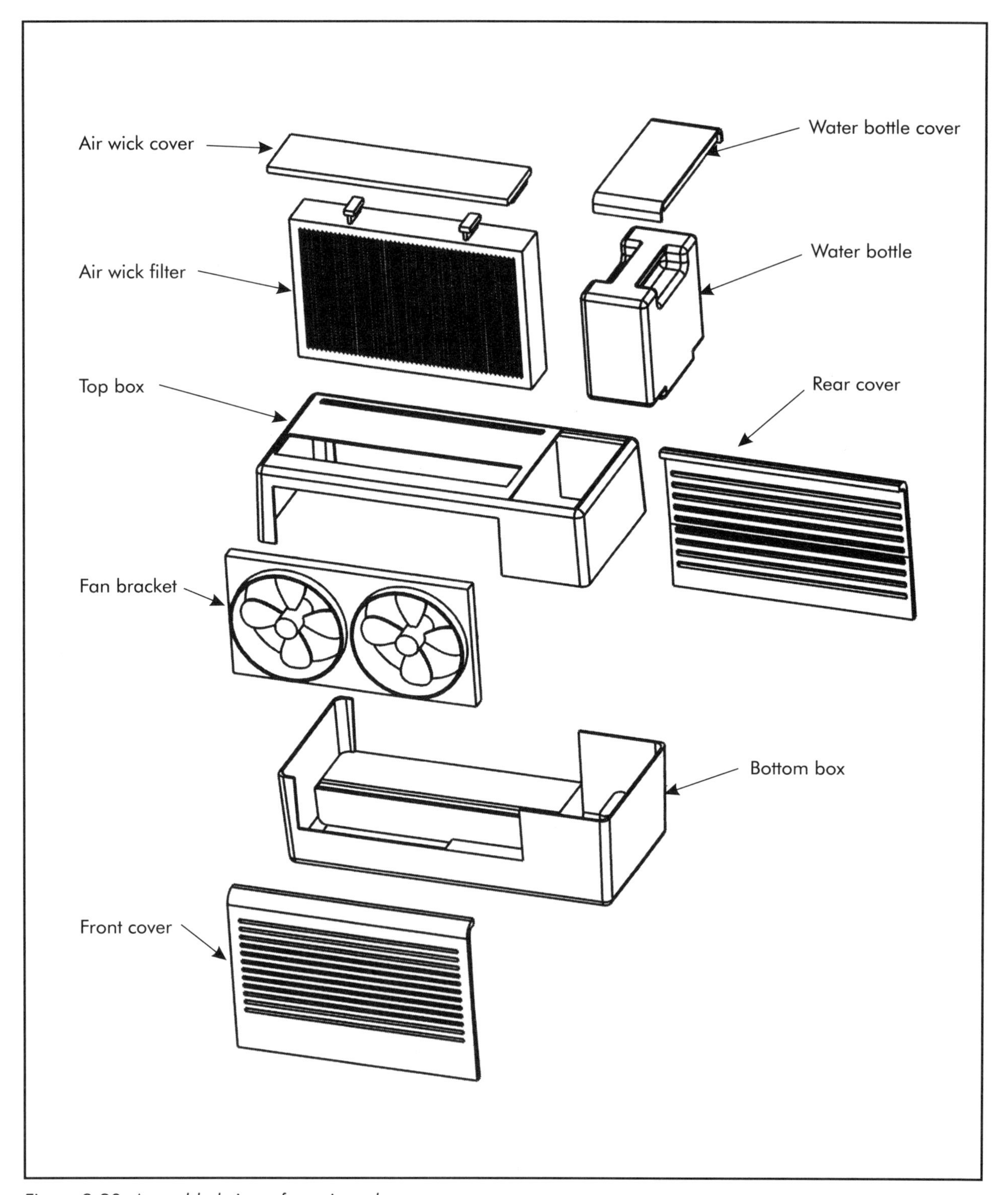

Figure 2-20. Assembled view of an air cooler.

requirements that characterize design objectives for a specific need.

- *Design parameters* (DPs) are variables that characterize the physical entity created by the design process to fulfill the FRs.

Axiomatic Principles

Design begins with a definition of the problem from an array of facts, which are formulated into a coherent statement of the questions. The objective of design is stated in the functional domain, while the physical solution is generated in the physical domain. Design involves what a planner wants to achieve and how he or she wants to do it. The design process links these two domains, which are independent of each other.

The next step in the design process is to determine the objectives of the design by defining it in terms of specific FRs. To satisfy these functional requirements, a physical embodiment is developed in terms of DPs. The design process relates FRs of the functional domain to DPs of the physical domain. This mapping feature between FRs and DPs is illustrated in Figure 2-21. Design axioms provide principles that aid the creative process by enabling good designs to be identified from an infinite number of designs.

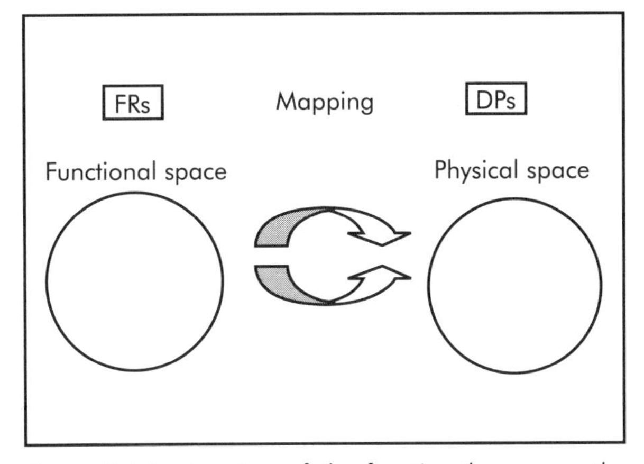

Figure 2-21. Mapping of the functional space to the physical space.

Two main axioms are:

1. The independence of the functional requirements is maintained.
2. The information content of the design is minimized.

The axioms provide an insight into questions like how design decisions are made, and why a particular design is better than others. Axiom 1 is related to the process of translation from the functional to the physical domain. Axiom 2 states that the complexity of a design, once Axiom 1 is satisfied, should be reduced. Questions like whether it is a rational decision and how many design parameters are needed to satisfy the functional requirements are answered. The same principles are used in all design situations, irrespective of whether they are product related, process related, or organization related.

In mathematical terms, the independence axiom can be represented as follows:

$$FR = [DM]\,[DP] \tag{2-1}$$

where:

$[FR]$ = vector of the functional space to the vector of the physical space
$[DM]$ = design matrix relating the functional and physical domains
$[DP]$ = vector of the design parameters

$$[DM] = \begin{bmatrix} x_{11} & x_{12} & x_{13} & \cdots & x_{1m} \\ \vdots & \vdots & \vdots & & \vdots \\ x_{n1} & x_{n2} & x_{n3} & \cdots & x_{nm} \end{bmatrix} \tag{2-2}$$

X_{ij} represents the relationship between each FR_i and DP_j. If the FR_i is affected by DP_j, then X_{ij} has a finite value. If FR_i is not affected by DP_j, then X_{ij} is zero. A design equation and design matrix can be written for each possible solution.

Implementation of the independence design axiom results in a case where every functional requirement is associated with a single design parameter. This is called the *uncoupled design* and is represented by a diagonal matrix.

$$\begin{bmatrix} FR_1 \\ FR_2 \\ \vdots \\ \vdots \\ FR_n \end{bmatrix} = \begin{bmatrix} X000..0 \\ 0X00..0 \\ 00X.. \\ .... \\ 0000X \end{bmatrix} \begin{bmatrix} DP_1 \\ DP_2 \\ \vdots \\ \vdots \\ DP_n \end{bmatrix} \qquad (2\text{-}3)$$

It can be observed from the first axiom that for a design to be uncoupled, it requires that the number of FRs and DPs be the same.

When the matrix is triangular (for example, $X_{nm} = 0$ when $n \neq m$ and $m > n$), the design is a decoupled design. Both uncoupled and decoupled designs satisfy the independence axiom. All other matrices, which do not satisfy Axiom 1, are called *coupled designs*.

Mathematical Relationships

Design parameters can be subdivided into (DP_1, DP_2...DP_n). Functional requirements are also broken down into subfunctional requirements (FR_1...FR_n). A matrix representation of FRs and DPs is shown in Figure 2-22.

Three types of design equations are used to represent the FR and DP relationships.

Type 1—uncoupled. The uncoupled design equation satisfies Axiom 1 (see Figure 2-23).

$$\begin{bmatrix} FR_1 \\ FR_2 \\ FR_3 \end{bmatrix} = \begin{bmatrix} a_{11} & 0 & 0 \\ 0 & a_{22} & 0 \\ 0 & 0 & a_{33} \end{bmatrix} \begin{bmatrix} DP_1 \\ DP_2 \\ DP_3 \end{bmatrix} \qquad (2\text{-}4)$$

Type 2—coupled. The coupled design equation always violates Axiom 1 (see Figure 2-24).

$$\begin{bmatrix} FR_1 \\ FR_2 \\ FR_3 \end{bmatrix} = \begin{bmatrix} a_{11} & a_{12} & a_{13} \\ a_{21} & a_{22} & a_{23} \\ a_{31} & a_{32} & a_{33} \end{bmatrix} \begin{bmatrix} DP_1 \\ DP_2 \\ DP_3 \end{bmatrix} \qquad (2\text{-}5)$$

Type 3—decoupled. Here, Axiom 1 is satisfied, and the independence of FRs is assured if DPs are arranged in a certain order (see Figure 2-25).

$$\begin{bmatrix} FR_1 \\ FR_2 \\ FR_3 \end{bmatrix} = \begin{bmatrix} a_{11} & 0 & 0 \\ a_{21} & a_{22} & 0 \\ a_{31} & a_{32} & a_{33} \end{bmatrix} \begin{bmatrix} DP_1 \\ DP_2 \\ DP_3 \end{bmatrix} \qquad (2\text{-}6)$$

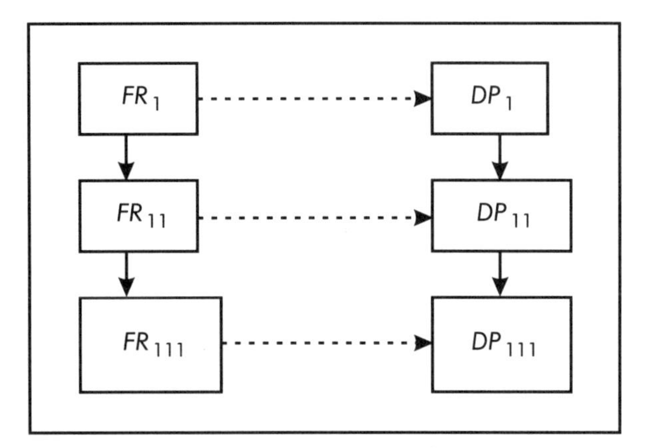

Figure 2-22. Matrix representation of FRs and DPs.

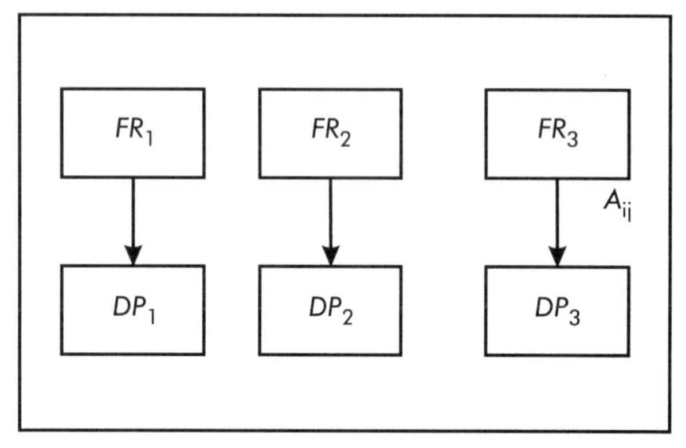

Figure 2-23. Graphical representation of an uncoupled design.

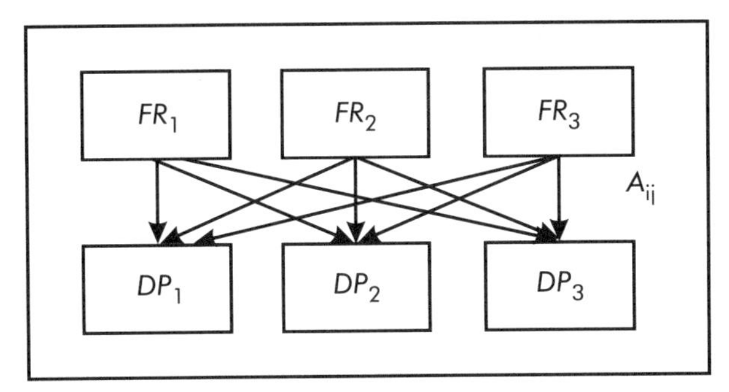

Figure 2-24. Graphical representation of a coupled design.

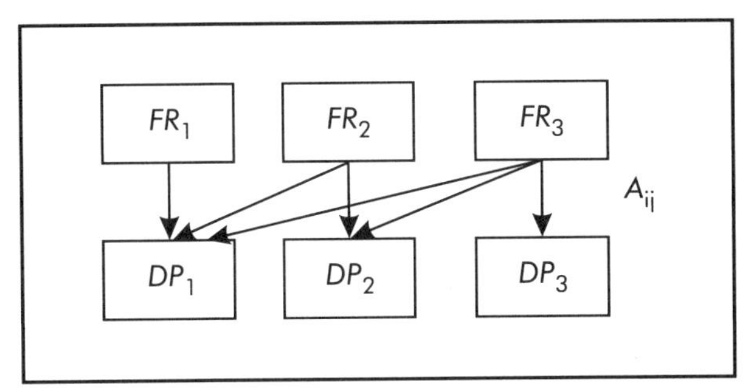

Figure 2-25. Graphic representation of a decoupled design.

Corollaries

There are a few modified versions of design axioms. These design rules or corollaries are derived from the basic axioms to facilitate applications. Some of the corollaries are as follows:

1. Decouple or separate parts or aspects of a solution if FRs are coupled or if they become interdependent in the designs proposed. Decoupling does not mean that a part has to be broken into parts, or that a new element has to be added to the design.
2. Minimize the number of FRs and constraints. Increasing these elements of design increases the information content. A designer should not produce a design that does more than what is intended. Such a design tends to be more expensive and may have less reliability.
3. Integrate the design features in a single physical part if the FRs can be independently satisfied in the proposed solution.
4. Use standardized or interchangeable parts if the use of these parts is consistent with the FRs and the constraints.
5. Use symmetrical shapes and/or arrangements if they are consistent with the FRs and constraints. Symmetrical parts require less information to manufacture and assemble.
6. Specify the largest allowable tolerance when stating the FRs.
7. Seek an uncoupled design that requires less information than coupled designs when satisfying a set of FRs. If a designer proposes an uncoupled design that has more information than a coupled design, then the design should be started as new because a better design lies somewhere.

In applying axiomatic design, Axiom 1 must be satisfied at all stages of mapping from the functional to the physical domain. Therefore, the matrix (DM) should be either triangular or diagonal. Axiom 2 is stated in terms of information and complexity. If a product is more complex, more information is needed to describe it. The major objective of product design is to determine the right combination of product/process parameters, and the material selection for getting the most economical solution of a product with the right quality. The axiomatic approach is intended to help the designer by choosing the right combination of information content to maximize the probability of achieving FRs.

Example of a Two-knob Water Faucet

Readers should consider the example of a two-knob water faucet. The basic objective is to provide continuous water at a desired flow rate and temperature. Hot and cold water are supplied separately. There are two functional requirements.

$$\begin{bmatrix} FR_1 \\ FR_2 \end{bmatrix} = \begin{bmatrix} X & X \\ X & X \end{bmatrix} \begin{bmatrix} DP_1 \\ DP_2 \end{bmatrix} \quad (2\text{-}7)$$

where:

FR_1 = obtain water flow rate
FR_2 = obtain water temperature
DP_1 = means to adjust cold water flow
DP_2 = means to adjust hot water flow

This equation illustrates that the needs of the two-handled faucet can be represented as a coupled system. This is because, to get the needed flow rate and temperature, the hot and cold water flow amounts have to be adjusted at the same time. In this representation, the water-flow temperature and the water flow are linked.

To obtain a different solution where the water temperature is maintained independent of the water flow, DPs can be reformulated without saying how the temperature is maintained. Uncoupling the DPs with the original FRs may result in alternate designs. One possible uncoupled design is:

$$\begin{bmatrix} FR_1 \\ FR_2 \end{bmatrix} = \begin{bmatrix} X & 0 \\ 0 & X \end{bmatrix} \begin{bmatrix} DP_1 \\ DP_2 \end{bmatrix} \quad (2\text{-}8)$$

where:

DP_1 = water-flow regulating device
DP_2 = water-temperature regulating device

Manufacturing Domain

Axiomatic procedures can be extended to the manufacturing domain. Figure 2-26 shows the mapping between the FRs (defined in functional space) of the product to DPs (defined in physical space). If the variables in the manufacturing space are defined as process variables (PVs), there can be further mapping of the manufacturing space between DPs and PVs.

Axiom 1 applies to both of these mapping situations as:

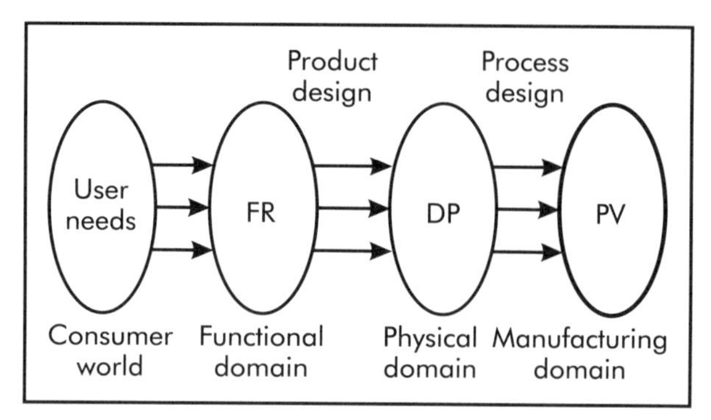

Figure 2-26. Domains of design for manufacturing.

$$\{FRs\} = [A]\{DPs\} \quad (2\text{-}9)$$

$$\{DPs\} = [B]\{PVs\} \quad (2\text{-}10)$$

where:

matrices [A] and [B] must be either uncoupled or decoupled to satisfy the independence axiom

The functional space is then related to the manufacturing space as:

$$\{FRs\} = [A][B]\{PVs\} = [C]\{PVs\} \quad (2\text{-}11)$$

Therefore, for the product to be manufacturable, the matrix [C] must also be an uncoupled or decoupled type. That means that a product design cannot be manufactured within specifications unless both the product and process design are either uncoupled or decoupled designs. As users search for a design solution to satisfy a given set of FRs, they know that the design matrix must be diagonal or triangular and that the number of DPs must be equal to the number of FRs in an ideal design.

The only unknowns in the equation are DPs. Thus, the user can proceed to conceptualize a design solution that consists of at least three DPs. To achieve good quality products, the functional requirements specified by the designer in terms of geometry, hardness, etc., must be satisfied by the manufacturing process and system. When the manufacturing process meets the design specifications, probability = 1, and the information required to achieve the task = 0, since the manufacturing process can produce good parts each and every time. When the probability < 1, additional information must be supplied by the operator or by some other source so that functional specifications can be met.

Example of Refrigerator Design

Functional requirements and design parameters can provide insight into the axiomatic-design application for a refrigerator. The main requirement of the refrigerator is to preserve food for long-term use and to keep some food at a cold temperature for short-term use without freezing (Suh 2001).

Functional domain.

FR_1 = freeze food for long-term preservation
FR_2 = maintain food at cold temperature for short-term preservation

A conventional refrigerator has a compressor, condenser, and evaporator with one fan to circulate the cold air. To satisfy the FRs, a refrigerator with two compartments is designed.

Physical domain.

DP_1 = freezer section
DP_2 = chiller section

The freezer section should affect the freezer area only and the chiller section should affect only the food to be chilled, but not frozen. The design matrix to satisfy this will be diagonal.

Second-level functional domain. Decomposition of FR_1:

FR_{11} = temperature control of freezer section in the range 18 to 25 °F (–8 to 4 °C)
FR_{12} = maintain the uniform temperature throughout the freezer section at the preset temperature
FR_{13} = control relative humidity to 50%

Decomposition of FR_2:

FR_{21} = control chilled section temperature in the range of 45 to 55 °F (7 to 13 °C).
FR_{22} = maintain uniform temperature in the chilled section to within 2 °F (1 °C)

To satisfy the second level FRs, a user has to design DPs in such a way that DP_{11}, DP_{12}, and DP_{13} satisfy FR_{11}, FR_{12}, FR_{13} and are independent from each other.

Second-level DP domain.

DP_{11} = turn on/off compressor when air temperature is higher or lower than set values
DP_{12} = blow air into freezer section and circulate it uniformly
DP_{13} = condense return-air moisture when the dew point is exceeded

The equations can be represented as:

$$\begin{bmatrix} FR_{12} \\ FR_{11} \\ FR_{13} \end{bmatrix} = \begin{bmatrix} X & 0 & 0 \\ X & X & 0 \\ X & 0 & X \end{bmatrix} \begin{bmatrix} DP_{12} \\ DP_{11} \\ DP_{13} \end{bmatrix} \tag{2-12}$$

The following equation indicates that the design is a decoupled design.

$$\begin{Bmatrix} FR_{21} \\ FR_{22} \end{Bmatrix} = \begin{bmatrix} X & 0 & 0 \\ X & X & 0 \end{bmatrix} \begin{Bmatrix} DP_{21} \\ DP_{22} \end{Bmatrix} \tag{2-13}$$

where:

DP_{21} = refrigerator fan
DP_{22} = vents

Using the Axiomatic Approach in Production-system Design

A manufacturing system is a complex arrangement of physical elements characterized by measurable parameters. A manufacturing-system design consists of the design of physical elements and operations required to produce a product. A production system provides supporting functions to the manufacturing system. It defines the performance measures of the manufacturing system. The production system consists of the design of all of the elements and functions that support the manufacturing system.

Designing production systems for a product is crucial to understanding the relationship between system design objectives and physical design implementation. It can provide a framework for explaining why low-level decisions tend to affect the viability of an entire production system. Designing a system requires an understanding of what variables have negatively impacted the operation of the manufacturing system. The lean production system represents a new production design and therefore requires a new set of performance and cost measurement criteria, which are inherently system-design based. Most production systems today are measured in a way that causes their design to move in the opposite direction of meeting lean production design objectives. Axiomatic design helps define what the design system objectives are, and how they are to be accomplished and implemented from a system-design perspective.

The development of production system design decomposition is based on the power of axiomatic design. Two elements of axiomatic design are design axioms and *zigzagging*. The idea of zigzagging means that any design, no matter how complex, may be decomposed into its constituent levels. Production system design decomposition provides a systematic means for designing production systems. Its scopes include the functional and physical domains of design. The functional requirements or objectives that are defined by the functional domain are measurable parameters of the production system design.

The benefits of the production system design decomposition are:

- an ability to concretely describe and distinguish between various production system design concepts;
- an adaptability to different products and manufacturing environments;
- an ability to design or create new system designs to meet new environments (for example, what happens when FRs or DPs change, as in lean versus mass production?);
- portability of a production system design methodology across industries;
- an impact of lower-level design decisions on the total system performance;
- providing a foundation for developing a new set of manufacturing performance measures from a system-design perspective; and
- making a connection between the machine-design requirements and the manufacturing-system objectives.

Figure 2-27 illustrates the difference between mass and lean production. The difference is the result of a change in design parameters, which affect the functional requirements of sales revenue, production costs, and production investment. In mass production, increasing sales revenue simply means making more products.

Figure 2-28 shows how axiomatic thinking can be applied to show the differences between mass production and lean manufacturing. As shown, if the business objective is to increase return on revenue, then FRs can be shown as sales revenue, production cost, and production investment. One component of lean manufacturing is increasing sales revenue. An increase in sales revenue can be mapped out to maximize customer satisfaction.

For mass production, the main aim is to produce products at a minimum cost by maximizing the production output and machine utilization. In lean manufacturing, an

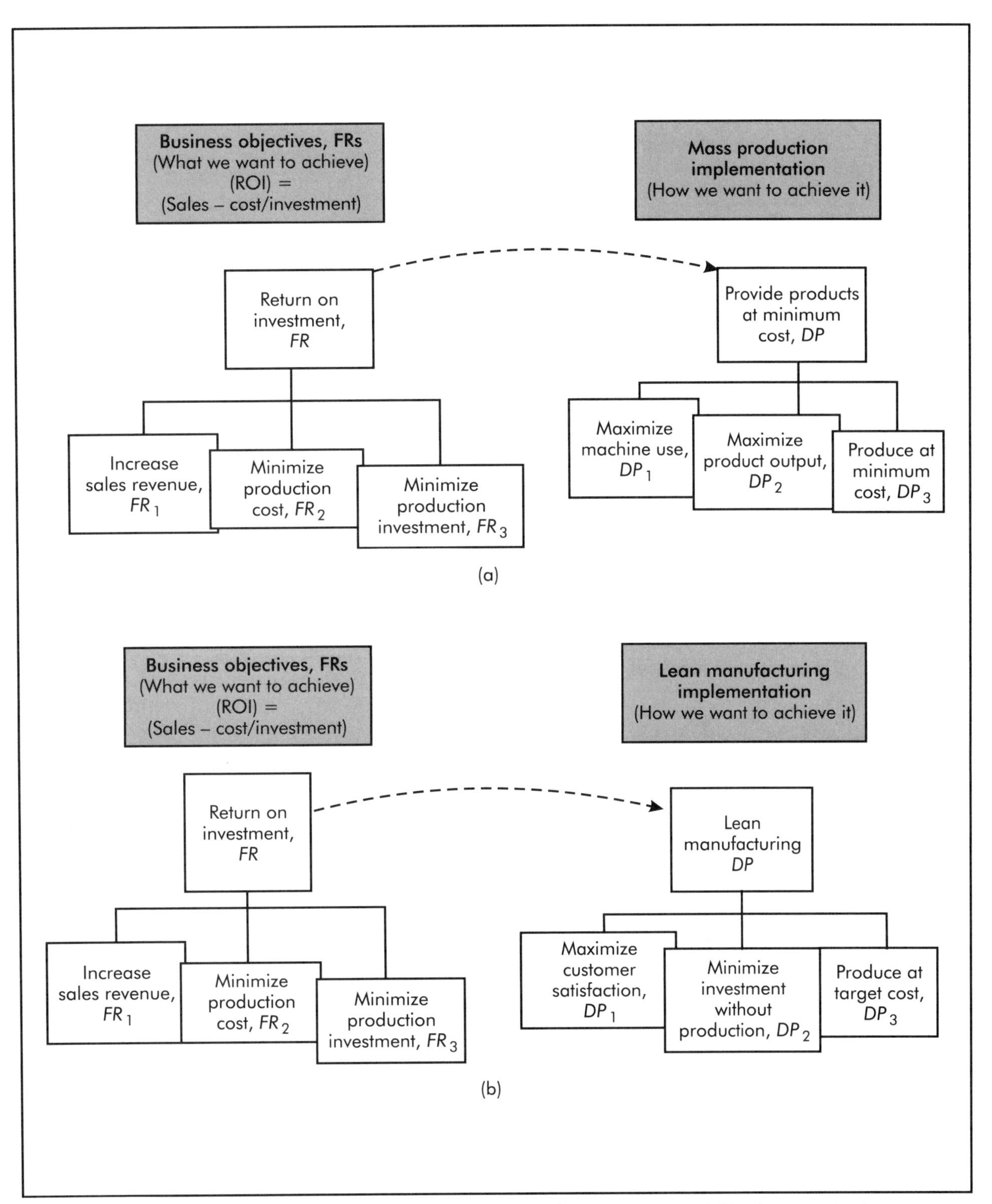

Figure 2-27. (a) Axiomatic thinking for the design of a mass production system and (b) a lean production system.

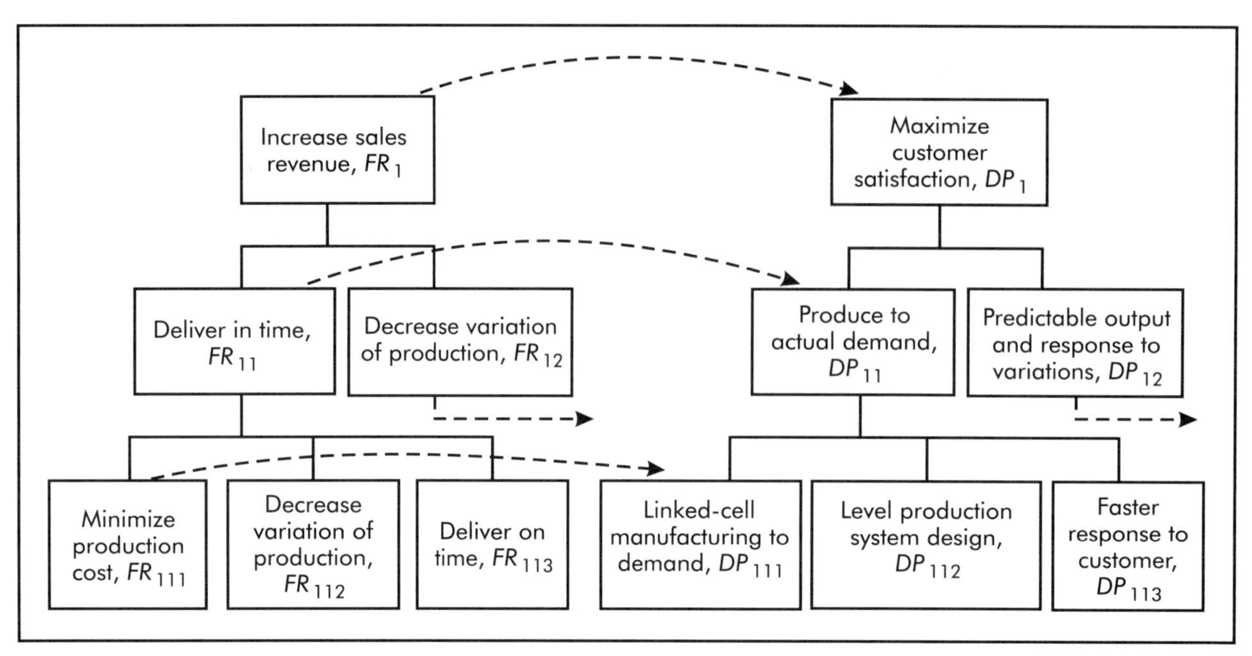

Figure 2-28. Axiomatic thinking applied to increasing sales revenue.

emphasis is placed on reducing defective products, increasing customer satisfaction, and achieving on-time delivery. In mass production, increased sales revenue (FR) means maximizing production output (DP). In lean production, increased sales revenue (FR) means satisfying the customer (DP).

Axiomatic Design Applied to Quality Function Deployment (QFD)

Axiomatic design methodology presents a structured approach to streamlining the design process. This allows products to be designed with their functional objectives in mind. Quality function deployment (QFD) is a means of understanding the customer's needs and translating those needs into requirements that will satisfy them. The objective of this section is to show how these two methods can be bridged together to further enhance and structure the design process. This section first examines some of the key similarities between axiomatic design (AD) and QFD. Following this, some basic rules to apply AD to QFD are explained. Finally, a case study of a receptacle is used to demonstrate the concept.

QFD deals with linking customer requirements (CRs) to design requirements (DRs). In axiomatic design, this linking takes place between functional requirements (FRs) and design parameters (DPs). These ideas are expressed in Figure 2-29. QFD achieves this link through matrices—QFD matrix or relationship matrix (RM)—and an axiomatic design referred to as a design matrix (DM). These matrices result from mapping one domain to the other, both in axiomatic design space and QFD space.

In QFD, an interdependency of design requirements is assessed using what is referred to as a correlation matrix (CM) (see Figure 2-30). In AD, interdependency is achieved through a particular DM, which may be in the form of a coupled (C), decoupled (DC), or uncoupled (UC) format.

Customer requirements are ranked within QFD. This ranking helps pinpoint which

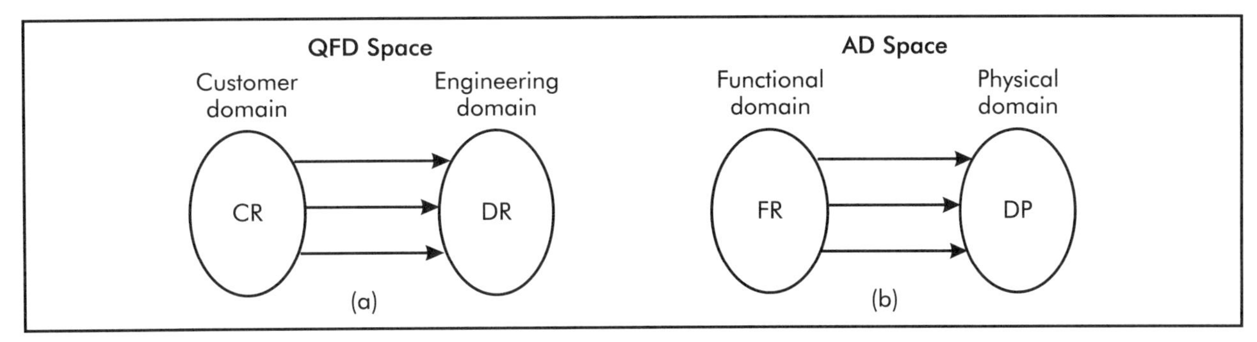

Figure 2-29. Linking representation.

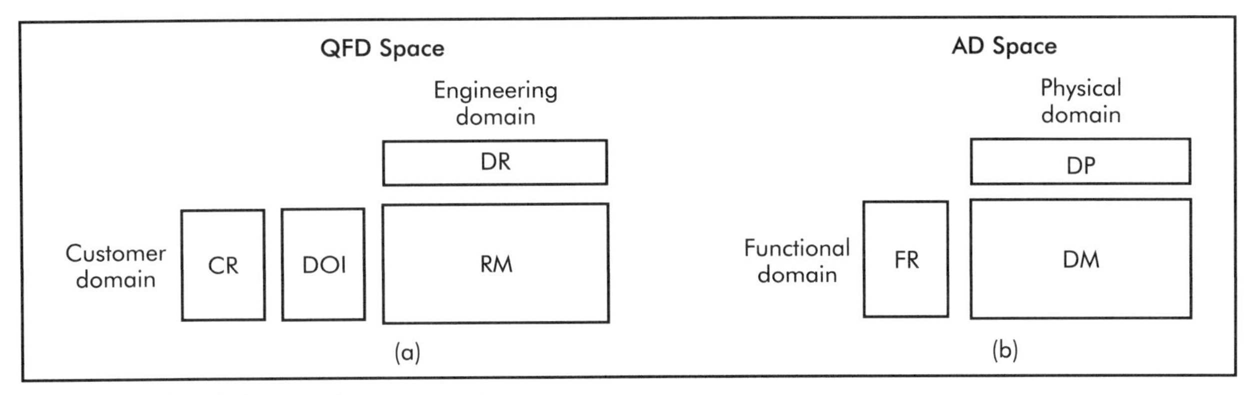

Figure 2-30. Correlation matrix representation.

design requirements (DRs) are most important with respect to the customers' needs. The ranking performed in QFD is achieved through what is referred as a degree of importance (DOI) matrix (see Figure 2-31).

For AD, an application of ranking is not so direct, and the ranking performed is primarily done for functional requirements as opposed to customer requirements. To perform ranking in AD, the two design axioms—functional independence and minimum information content—are applied. The less information needed to implement a design, the better the design; and the more independent its FRs, the better the design.

AD methodology uses design equations (DEQ) to show the relationship between FR and DP. In the same fashion, QFD can use similar matrices to establish a set of prioritization equations (PEQ). They are represented in Figure 2-32. These equations represent the essence of both methodologies. They are the mathematical embodiment of the entire AD and QFD processes at all levels.

The goal of the matrix in QFD is to formulate, capture, and identify those design requirements (DRs) most significant to satisfying customer needs. For AD, this goal is to capture and identify the nature of a particular design (coupled, uncoupled, or decoupled) with respect to its ability to satisfy certain design objectives.

To start applying AD to QFD, a link first must be established between the FRs and DRs. To establish this link, the framework depicted in Figures 2-33 and 2-34 is applied. This allows restructuring and mapping QFD-space DRs to AD-space FRs.

Action is defined as being that singular task required to achieve a particular result;

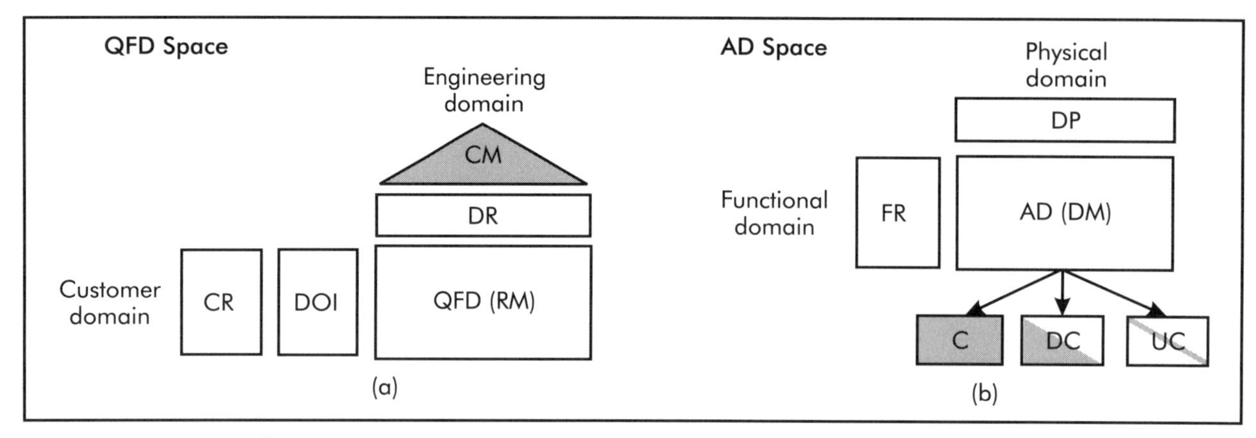

Figure 2-31. Degree of importance matrices representing the QFD and AD spaces.

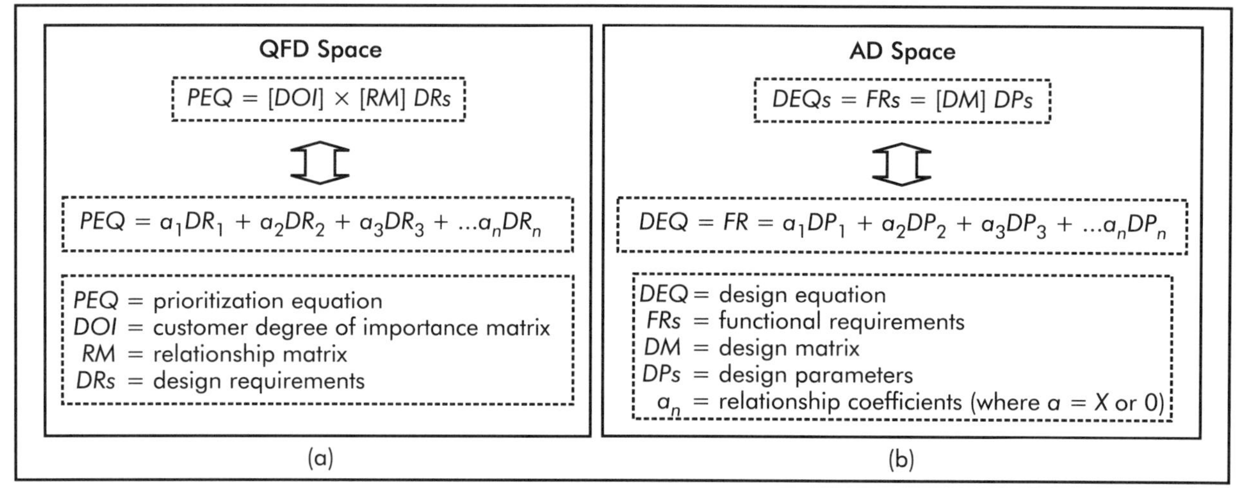

Figure 2-32. Prioritization equation and design equation for QFD and AD space.

object is defined as the entity requiring the task; and *criteria* are the standards against which the action will be measured.

QFD and AD share similar traits. Because of this, both methods can be applied to enhance the design process, thus taking advantage of both the strengths of AD and of QFD. Figure 2-35 summarizes this framework.

Electrical Receptacle Example

In the QFD receptacle example shown in Figure 2-36, assume that the QFD team has now decided to incorporate AD into its QFD analysis. The following details show how this can be done.

QFD space. To start, the team first formulates and captures CRs and DRs as follows:

CR_1 = wiring too time consuming
CR_2 = modern look
CR_3 = low cost
CR_4 = meets UL tests
DR_1 = reduce time to wire
DR_2 = reduce number of process steps
DR_3 = use high-ranking jury
DR_4 = reduce cost to under \$2.50
DR_5 = pass UL 498 test

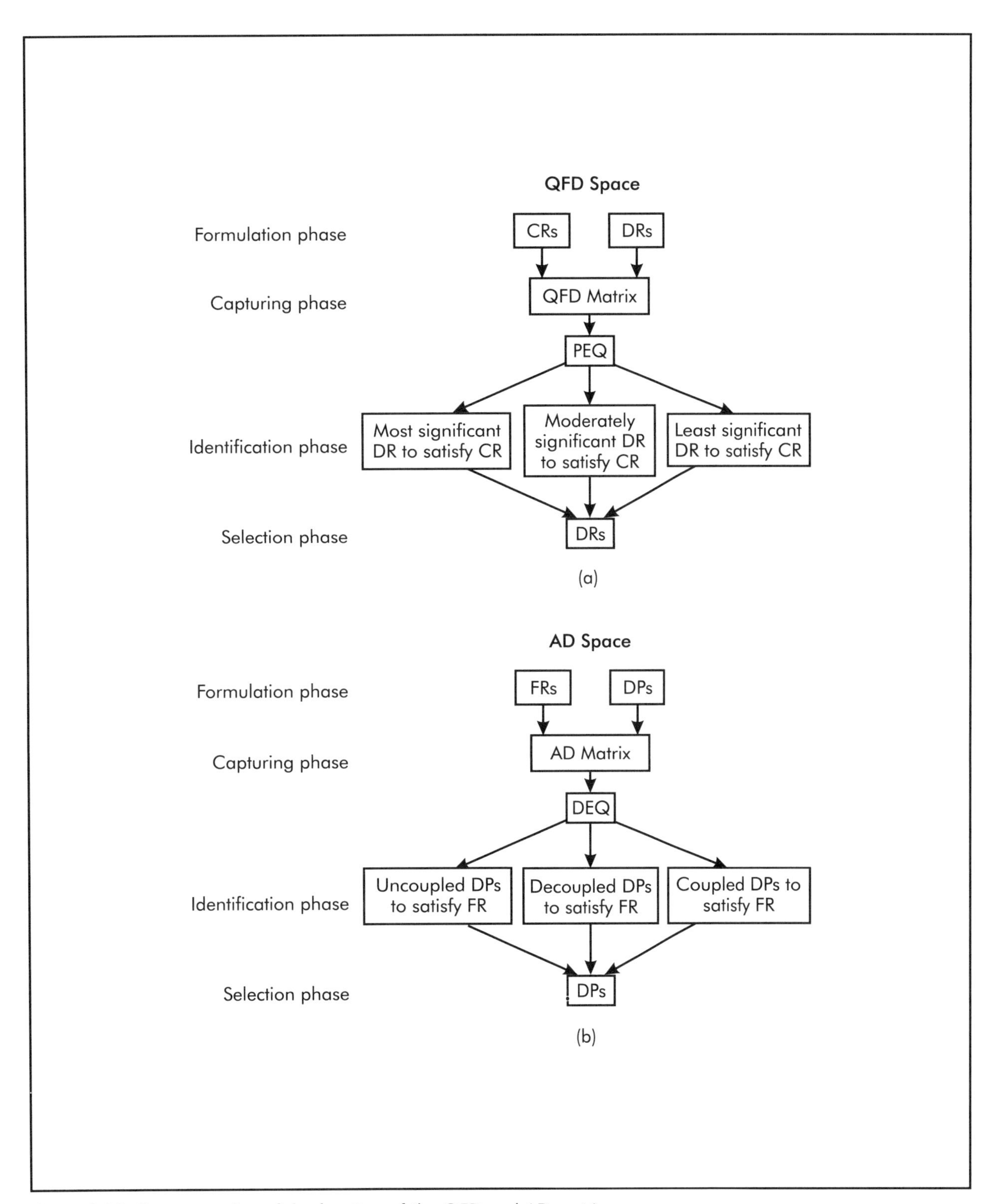

Figure 2-33. Representation of the function of the QFD and AD matrix.

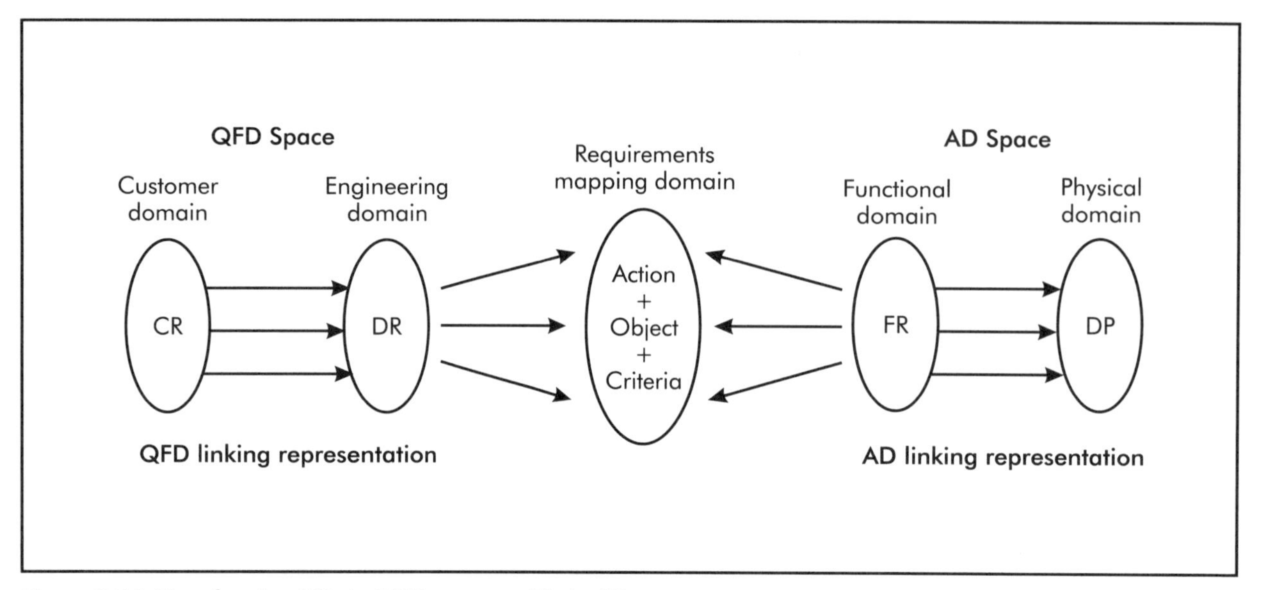

Figure 2-34. Transforming DRs in QFD space to FRs in AD space.

The team then identifies the DRs that are at least moderately satisfying the CRs. To do this, it applies the following PEQ and then picks the DR terms with the largest a_i or coefficient value.

$$PEQ = [DOI]\ [RM]\ DRs \tag{2-14}$$

$$PEQ = a_1DR_1 + a_2DR_2 + a_3DR_3 + \ldots + a_nDR_n \tag{2-15}$$

where:

PEQ = prioritization equation
DOI = customer degree-of-importance matrix
RM = relationship matrix
DRs = design requirements
a_n = relative weight of design requirements

The house of quality shown in Figure 2-36 lists the relationships between the "whats" and "hows." The quick-wiring requirement has a positive ranking of 9, while meeting UL-test standards has a moderate relationship of 3. The degree of importance is shown to be (4, 4, 5, 5). Using Equations 2-14 and 2-15, the PEQ can be found.

$$DOI = [4, 4, 5, 5],$$

$$RM = \begin{bmatrix} 99000 \\ 00900 \\ 00090 \\ 03009 \end{bmatrix} \text{ and } DRs = \begin{bmatrix} DR_1 \\ DR_2 \\ DR_3 \\ DR_4 \\ DR_5 \end{bmatrix}$$

$$PEQ = 36DR_1 + 51DR_2 + 36DR_3 + 45DR_4 + 45DR_5$$

DRs that are found to have very large coefficients should be chosen. The ones that appear to satisfy this criteria are DR_2, DR_4, DR_5 and should included in the AD analysis. These DRs are the most significant DRs to satisfy customer requirements. Since both DR_1 and DR_3 are found to have moderately large coefficients, these are selected as well for inclusion into the AD analysis (see Figure 2-37).

Finally, the DRs are converted into FRs. Once done, the team is ready to proceed over to its AD space to start analyzing design requirements as functional requirements.

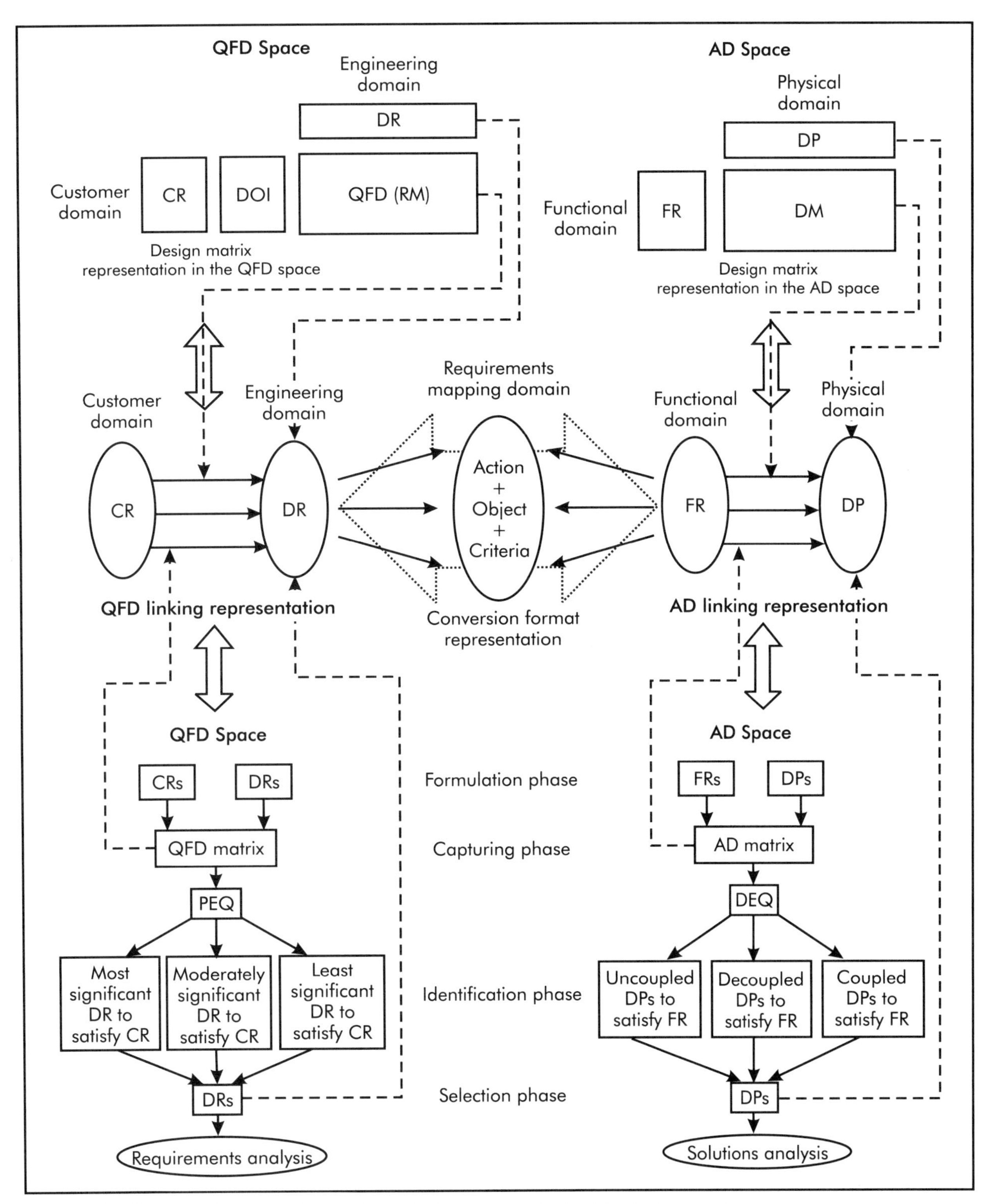

Figure 2-35. Frameworks for applying axiomatic design analysis to QFD design requirements.

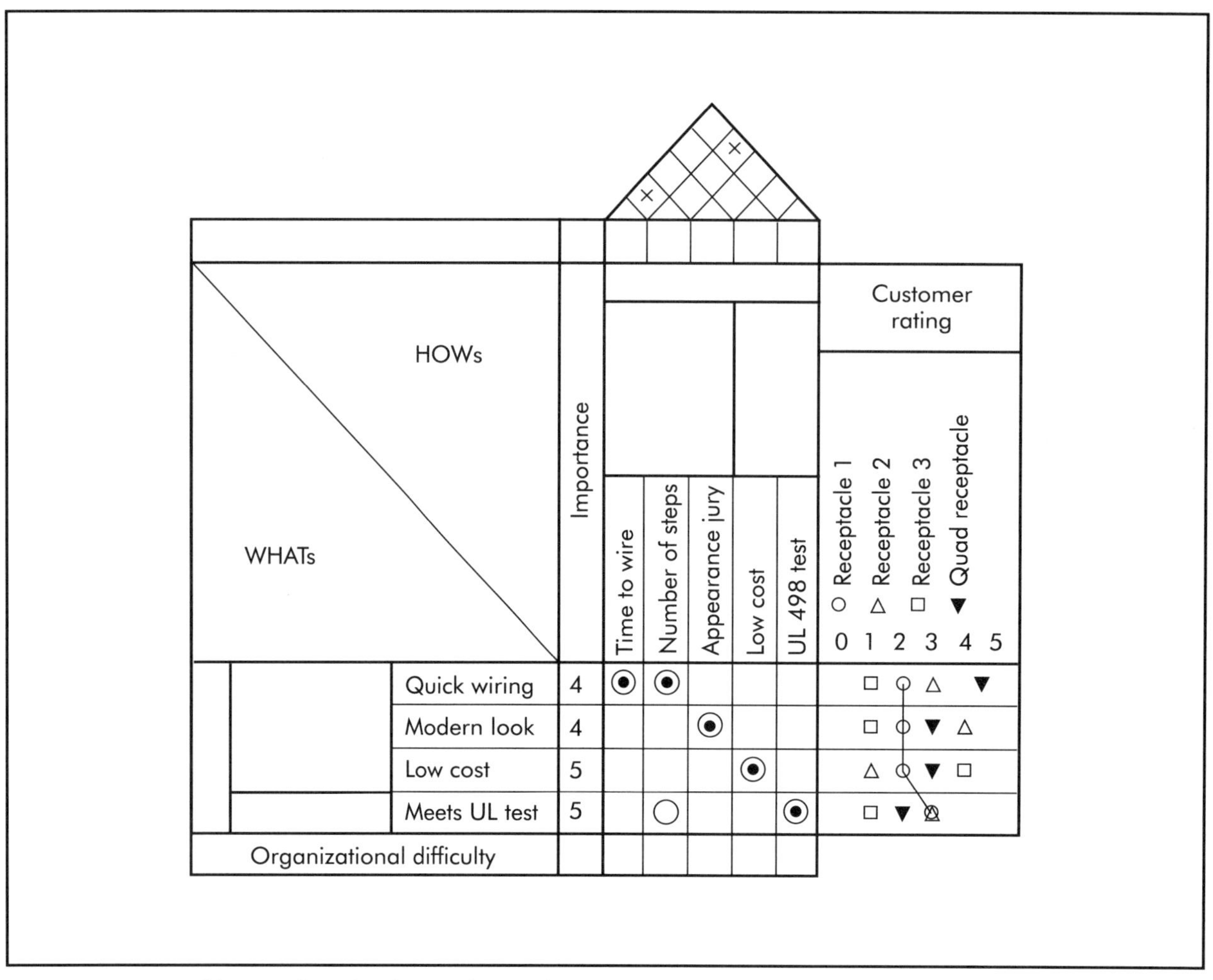

Figure 2-36. Partial house of quality for an electrical receptacle.

AD space. The team must express its FRs as DPs using the format specified.

Note: Typically, the goal is to find the best means possible to satisfy a particular FR. The term *means* refers to any solution; this could be a device, a material type, a parameter change, a software type, etc. Depending on the nature of a particular design, arriving at an appropriate means may require the use of an experimental design or some formal application of decision analysis. In any event, whatever the solution, it should be consistent with Axiom 2—the information-content axiom.

INVENTIVE PROBLEM-SOLVING TECHNIQUES

There have been a number of design methodologies used to create new products and processes. Some of these methods overlap and use different definitions of key concepts, such as function. While these methodologies emphasize the importance of knowing physical effects to solve technical problems, they were developed independently and have different backgrounds.

TRIZ is the acronym for a Russian term that translates to theory of inventive prob-

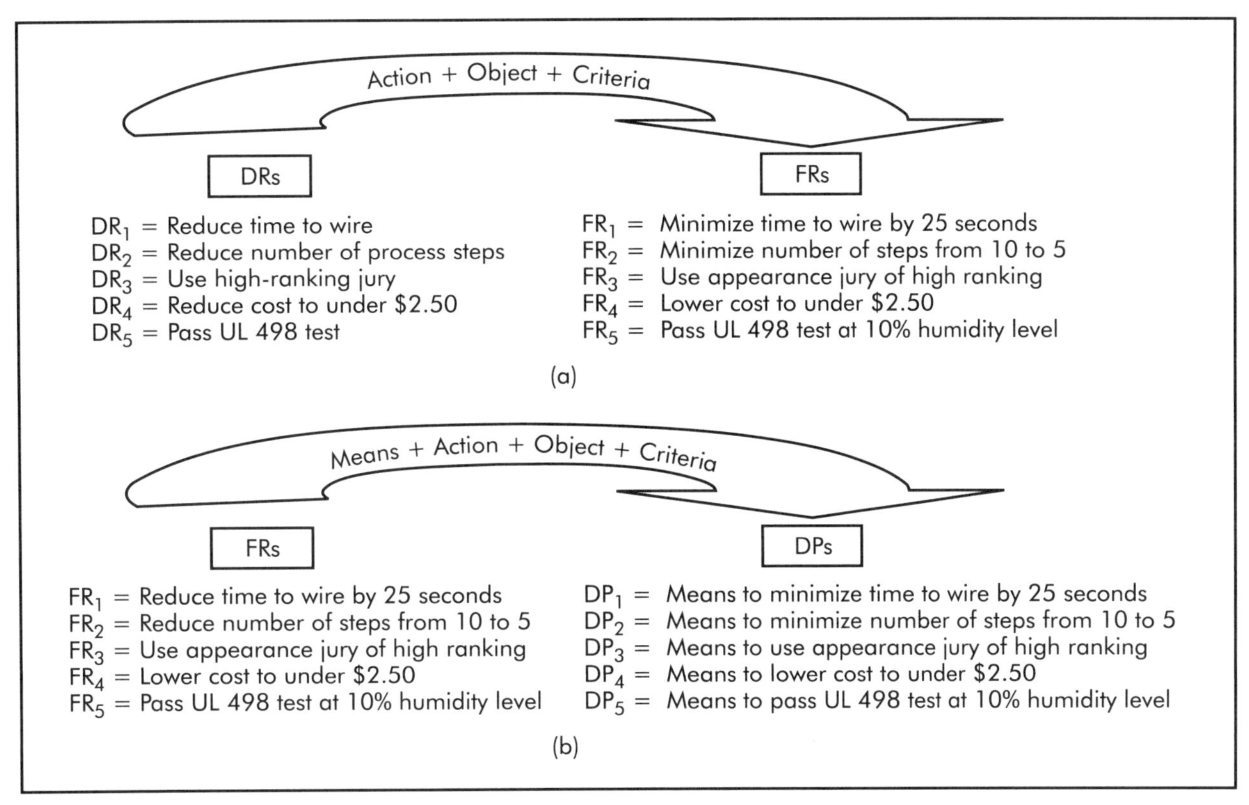

Figure 2-37. Action, object, and criteria analysis.

lem solving (TIPS). Genrich Altshuller developed the theory in 1946. He began with the hypothesis that universal principles serve as the basis for creative innovation across all scientific fields. If these principles were codified and taught, it would be possible to make innovation more predictable. To test this theory, he reviewed about 200,000 patents submitted at that time in the Soviet Union. The analysis showed that most patents suggested a means for eliminating system conflicts.

For a problem to be inventive, it has to pose at least one contradiction. Such contradictions arise when a certain parameter cannot be improved without causing another parameter to deteriorate. A contradiction between speed and sturdiness is one example. A sturdy automobile means more weight. More weight generally results in less speed. How can the same vehicle be designed to run faster? TRIZ researchers found about 39 parameters, each of which could be in contradiction with one another. The initial step in using TRIZ is to find out which design parameters contradict one another.

Another technique, called the *systematic approach to engineering design* (SAPB), has a European origin and was developed by Gerhard Pahl and Wolfgang Beitz (Pahl and Beitz 1988). SAPB states that design problem solving is a variant of general problem solving. When designing, the designer usually follows a path consisting of certain fundamental activities—problem and requirements formulation, a search for alternative solutions, evaluation and documentation, and communication of results. Design methodologies support this process by providing specific design methods and design knowledge.

TRIZ methodology systematically investigates the problem as an innovative solution and applies step-by-step guidelines to generate solution alternatives—improving product parameters while maximizing product changes and costs. This procedure was developed with a very limited knowledge of other methodologies, but is based on a large empirical knowledge base of patents. The concept has been adopted by many organizations as an effective concept-generating tool. Apart from solving technological issues, it is capable of affecting key management functions. The different elements of TRIZ/TIPS are shown in Figure 2-38.

For example, the engineering principle used to split gems is also used to extract seeds from bell peppers so that they can be canned. In both cases, the objective is to break something apart without breaking it. The procedure involves placing the object in an airtight container, gradually applying pressure, and then suddenly releasing the pressure. A sudden pressure drop creates an explosion, which splits the object along a fracture line. TRIZ may assist in generating ideas and looking at correct conditions that may satisfy the manufacturing process. In the case of diamond cracking, the splitting pressure will be much higher than the one used for bell peppers. This has to be determined experimentally. Systematically codifying which principles can help solve given combinations of controlling parameters speeds up idea generation, instead of leaving it to trial and error. The process of splitting and canning peppers was patented in 1968. The patent given for crystal splitting was issued 20 years later.

Step-by-step TRIZ Process

1. Identify the problem. The engineering system under consideration should be identified—including its operating environment, resource requirements, primary useful function, harmful effects, and ideal result.
2. Formulate the problem in terms of TRIZ and contradictions. The problem should be restated in terms of physical contradictions, identifying problems that could occur, and analyzing for tech-

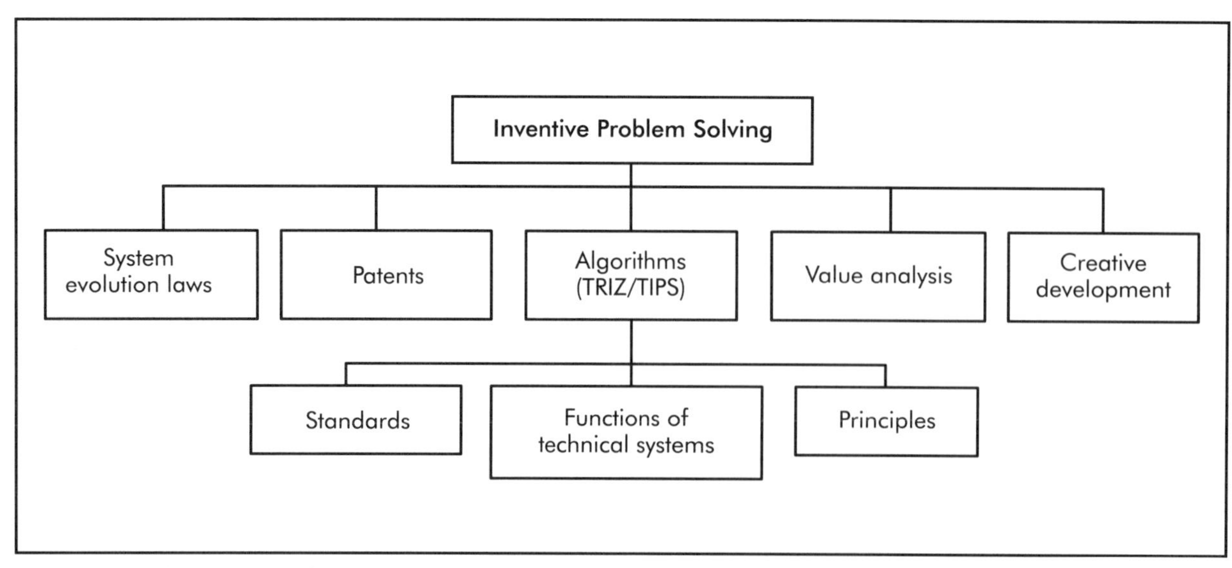

Figure 2-38. The structure of TRIZ/TIPS.

nical conflicts in problems that may force a compromise solution.

3. Search for a previously solved problem. By TRIZ functioning, there are 39 standard technical characteristics that could cause conflict. Users should find the contradictory engineering principles.
4. Examine TRIZ inventive principles. Users should first find the principle that needs to be changed, as well as the principle that has an undesirable secondary effect. They should identify any technical conflicts and apply the necessary algorithms to solve the contradictions.
5. Examine the effects. Examine whether there are any harmful actions that need to be eliminated. Is it necessary to maximize useful actions? If there are harmful actions, what is the best way to reduce them?
6. Specify the solution. The designed system should be reduced in such a way that the function will be fulfilled satisfactorily, but so the system will be reduced or eliminated (see Figure 2-39).

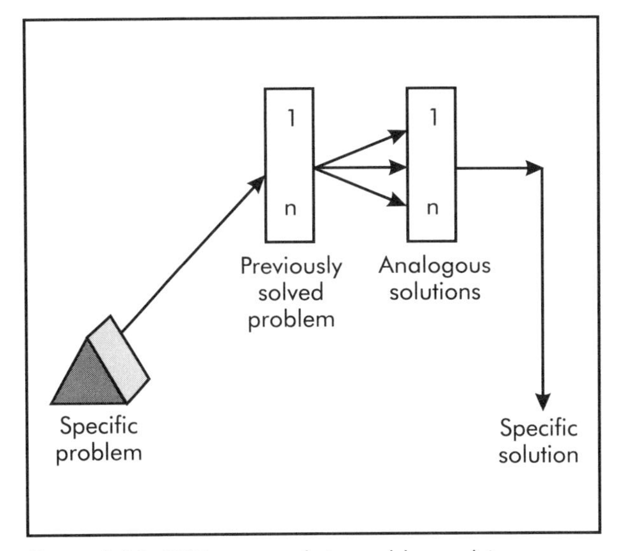

Figure 2-39. TRIZ approach to problem solving.

Resolution of Technical Contradictions

As stated earlier, physical contradictions arise when a certain parameter cannot be improved without causing another to deteriorate. The basic concept of TRIZ/TIPS is the resolution of a contradiction. A contradiction arises from placing mutually exclusive demands on the same system. Improvement of one system's parameters leads to the deterioration of others. To resolve this, it is important to find physical contradictions that are the hidden root of the technical problem. The most effective solutions are achieved when a designer solves a technical problem that contains a contradiction. This generally occurs when the designer tries to improve on specific parameters. The physical contradictions and principles are combined in a matrix (see Table 2-12), the rows and columns of which contain 39 generalized parameters corresponding to the most common parameters engineers try to improve.

Table 2-13 provides a list of technical contradictions. After reviewing the table, the designer can obtain some idea about where to start looking for solutions. The contradiction idea should not be a reflection of the whole design problem, but deal with conflicting elements of weak spots of the project.

TRIZ/TIPS has a collection of 39 inventive principles for resolving physical contradictions. The full list of principles appears in the Appendix to this chapter.

TRIZ principles do not constitute final solutions to problems, but rather, they are high-level strategies for finding ideas. The principles of TRIZ will assist the designer in finding a highly inventive solution. They force the designer to pre-formulate the problem in terms of standard engineering parameters.

One TRIZ design principle is segmentation or division. This technique involves dividing up an object usually seen as a whole. As an example of segmentation, consider the detachable car radio. The removable nature

Table 2-13. Technical contradictions

Technical Characteristic	
1. Weight of moving object	21. Energy spent by nonmoving object
2. Weight of nonmoving object	22. Power
3. Length of moving object	23. Waste of energy
4. Length of nonmoving object	24. Waste of substance
5. Area of moving object	25. Loss of information
6. Area of nonmoving object	26. Waste of time
7. Volume of moving object	27. Amount of substance
8. Volume of nonmoving object	28. Reliability
9. Speed	29. Accuracy of measurement
10. Force	30. Accuracy of manufacturing
11. Tension, pressure	31. Harmful factors acting on object
12. Shape	32. Harmful side effects
13. Stability of object	33. Manufacturability
14. Strength	34. Convenience of use
15. Durability of moving object	35. Repairability
16. Durability of nonmoving object	36. Adaptability
17. Temperature	37. Complexity of device
18. Brightness	38. Complexity of control
19. Energy spent by moving object	39. Level of automation
20. Productivity	

of the radio component prevents theft. Contradicting design principles are reliable use and, at the same time, the least harmful side effect. In this case, reliable use can be interpreted as using the radio whenever the motorist wants to. The harmful side effect is the theft of the car radio.

Additional TRIZ Tools

Problems of a more difficult nature are solved with the following precise TRIZ tools:

- algorithm for inventive problem solving (ARIZ);
- SU-field analysis;
- directed product evolution (DPE); and
- anticipatory failure determination (AFD).

ARIZ

ARIZ is an analytical tool of TRIZ. It provides specific sequential steps for developing a solution to complex problems without apparent contradictions. Depending on the nature of the problem, anywhere from five to 60 steps may be involved. From an unclear technical problem, the underlying technical problem can be revealed. Below is a brief description of eight steps.

1. Analysis of problem: begin by making the transition from vaguely defined statements of the problem to a simply stated mini-problem. This step also provides for the analysis of conflicting situations (technical contradictions). A decision is made as to which contradiction is considered for further resolution. Once decided, a model of the problem is formulated.
2. Analysis of problem model: a simplified diagram that models conflict in the operating zone is drawn—the operating zone is a specified narrow area of conflict. Then, an assessment of all the available resources is made.

3. Formulation of ideal final result (IFR): usually the statement of the IFR reveals contradictory requirements to the critical component of the system in the operating zone. This is called the *physical contradiction*.
4. Utilization of outside substances and field resources: consider solving a problem by applying standards in conjunction with a database of physical effects.
5. Reformulation of problem: if the problem still remains unsolved until this stage, ARIZ recommends returning to the starting point and reformulating the problem in respect to the super system. This looping process can be done several times.
6. Analysis of the method that removed the physical contradiction: the main goal of this step is to check the quality of the solution (with the physical contradiction removed).
7. Utilization of current found solution: this step guides users through an analysis of the effects the new system may have on adjacent systems. It also forces a search for applications to other technical problems.
8. Analysis of steps that lead to solution: this is a checkpoint where the real process used to solve a problem is compared with that suggested by ARIZ. Deviations are analyzed for possible further use.

SU-field Analysis

SU-field analysis is a tool for expressing function statements in terms of one subject acting on another subject. The objects are called *substances* and the action is a *field*. SU-field analysis is helpful in identifying functional failures. By looking at actions as fields, undesirable or insufficient actions can be countered by applying opposite or intensified fields. There are 76 standard solutions that permit quick modeling of simple structures for SU-field analysis.

There are essentially four steps to follow in making the SU-field model.

1. Identify the elements.
2. Construct the model. (After completing these two steps, stop to evaluate the completeness and effectiveness of the system. If an element is missing, identify that element.)
3. Consider the solutions from the standard solutions.
4. Develop a concept to support the solution.

Directed Product Evolution

Future characteristics of machines, procedures, or techniques are subject to prediction attempts using traditional forecasting techniques. They rely on surveys, simulations, and trends to create a probabilistic model of future developments. Directed product evolution (DPE) gives a forecast, but does not precisely detail the technology being forecast.

DPE is essentially a prediction based on a level of confidence in a technological achievement during a given time frame with a specified level of support. Most innovations of the next 20 years will be based upon scientific and technological knowledge existing now. Difficulty lies in identifying what is of real significance. The role of DPE is to evaluate today's knowledge systematically, thereby identifying how an achievable technological advance can fulfill a human need.

The principle of DPE basically works on the following eight patterns of evolution:

1. stages of evolution;
2. evolution toward increased ideality;
3. non-uniform development of system elements;
4. evolution toward increased dynamism and controllability;

5. evolution toward increased complexity and then simplification;
6. evolution with matching and mismatching elements;
7. evolution toward the micro level and increased use of fields; and
8. evolution toward decreased human involvement.

By analyzing the current technology level and contradictions in products, TRIZ enables users to analyze evolutionary processes and create the future. Using the eight patterns of evolution, the DPE method looks at the past and present scene and recommends directions of innovation.

Anticipatory Failure Determination

Anticipatory failure determination (AFD) is an efficient and effective method for analyzing, predicting, and eliminating failures in systems, products, and processes. AFD guides designers in documenting situations, formulating related problems, developing hypotheses, verifying potential failure scenarios, and finding solutions to eliminate problems.

AFD is a powerful approach that favorably impacts costs associated with quality, safety, reliability, recalls, and warranty claims. The prevention of unanticipated failures is important in new product development. AFD, in effect, invents failure mechanisms and then examines the possibilities of their actually occurring. As a result, factors contributing to failures can be eliminated with this highly proactive technique.

The AFD system consists of two modules:

1. analysis of previous failures; and
2. prediction of failures that can occur in the future.

The AFD system supports these applications by providing a disciplined, rigorous process where the designer can:

- thoroughly analyze given failure mechanisms;
- obtain an exhaustive set of potential failure scenarios; and
- develop inventive solutions to prevent, counteract, or minimize the impact of failure scenarios.

More Principles of TRIZ

TRIZ states that the evolution of engineering systems is not a random process, but is governed by certain objective laws. These laws are used to predict how a certain system will develop in the next phase. The laws can be a useful tool in product planning by providing support for technological forecasting.

Principle of technical systems. In TRIZ methodology, anything that performs a function is a technical system and consists of one or more subsystems. The hierarchy of a technical system spans from the least complex, with only two elements, to the most complex with many interacting elements. When a technical system produces inadequate functions, it may need to be improved. This requires an imaginative reduction of the system to its simplest state. In TRIZ, the simplest technical system consists of two elements with energy passing from one element to another.

There are two groups of problems people face: those with generally known solutions and those with unknown solutions. In the case of known solutions, they follow a general pattern of problem solving. In Figure 2-40, the specific problem is considered as a standard problem of a similar or analogous nature. A standard solution results in a specific solution. The functions of technical systems are realized by using physical, chemical, and geometrical effects. It follows that knowledge of such effects is crucial in inventive situations.

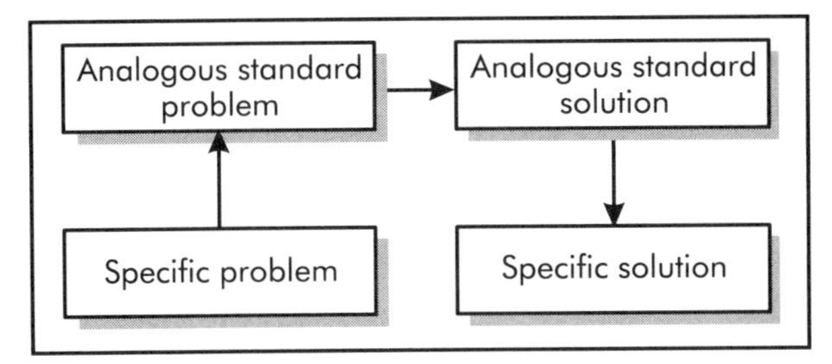

Figure 2-40. General model of technical systems.

Law of Increasing Ideality. The most fundamental law is that of the *ideal system*, one in which the given function is realized, but where no resources are consumed. This ideal solution may never be found, but the ratio function divided by the resources is likely to increase over time. The ideal-system law states that any technical system, throughout its lifetime, tends to become more reliable and effective. Ideality always reflects the maximum utilization of existing resources, both internal and external to the system. Using readily available resources makes the system more ideal. The *Law of Increasing Ideality* means that technical systems evolve to increasing degrees of ideality.

Psychological inertia. If the problem being explored has no solution forthcoming or if it lies beyond the designer's experience, a number of trials could vary, depending on the designer's intuition and creativity. But the drawback is that it could be difficult to extrapolate these psychological tools to other people in the organization. This leads to *psychological inertia* in which solutions considered are only within an individual's domain of expertise and do not consider alternative means. An ideal solution to a problem may lie outside the designer's field of expertise. The limitation of psychological inertia leads to only looking where there is personal experience. It does not consider that the solution could occur randomly within a solution space.

Combination of TRIZ and QFD. Since TRIZ can help engineers and developers solve technical contradictions and invent new technologies, its use in new product development is very important. Combined with QFD, a company should be able to identify important customer requirements and then solve any technical bottlenecks that arise. TRIZ can also help identify new functions and performance levels to achieve truly exciting levels of quality. Table 2-11 showed areas where QFD and TRIZ can complement each other.

QFD identifies and prioritizes the voice of the customer and the capabilities of a company's technologies. It then helps prioritize new concepts for design and production. TRIZ helps create new concepts.

USING A SYSTEMATIC APPROACH TO PROBLEM SOLVING

A *systematic approach to problem solving* (SAPB) assumes that design problem solving is a variant of general problem solving. When designing, the designer usually follows a path consisting of certain fundamental activities—problem and requirement formulation, a search for alternative solutions, evaluation and documentation, and communication of the results. Design methodologies support this process by providing specific design methods and knowledge. SAPB divides the design process into a number of phases where decisions are made (see Figure 2-41). In using SAPB, delimitations between phases are approximate and therefore there is need for iteration and recursion. The main phases of SAPB are:

1. Clarify the task.
2. Conceptualize the design.
3. Embody the design.
4. Detail the design.

Clarify the Task

Clarifying the task involves collecting information about requirements that products

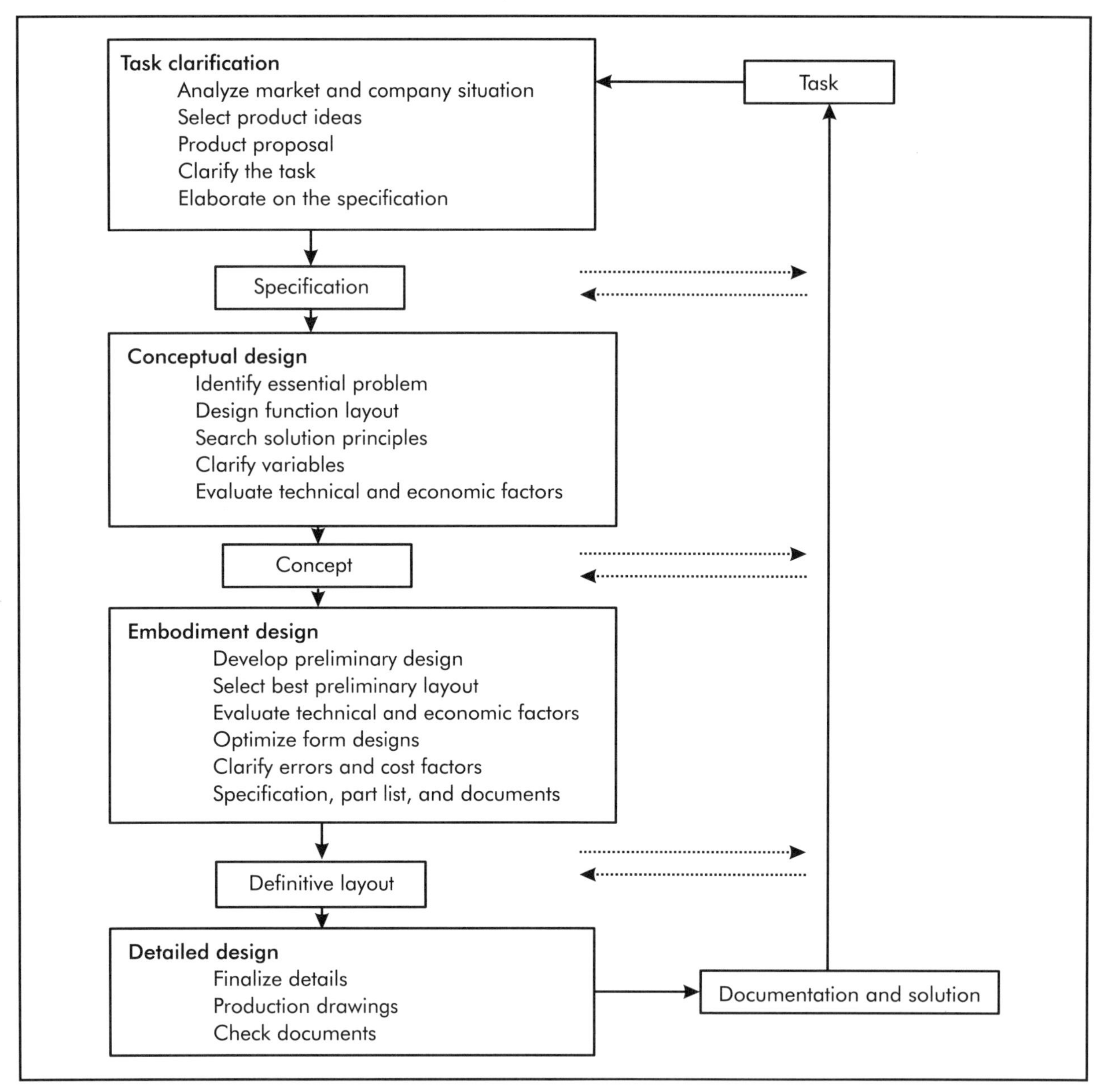

Figure 2-41. The design process according to SAPB (Pahl and Beitz 1988).

should meet and about constraints. This phase results in a detailed design specification.

Conceptualize the Design

The conceptual design phase starts with an analysis of the specifications to identify essential problem(s) to be solved. The design problem is then formulated in an abstract, solution-neutral form. This makes the solution space as wide as possible, which serves to dispel prejudices that may tempt the designer to decide on a certain solution before other alternatives have been considered. The problem may then be decomposed into subproblems and a function structure is established. Solutions to subfunctions are then sought. This process is supported by creative methods (brainstorming), conventional methods (patent searches, etc.), and systematic methods.

The systematic methods make use of design catalogues with physical and chemical effects and machine elements. Morphological matrices are used to combine subfunction solutions into system solutions. Promising system solutions are then developed further into concept variants. Finally, use-value analysis is used to evaluate concept variants, and the best concept is selected for further development. Here, use-value analysis helps ensure that a rational, objective decision is made. SAPB emphasizes the importance of the decisions made in the conceptual design phase, since it is very difficult to correct fundamental shortcomings of a concept in the later embodiment and detail design phases.

Embody the Design

To embody the design, the designer develops the layout and form of the final system using the concept as a starting point. Also, several alternative designs may be considered, for example layout variants. The system in this phase will be developed to the point that a clear check of function, durability, production, assembly, and other requirements can be carried out. SAPB supports these activities by providing the following rules, principles, and guidelines:

- The rules state that three important conditions must be fulfilled if the design is to meet the requirements: clarity, simplicity, and safety.
- The principles relate to the fundamental engineering design knowledge. Examples are principles of subdivision of tasks and the use of self-reinforcing solutions.
- Guidelines are more domain-specific, such as design for assembly guidelines.

Detail the Design

Finally, in the detail design plane, detail drawings and production documents are completed. While most high-level decisions will have been made at this point, SAPB warns the designer not to relax too early; even the best concept can be ruined by a lack of attention during the detail design phase.

Comparing TRIZ and SAPB

This section presents a systematic comparative analysis of TRIZ and SAPB design methodology. Similarities and differences are discussed with respect to 14 different aspects. These aspects have been selected to cover task clarification and the conceptual design stages of the design process. For these stages, this section compares the steps, methods, design knowledge, and product models included in the methodologies. Two other aspects are also discussed: learning time and available computer support. Results are summarized in Table 2-14.

The SAPB methodology covers the entire design process; from task clarification through the conceptual, embodiment, and detail design phases. By contrast, TRIZ focuses on the activities that help to specify the problem, finding a physical, chemical, or geometrical effect to solve it, and then giving the solution some initial form of a conceptual design. This is the crucial step in the design process, and also the most poorly understood. SAPB is applicable to a wide range of design problems, simple as well as difficult, and engineering as well as other types of problems. Perhaps because of its emphasis on invention, TRIZ does not provide much support for solving simple problems. For difficult problems, however, TRIZ provides some tools that do not have SAPB correspondents, such as the principles.

Table 2-14. TRIZ versus SAPB

Aspect	TRIZ	SAPB
Scope	Inventive problem Emphasis on difficult problems Component design	Entire design process Simple and difficult problems Systems design
Task clarification	Laws of engineering systems evolution	General procedure
Problem formulation	Identified physical contradiction	Abstraction of essential problem
Systematic methods for generating solutions	Functions coupled to physical effects Standard principles	Functions coupled to physical effects Design catalogues
Creative methods	Not included	Brainstorming Synectics Delphi method
Function vocabulary	30 basic functions	Five generally valid functions
Solution space	Focused—only promising directions Minimal change of system	All possible solutions considered
Product models	Su-field model	Design specification Function structure Component structure
Principles	Principles for resolving contradictions Standards	Principles and guidelines for all phases of the design process
Knowledge base	Effects	Effects Design catalogues Engineering knowledge
Evaluation	Checklist	Use-value analysis
Learning time	Long time to team	Short time to learn
Computer support	Commercial	Research prototypes

Example of Using Contradictions

The example chosen can be described as, "Design of a comfortable bicycle seat." Bicycle seats are generally uncomfortable. The bicycle industry has developed a range of designs. There are literally hundreds of tiny variations on this same set of design principles. In each design, there is a subtle balance between tradeoffs that are inherent to a product required to give both weight support and the freedom to pedal.

Contradictions. The elimination of physical contradictions is a fundamental principle of the TRIZ method. In the case of the bicycle seat, the fundamental design trade-off is a compromise between a wide seat to achieve comfort for the cyclist, and a narrow seat to provide the freedom of movement for the legs during pedaling.

Seeking out the best compromise between these two extremes is not the correct problem that needs to be solved here. The right problem is more likely to be how to achieve a bicycle seat that is both wide and narrow. Expressed in terms of TRIZ contradiction, the situation that a designer is trying to improve about the bicycle seat is the *width of a stationary object.*

The factor that gets worse as the improvement of the seat's width is sought is the shape of the seat. For such a width/shape technical contradiction, the matrix recommends the following:

- principle of inversion,
- principle of dynamics, and
- principle of nesting.

For the inversion principle—or the other way around—the suggestion is: make movable parts fixed, and make fixed parts movable.

For the dynamic principle, the suggestions are:

- Divide the object into parts capable of moving relative to each other.
- Make the object rigid or inflexible so it becomes movable or adaptable.

For the nesting principle, the suggestion is to contain the object inside another, which in turn is placed inside of another object.

It is obvious that here is a solution to the bicycle-seat problem that not only uses the two inventive principles recommended by TRIZ, but that also is fundamentally right. This design gives cyclists support where the body desires support. Using the contradiction matrix is a good means of finding a solution to the right problem.

APPENDIX

The following are the 39 TRIZ/TIPS inventive principles for resolving physical contradictions.

Anti-weight

- To compensate for the weight of an object, merge it with other objects that provide lift.
- To compensate for the weight of an object, make it interact with the environment.

Asymmetry

- Replace a symmetrical form with an asymmetrical form.
- If an object is asymmetrical, increase its degree of symmetry.

Change the Color

- Change the color of an object or its surroundings.
- Change the translucency of objects or processes that are difficult to see.
- Use colored additives to observe objects or processes that are difficult to see.
- If such additives are already used, employ luminescent traces or tracer elements.

Cheap Short-lived Objects

Replace an expensive object with multiples of inexpensive objects, comprising certain qualities.

Combining

- Combine in-space homogeneous objects; assemble identical/similar parts to perform parallel operations.
- Make operations contiguous; bring them together in time.

Composite Materials

Replace a homogeneous material with a composite one.

Continuity of Useful Action

- Carry out an action continuously, where all parts of an object operate at full capacity.
- Remove idle and intermediate motions.

Convert Harm to Benefit

- Use harmful factors to achieve a positive effect.
- Eliminate a primary harmful action by adding to it another harmful action to resolve the problem.

Copying

- Use a simple and inexpensive copy, instead of an object that is complex, expensive, fragile, or convenient to operate.
- Replace an object with its optical copy or image. A scale can be used to reduce or enlarge the image.
- If visible optical copies are used, replace them with infrared or UV copies.

Dynamics

- Allow the characteristics of an object, the external environment, or the process to change to be optimal, or find an optimal operating condition.
- Divide an object into elements that can change their position relative to each other.
- If an object (or process) is rigid or inflexible, make it movable or adaptive.

Equipotentiality

Change the working conditions so that an object need not be raised or lowered; limit the position changes.

Extraction

- Extract (remove or separate) a disturbing part or property from an object.
- Extract only the necessary part or property.

Feedback

- Introduce feedback to improve a process or action.
- If feedback already exists, reverse it.

Flexible Membranes or Thin Film

- Replace traditional construction with products made from flexible membranes or thin film.
- Isolate the object from the external environment using flexible shells and thin films.

Homogeneity

Make objects interact with a given object of the same material (or material with identical properties).

Inert Atmosphere

- Replace a normal environment with an inert one.
- Add neutral parts or inert additives to an object.

Intermediary

- Use an intermediary carrier article or intermediary process.
- Temporarily connect one object to another that is easy to remove.

Inversion

- Implement an opposite action, instead of an action dictated by the specifications of the problem.
- Make movable parts fixed, and fixed parts movable.
- Turn an object upside down.

Local Quality

- Change the structure of an object from uniform to non-uniform; change an external environment (or external influence) from uniform to non-uniform.

- Have different parts of the body perform different functions.
- Make each part of an object function in conditions most suitable for its operation.

Mechanical Vibration

- Set an object into oscillation.
- If oscillation exists, increase its frequency, even as far as ultrasonic.
- Use resonant frequency.
- Instead of mechanical vibrations, use piezo-vibrators.
- Use ultrasonic vibrations in conjunction with an electromagnetic field.

Moving to a New Dimension

- Move an object in two- or three-dimensional space.
- Use a multistory arrangement of objects instead of a single-story arrangement.
- Tilt or reorient an object; lay it on its side.
- Use another side of a given area.

Nesting

- Contain the object inside another, which in turn is placed inside another object.
- Pass an object through the cavity of another object.

Parameter Changes

- Change an object's physical state.
- Change the concentration or consistency.
- Change the degree of flexibility.
- Change the temperature.

Partial or Excessive Actions

If it is difficult to obtain 100% of a desired effect, achieve somewhat more or less to simplify the problem.

Periodic Action

- Instead of continuous action, use periodic or pulsating actions.
- Change the periodic magnitude or frequency if an action is already periodic.
- Use pauses between impulses to perform a different action.

Phase Transformation

Implement an effect developed during phase transition of a substance.

Pneumatics and Hydraulics

Replace solid parts of an object by gas or liquid. These parts can use air or water for inflation, or use air or hydrostatic conditions.

Porous Materials

- Make an object porous or add porous elements (inserts, coatings, etc.).
- If an object is already porous, use pores to introduce a useful substance or function.

Prior Action

- Perform in advance the required change of an object (either fully or partially).
- Arrange objects beforehand, so they can come into action from the most convenient place without losing time for delivery.
- Compensate for the relatively low reliability of an object by taking countermeasures in advance.

Prior Anti-action

- If it is necessary to do an action with both harmful and useful effects, this action should be replaced with anti-actions to control harmful effects.
- If an object is under tension, provide anti-tension in advance.

Rejecting and Regenerating Parts

- After it has completed its function or become useless, reject or modify an element of an object.
- Immediately restore any part of an object that is exhausted or depleted.

Replacement of a Mechanical System

- Replace a mechanical means with a sensory (optical, acoustic) means.
- Use an electrical, magnetic, and electromagnetic field to interact with the object.
- Change from static to movable fields, and from unstructured fields to those having structures.
- Use fields in conjunction with field-activated particles (for example, ferromagnetic).

Segmentation

- Divide an object into independent parts.
- Make an object sectional.
- Increase the degree of object segmentation.

Skipping

Conduct a process, or certain stages, at high speed.

Self-service

- Make an object serve itself by performing auxiliary helpful functions.
- Use waste resources, energy, or substances.

Spheroidality

- Replace rectilinear part surfaces with curved ones; replace cubical shapes with spherical shapes.
- Use spirals, rollers, and balls.
- Go from linear to rotary motion; use centrifugal forces.

Thermal Expansion

- Use thermal expansion (or contraction) of materials.
- If thermal expansion is being used, use multiple materials with different coefficients of thermal expansion.

Use Strong Oxidizers

- Replace normal air with enriched air.
- Replace air with oxygen.
- Treat an object in air or in oxygen with ionizing radiation.
- Use ionized oxygen.

Universality

Make parts or objects perform multiple functions; eliminate the need for other parts.

REFERENCES

Adams, James L. 1976. *Conceptual Block Busting: A Pleasurable Guide to Better Problem Solving*. San Francisco, CA: San Francisco Book Co.

Akiyama, Kaneo. 1991. *Function Analysis: Systematic Improvement of Quality and Performance.* Cambridge, UK: Cambridge Productivity Press.

Blair, Brian and Shetty, Devdas. 2002. *Integrating and Applying QFD to Axiomatic Design*. 14th International ASME Conference on Design Methodology. New York: American Society of Mechanical Engineers.

Cross, Nigel. 1994. *Engineering Design Methods— Strategies for Product Design*, 2nd Ed. Chichester, UK: John Wiley and Sons.

DeBono, E. 1970. *Lateral Thinking: Creativity Step by Step*. New York: Harper & Row.

Dominick, Peter, Demel, John, Lawbaugh, William, Freuler, Richard, Kinzel, Gary, and Fromm, Eli, 2001. *Tools and Tactics of Design*. New York: John Wiley & Sons, Inc.

Fogler, Scott and LeBlanc, Steven E. 1995. *Strategies for Creative Problem Solving*. Upper Saddle River, NJ: Prentice Hall.

Koestler, Arthur. 1969. *The Act of Creation*. New York: MacMillan.

Miles, Lawrence D. 1972. *Techniques of Value Analysis and Engineering,* 2nd Ed. New York: McGraw Hill, Inc.

Nierenberg, G.I. 1986. *The Art of Creative Thinking*. New York: Simon & Schuster, Inc.

Osborn, A.F. 1957. *Applied Imagination*. New York: Scribners.

Pahl, Gerhard and Beitz, Wolfgang. 1988. *Engineering Design: A Systematic Approach*. London, UK: Springer-Verlag.

Pugh, Stuart. 1981. "Concept Selection—A Method that Works." Proceedings of the International Conference on Engineering Design (ICED 81). Zurich, Switzerland: International Society for the Science of Engineering Design, WDK: March.

——. 1991. *Total Design—Integrated Methods for Successful Product Engineering*. London, UK: Addison-Wesley Publishing Co.

Stein, Morris I. and Heinze, S.J. 1960. *Creativity and the Individual*. New York: Free Press.

Suh, Nam P. 2001. *Axiomatic Design Advances and Applications*. Oxford, UK: Oxford University Press.

——. 1990. *The Principles of Design*. Oxford, UK: Oxford University Press.

——. 1989. "Design Axioms and Quality Control." Report 6-22-89. Boston, MA: Massachusetts Institute of Technology Industrial Liaison Program.

Ternicko, J. Zusman and Zlotin, B. 1998. *Systematic Innovation: An Introduction to TRIZ*. Boca Raton, FL: St. Lucie Press.

Ulrich, Karl and Eppinger, Steven. 1995. *Product Design and Development*. New York: McGraw-Hill, Inc.

Chapter 3

The Impact of Product Design

INTRODUCTION

The cost of a product is very much decided at the design stage, where costs of fabrication, assembly, and inspection are determined. The designer should be aware of the nature of assembly processes. He or she should always have sound reasons for requiring separate parts, and hence, higher assembly costs. Each combination of two parts into one will eliminate at least one operation in manual assembly or an entire section of an automatic assembly machine.

This section deals with different approaches to product design and highlights their weaknesses and strong points. Several techniques and tools enable designers to deal with design and avoid pitfalls in developing a new product. The axiomatic, Hitachi, Boothroyd-Dewhurst, and Lucas methods are used in many industries for product design (Boothroyd-Dewhurst 1987). The axiomatic method, based on a scientific approach to design, was discussed in Chapter 2. This method is based on an attempt to identify common properties of successful designs. These common properties, such as how proposed design satisfies functional requirements, were proposed as axioms of good design. Design axioms have to be satisfied to design a successful product. They can be viewed as global product guidelines that coexist with component guidelines for details of a product.

The remaining methods are used to analyze a design and investigate how design components are processed and assembled. The goal is to increase product quality and reduce cost and time to market. Most of these methods evaluate the efficiency of current design. Using comparison charts, they determine design validity.

This section also discusses choices of assembly methods and recent trends in automatic assembly.

Definitions of Relevant Terms

Easy to align and position—A part is easy to align and position if the position of the part is established by locating features on the part or on its mating parts, and if insertion is facilitated by well-designed chamfers or similar features.

Resistance to insertion—The resistance felt by a part during insertion can be due to small clearances, jamming, wedging, or acting against a large force. Examples are press fits or self-tapping screws.

Tangle—Components may tangle if reorientation is required to separate them from the bulk layout.

Severely tangle—Components are said to severely tangle if they require manipulation for specific orientations and if force is required to separate them. Operators may use both hands to untangle components.

Flexible—Flexible parts deform substantially during assembly and manipulation. Operators may use two hands when handling paper, belts, felt gaskets, and cable assemblies.

Handling difficulties—Components can present handling difficulties if they nest, tangle, or stick together. This happens because of magnetic attraction, grease coatings, or parts being slippery, delicate, or hot or cold.

Obstructed access—Obstructed access means that space available for assembly causes a significant increase in the assembly time. The access could be for fingers to reach or for tools used.

Restricted vision—If the assembly environment has restricted vision, an operator has to depend on tactile sensors during assembly.

Holding down required—If a part is unstable after placement or insertion, it will require gripping, realignment, or holding down before it is finally secured. Holding down is also an operation intended for maintaining the position and orientation of a part either during its assembly or its pre- or post-assembly operations.

Located—A part is said to be located if it is partially located, and if it does not require holding down or realignment for the next activity.

Envelope—The envelope is the smallest cylinder or rectangular prism that can completely enclose the part. The size is the length of the longest side of the envelope. The thickness is the length of the shortest side of the envelope.

DESIGN FOR MANUFACTURABILITY METHODOLOGY

The importance of design on manufacturing cannot be overstated. The ability to generate profit can be limited by specifying parts requiring secondary and other miscellaneous operations. Design simplification and designing a product for easy manufacturing should be part of every product's life cycle. The central issue in the design-for-manufacture system includes design guidelines that help designers optimize the number of parts. The ability of the designer to apply these rules is a key factor in superior product design.

Design for Assembly Guidelines

Common design for assembly guidelines used by designers are listed in Table 3-1.

Part Count Reduction

The designer should go through the assembly process part by part and evaluate whether the part can be eliminated, combined with another part, or if the job can be performed in another way. To determine a theoretical minimum number of parts, a designer should ask the following questions:

- Does the part move relative to all other moving parts?
- Must the part absolutely be of a different material from the other parts?
- Must the part be different to allow for possible disassembly?

The designer should simplify and reduce the number of parts because, for each part, there is an opportunity for a defective part and an assembly error. The probability of a perfect product goes down exponentially as the number of parts increases. As the number of parts goes up, the cost of fabricating and assembling the product goes up. As parts are reduced, the cost of inventory and purchasing are reduced. The use of manufacturing processes such as injection molding, extrusion, and metallurgy can bring about part-count reduction. Table 3-2 gives examples of how to reduce the number of parts in design for assembly.

Table 3-1. Design for assembly (DFA) guidelines

Reduction in part count	Simplify and reduce the number of parts. Reduction of parts in the product design stage reduces inventory, handling time, processing time, and manufacturing costs.
Standardization	Standardize and use common parts and materials. This assists in minimizing the amount of inventory. Standardize handling and assembly operations.
Design for ease of fabrication	Optimum combination should occur between the materials and processes. The manufacturing process should be compatible with the materials used and the volume of production.
Robustness in product design	Mistake-proof product design and assembly so that the assembly process is direct and easy. All parts should be designed so they can be assembled in only one way.
Design parts for easy orientation	Use the principles of design for ease of part handling and orientation. Minimize non-value-added efforts and uncertainty in orienting and merging parts. Design parts should be multi-functional.
Minimize flexible parts	Avoid using flexible parts and interconnections. If cables have to be used, use a dummy connector to plug the cable so it is easily located.
Design for ease of assembly	Use simple patterns of movement that minimize the axes of assembly during the assembly process. Design modular products that use components as building blocks.
Design for efficient joining and fastening	Design joints and connections for efficient fastening and removal.

Standardization

Standard components are less expensive than customized components. Common parts will result in lower inventories, reduced costs, and higher quality. The following guidelines apply to standard components:

- Standardize and use common parts and materials to facilitate design activities (see Table 3-3).
- Minimize the amount of inventory in the system, and standardize handling and assembly operations.

Operator learning is simplified and there is greater opportunity for automation as the result of high-production volumes and operation standardization. The designer should limit unique components because suppliers are less likely to compete on quality or cost for these components. Group technology can guide selection and development of manufacturing cells for common parts or product families.

Design for Ease of Fabrication

There should be an optimum combination of material and production processes to minimize manufacturing costs.

Designers should use net shape for molded and forged parts to minimize machining and processing as follows:

Table 3-2. Examples of design for assembly

Guideline	Poor Design	Better Design
Reduce number of parts	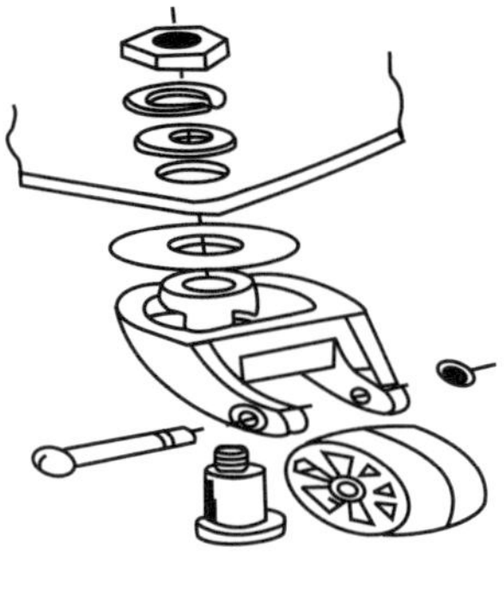 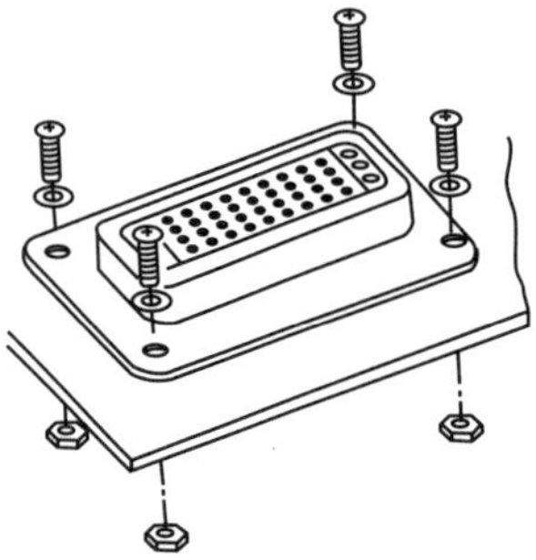 (1) (2) Pins 0.07 in. (1.7 mm) Ø × 2	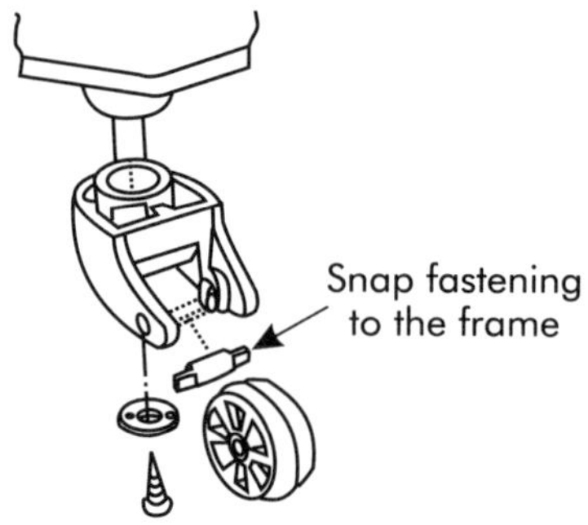 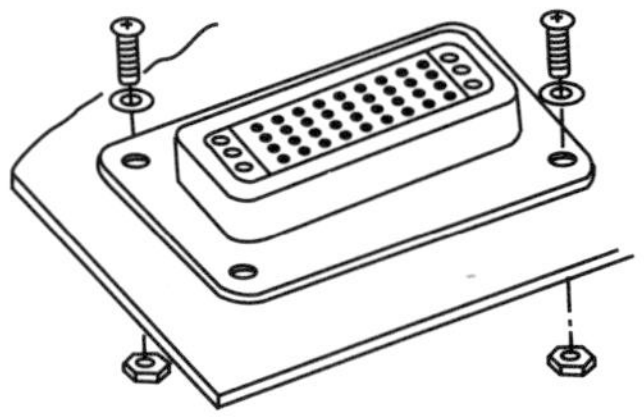 Snap fastening to the frame 0.08 in. (2 mm) Ø boss

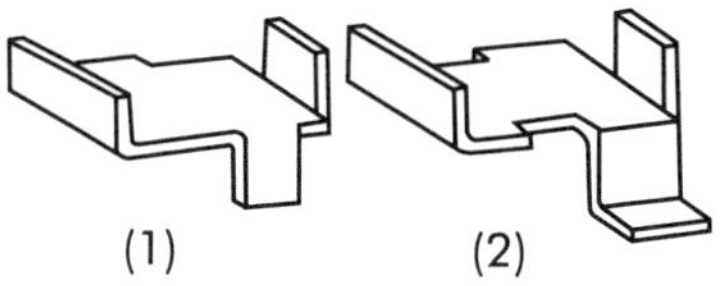

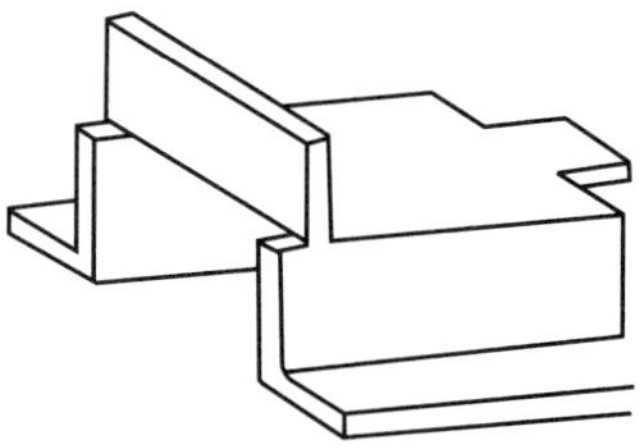

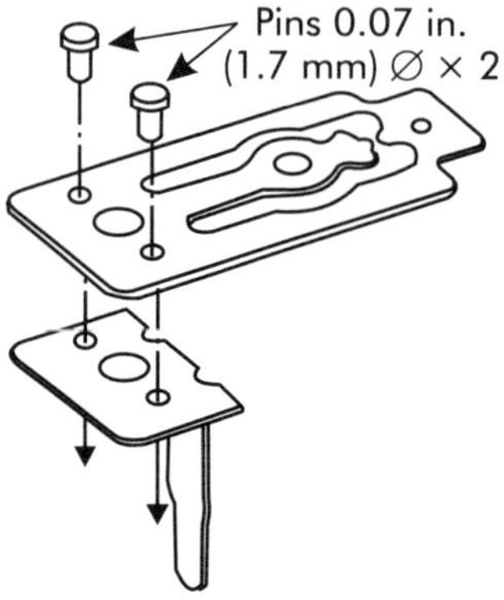

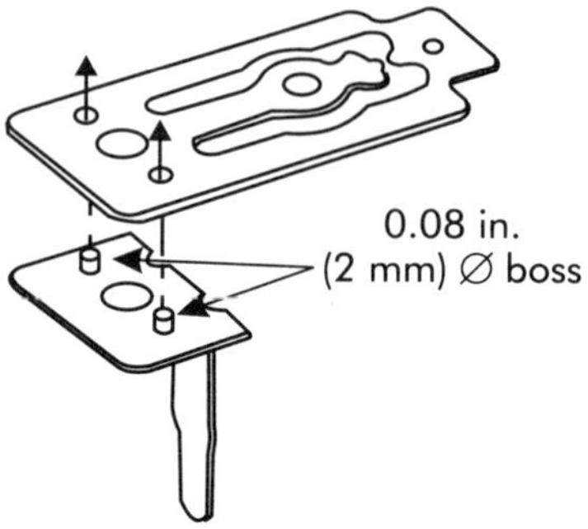

Table 3-2. (continued)

Guideline	Poor Design	Better Design
Reduce number of parts	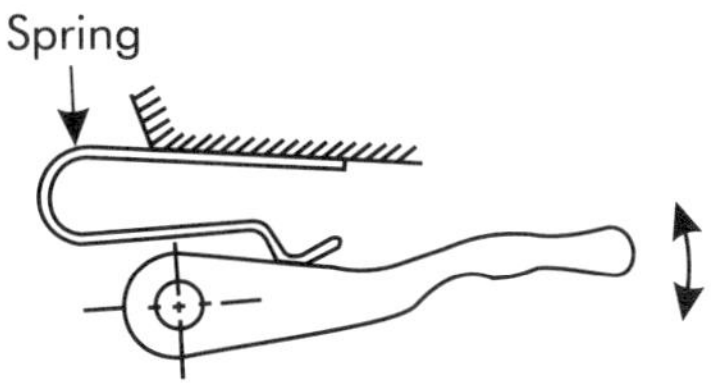 Spring	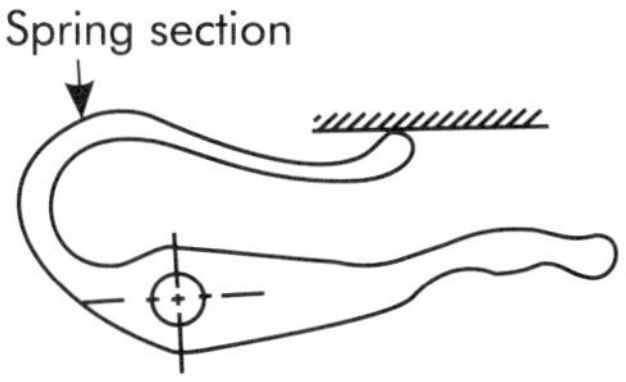 Spring section
	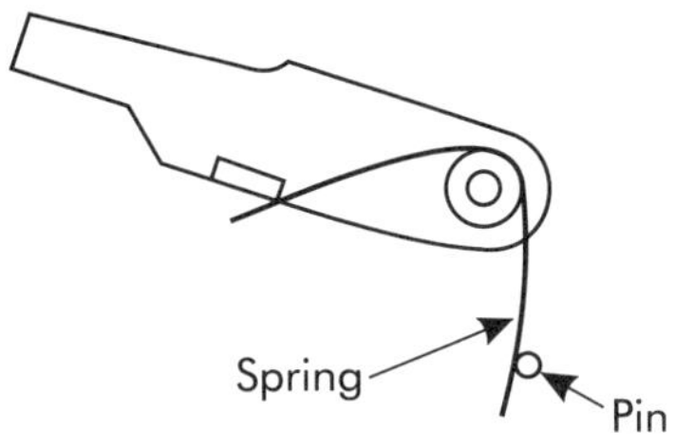 Spring Pin	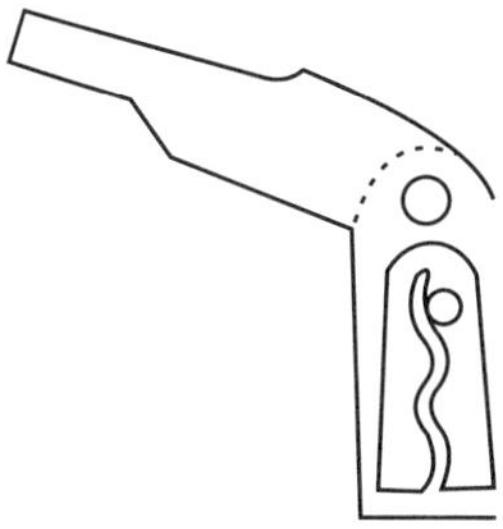
	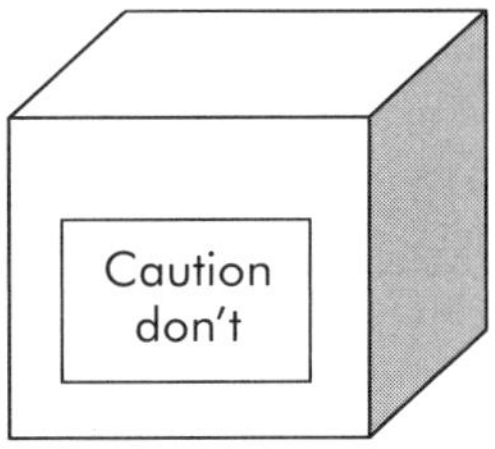 Caution don't Label	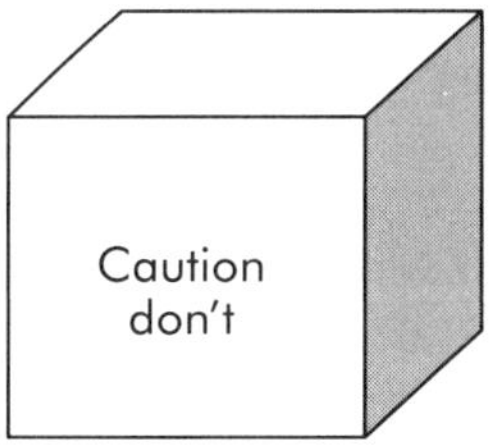 Caution don't Letter molded
	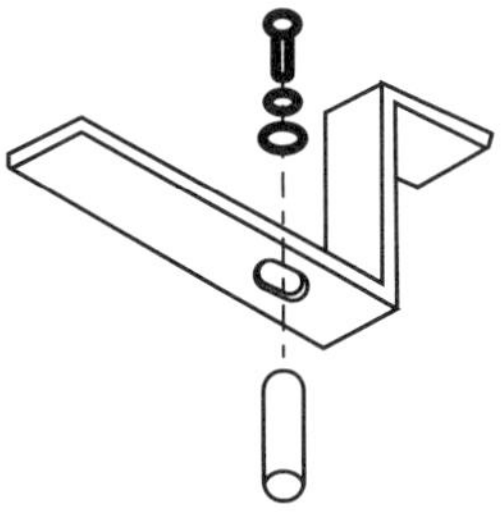 Five parts	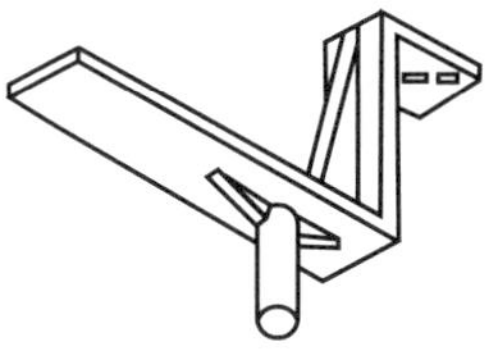 One part
	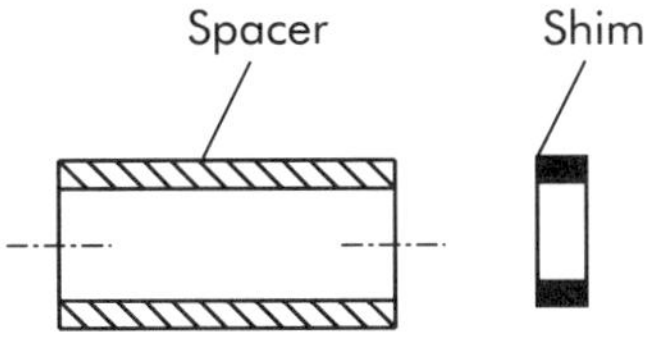 Spacer Shim	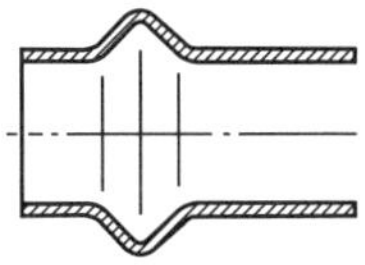

Table 3-3. Standardization in design for assembly

Guideline	Poor Design	Better Design
Specify standard parts: • reduced need for unique tools, • reduce assembly time, and • improve inventory control.	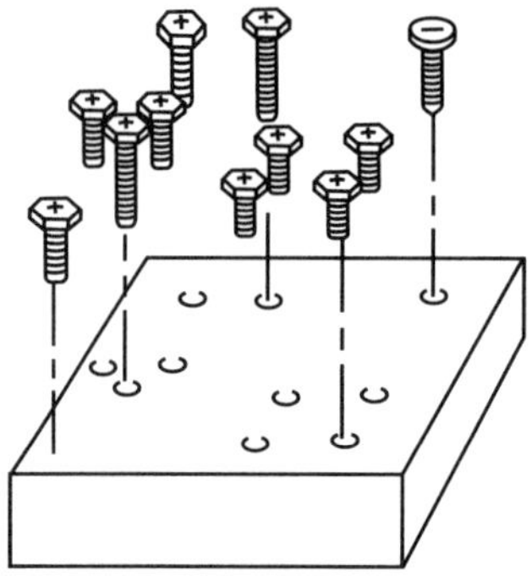	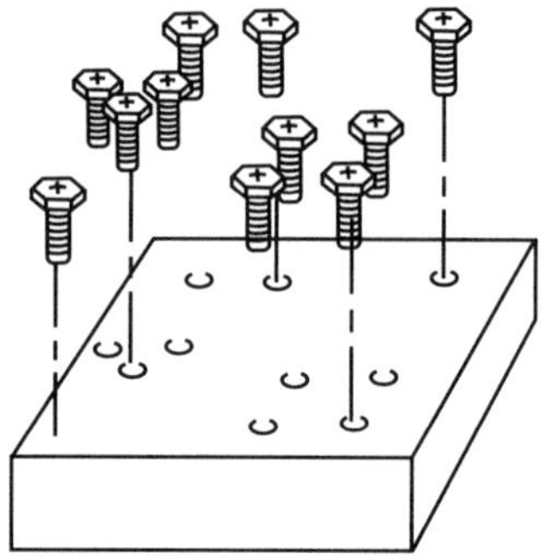
Eliminate post-assembly adjustments: • easier positioning, • reduced assembly time, and • avoid tolerance demands on mating parts.	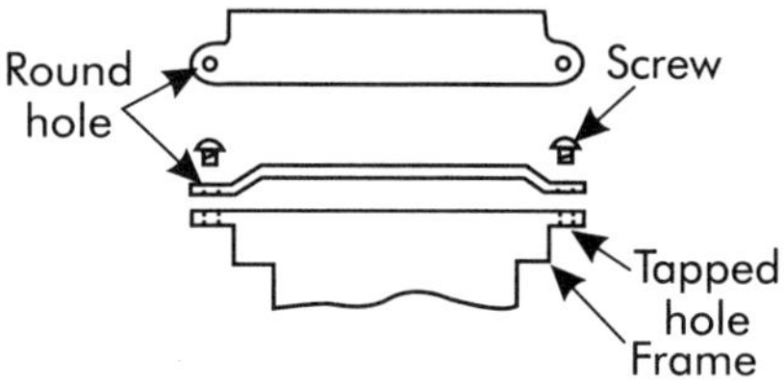 Round hole Screw Tapped hole Frame	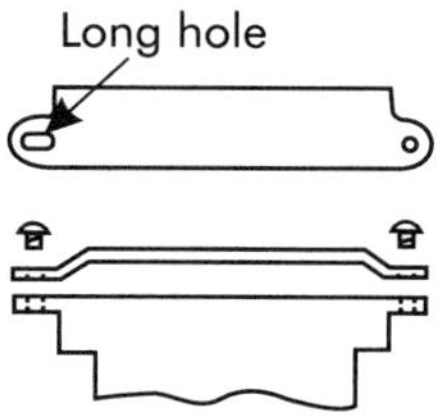 Long hole
Use screws with length-to-diameter ratio greater than 1.5 (so screw will not jam if fed automatically by being blown through a tube).	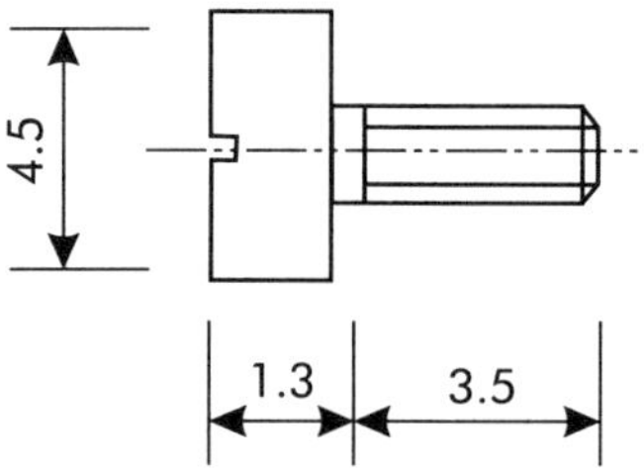 4.5 1.3 3.5 Short screw (L/D <1.5)	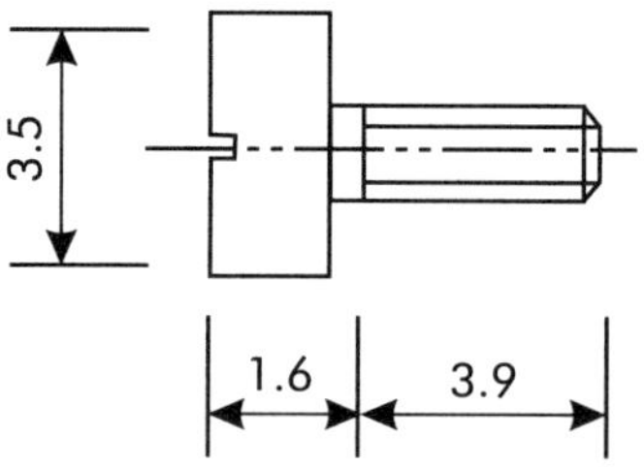 3.5 1.6 3.9 Longer screw (L/D >1.5)

- Avoid unnecessarily high tolerances beyond the natural capability of the manufacturing processes.
- Avoid tight tolerances on multiple interconnected parts. Tolerances on multiple assembled parts can stack up undesirably.

Robustness in Product Design

Design individual components so that part variation in the product does not compromise total performance. For mechanical products, verifiability can be achieved with simple go/no-go tools. Electronic products can be designed to contain self-test and/or diagnostic capabilities. A designer should provide mistake-proof assembly so assembly is consistent. He or she should design parts for multi-use and develop a modular design approach as follows:

- As far as possible, products should be designed to avoid post-assembly adjustments (see Table 3-3).
- A product and its components should incorporate design verifiability.

Design Parts for Easy Orientation

The designer should design parts for easy orientation so that the components can be handled with a minimum of difficulty and time. He or she should minimize non-value-added manual effort in orienting and merging parts. Adding notches, creating asymmetrical holes, and taking steps such as the following can prevent assembly mistakes:

- Parts must be designed to consistently orient themselves when fed into a process.
- Design must avoid parts that can become tangled, wedged, or disoriented.
- Design should incorporate symmetry in components, low centers of gravity, easily identifiable features, guide surfaces, and tools for pick-up and handling.

Design of parts for easy orientation will allow automation in parts handling and assembly—such as vibratory bowls, tubes, magazines, pick-and-place robots, and vision systems. When components are purchased, look for components that are already oriented in magazines, bonds, tape, or strips.

Orientation involves proper alignment of the part for insertion relative to its corresponding member (see Table 3-4). This is achieved by aligning the axis of the part corresponding to the axis of insertion.

Alpha and beta symmetry. There are two types of symmetry definitions for a part. Alpha symmetry (α-*symmetry*) is the rotational symmetry of a part about an axis that is perpendicular to the axis of insertion. For parts with one axis of insertion, end-to-end orientation is necessary when $\alpha = 360°$, otherwise $\alpha = 180°$. Beta symmetry (β-*symmetry*) is the rotational symmetry of a part about its axis of insertion, or equivalently about an axis that is perpendicular to the surface on which the part is placed during assembly. The magnitude of rotational symmetry is the smallest angle through which the part can be rotated and repeat its orientation. For a cylinder inserted into a circular hole, $\beta = 0$; for a square-section part inserted into a square hole, $\beta = 90°$, etc. The thickness is the length of the shortest side of the smallest rectangular prism that encloses the part. However, if the part is cylindrical, or has a regular polygonal cross-section with five or more sides, then the thickness is defined as the radius of the smallest cylinder that can enclose the part. The size is the length of the longest side of the smallest rectangular prism that can enclose the part (see Table 3-5).

Minimize Flexible Parts

Minimize flexible parts and interconnections as follows:

- Avoid flexible and flimsy parts such as belts, gaskets, tubing, cables, and wire harnesses. Their flexibility makes material handling and assembly more difficult and these parts are more susceptible to damage.
- Use plug-in boards and back planes to minimize wire harnesses. Interconnections such as wire harnesses, hydraulic lines, and piping are expensive to fabricate, assemble, and service.

Design for Ease of Assembly

Designers should plan for ease of assembly utilizing simple patterns of movement

Table 3-4. Design for ease of part orientation

Guideline	Poor Design	Better Design
Avoid reorientation, use symmetry	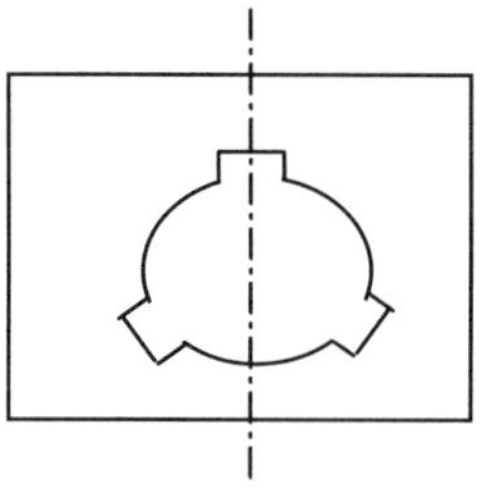 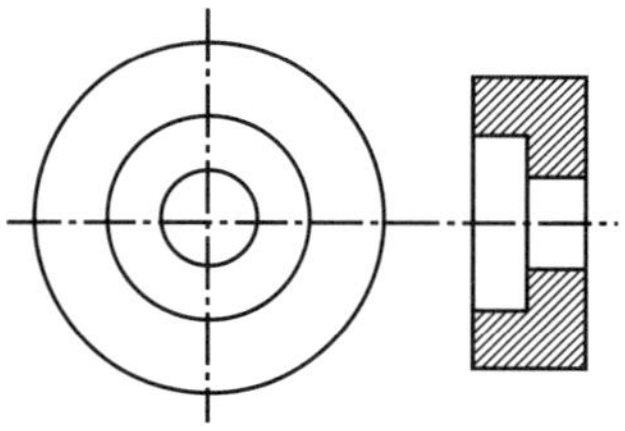 25 2 60 L_1 L_2 $L_2 = L_1$	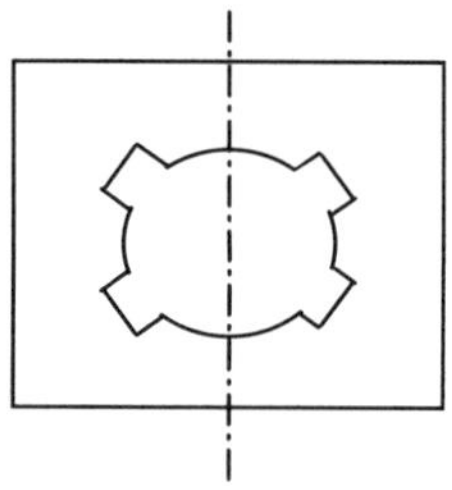 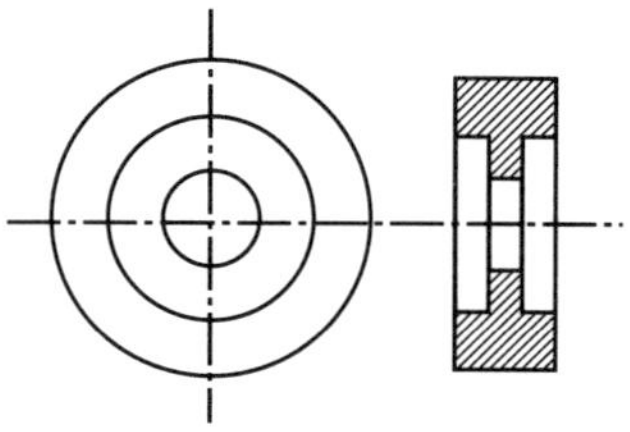 L_1 L_2 $L_2 > L_1$

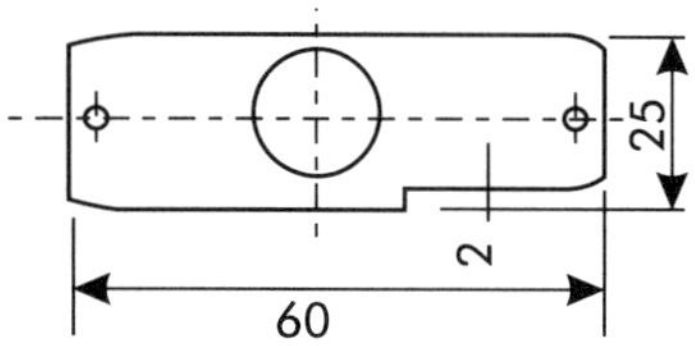

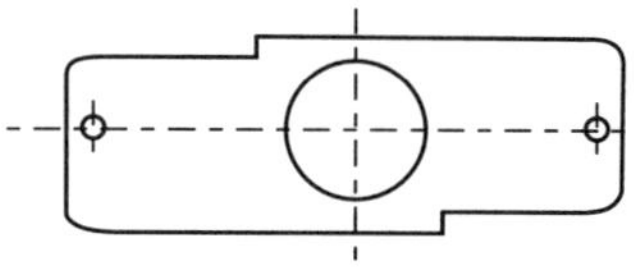

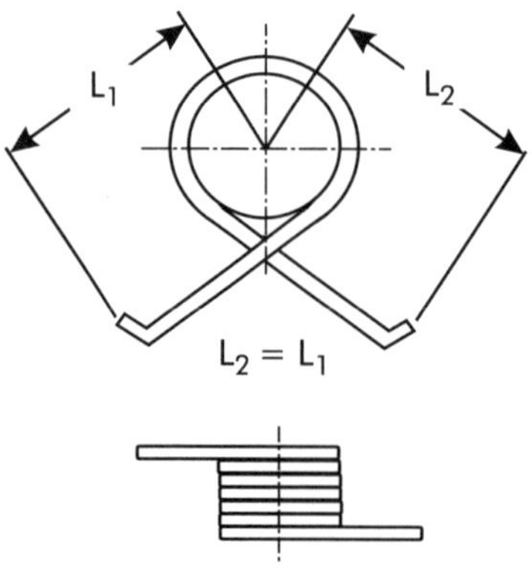

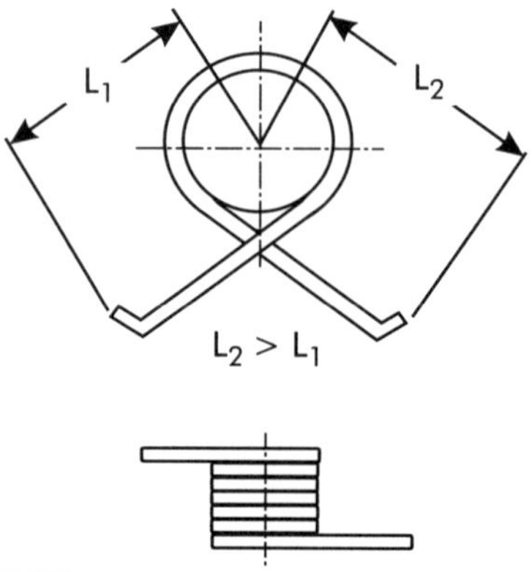

Table 3-4. (continued)

Guideline	Poor Design	Better Design
Define asymmetry by one main feature	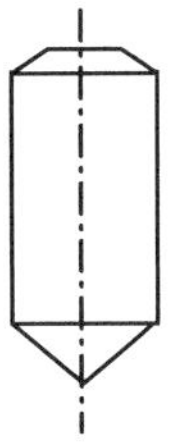	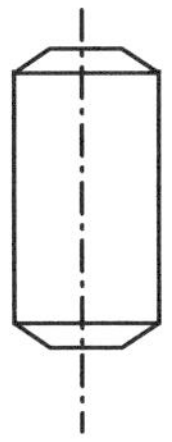
Exaggerate symmetry	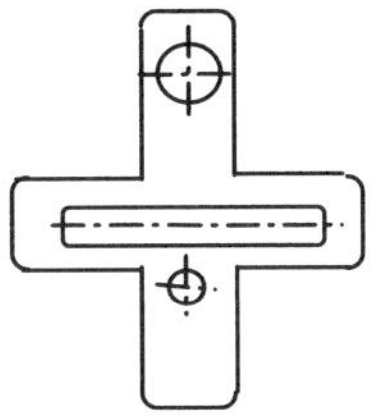	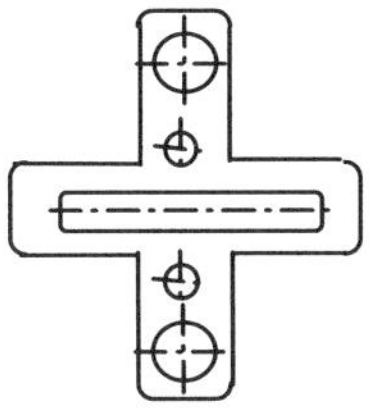
Provide a means to easily grip and hold the part	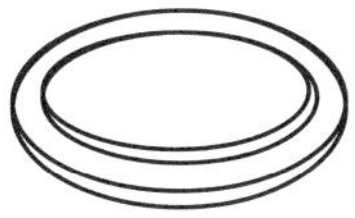	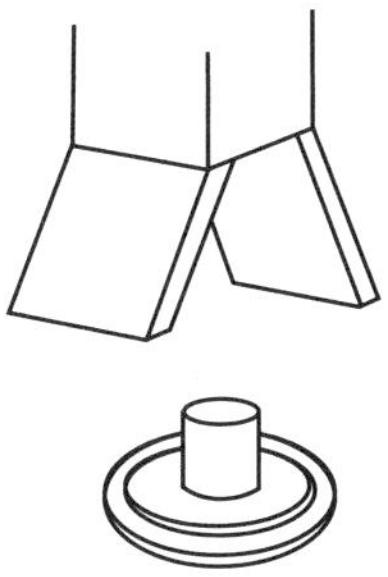
Technique to simplify part insertion	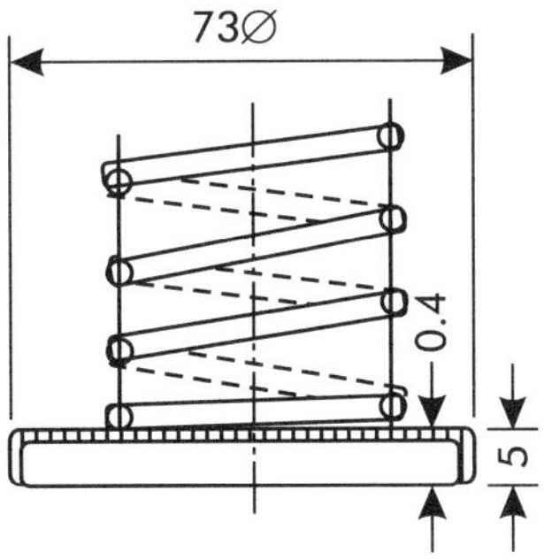 73∅ 0.4 5 Locating feature is subtle and inefficient	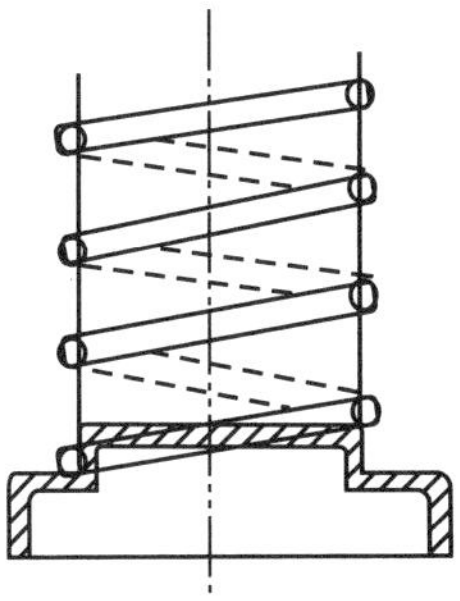 Locating protrusion provides a positive position for mating parts

Table 3-5. Alpha and beta symmetry

Alpha symmetry

α symmetry—is the rotational symmetry of a part about an axis perpendicular to the axis of insertion. Alpha symmetry depends on the angle through which a part must be rotated about an axis, perpendicular to the axis of insertion to repeat its orientation.

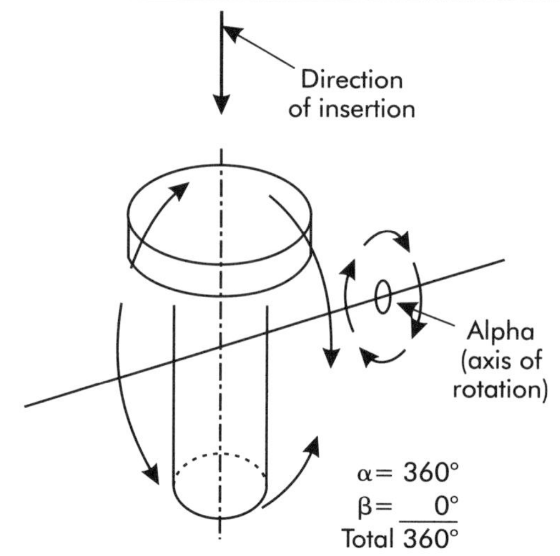

Beta symmetry

β symmetry—is the rotational symmetry of a part about its axis of insertion, or about an axis that is perpendicular to the surface on which the part is placed. β-symmetry depends on the angle through which a part is rotated about the axis of insertion to repeat its orientation. For a cylinder inserted into a circular hole, $\beta = 0°$; for a square section inserted into a square hole, $\beta = 90°$.

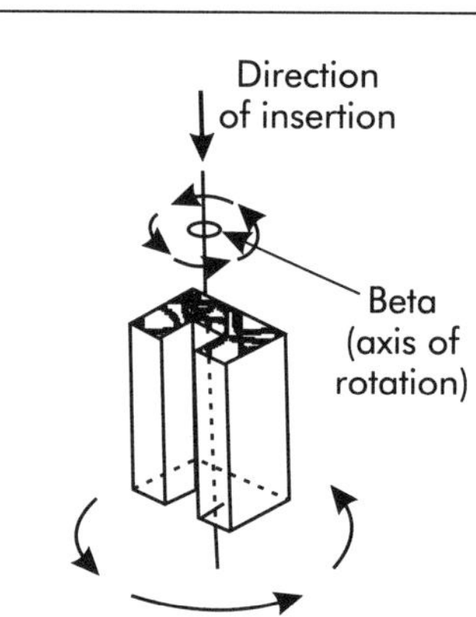

Examples of alpha and beta symmetry:

Alpha symmetry

An alpha-symmetric part is one that does not require end-to-end orientation.

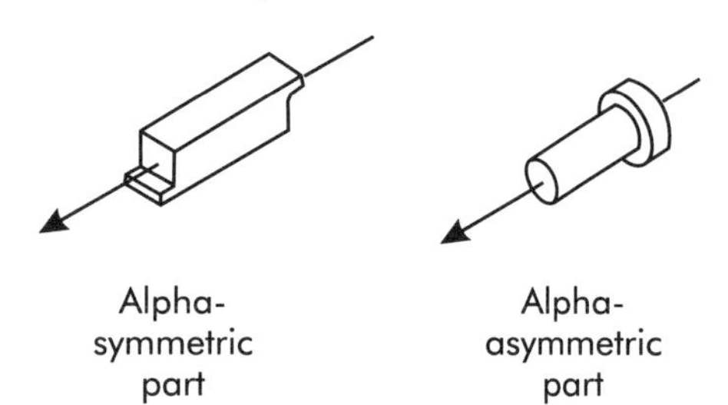

Beta symmetry

A beta-symmetric part is one that does not require orientation about the axis of insertion.

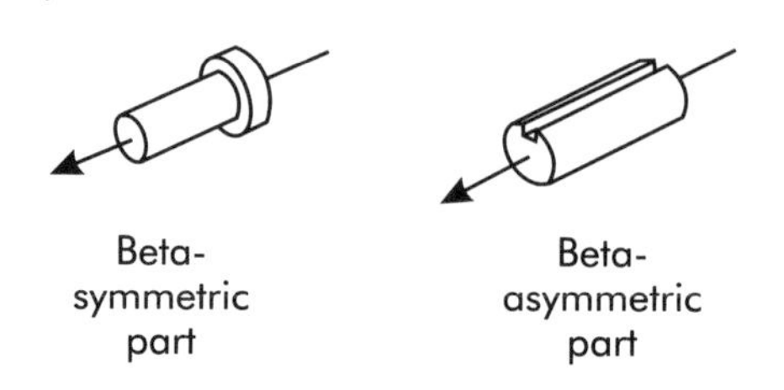

Table 3-5. (continued)

Basic shapes										
Alpha	0°	180°	360°	180°	90°	360°	180°	180°	360°	360°
Beta	0°	0°	0°	90°	180°	360°	0°	90°	0°	360°

and by minimizing the axes of assembly. They should avoid complex orientation and assembly movements made in various directions in the following ways:

- Part features should be provided such as chamfers and tapers.
- The product's design should enable assembly to begin with a base component with a large relative mass and a low center of gravity upon which other parts are added.

Assembly should proceed vertically with other parts added on top and positioned with the aid of gravity. This will minimize the need to reorient assembly and reduce the need for temporary fastenings and more complex fixturing. Table 3-6 provides examples of design for ease of assembly. A product that is easy to assemble manually will normally be easily assembled with automation. Automated assembly will be more uniform, more reliable, and of higher quality.

Design for Efficient Joining and Fastening

Threaded fasteners (screws, bolts, nuts, and washers) are time consuming to assemble and difficult to automate. Designers should:

- Use standardization and minimize variety when using fasteners such as self-threading screws and washers.
- Consider the use of a snap-on fit to replace welded joints.
- Evaluate other bonding techniques such as adhesives. Match fastening techniques to materials and product requirements.

Part Handling

The manual handling process involves grasping, transportation, and orientation of parts or subassemblies before they are inserted into or added to the work fixture or partially built-up assembly. Guidelines for design for ease of part handling are given in Table 3-7. When design function permits, make parts with functionally superfluous features that facilitate handling during assembly.

A subassembly is considered a part if it is added during assembly. However, adhesives, fluxes, fillers, etc., used for joining parts are not considered parts. Time spent on assembly increases if the part requires holding down. Holding down is required if the part is unstable after placement, insertion, or during subsequent operations. A part may require gripping, realignment, or holding down before it is finally secured. Holding down also refers to a situation that maintains the position and orientation of a part already in place.

Parts can present handling difficulties if they nest or tangle, stick together because

Table 3-6. Design for ease of assembly

Guideline	Poor Design	Better Design
Simplify part insertion	Original: design is not self-locating. Spring hangs up.	Improved: chamfer guides spring into appropriate position.
	Avoid mating two unchamfered parts. Retainer Cup	Always have at least one (preferably both) chamfered.
	Original: difficult to locate and align.	Improved: chamfered threaded fastener greatly improves centering and starting.
Self-fastening part	Original: insert and screw 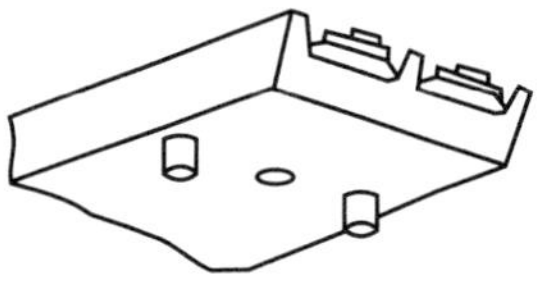	Improved: snap-fit feature 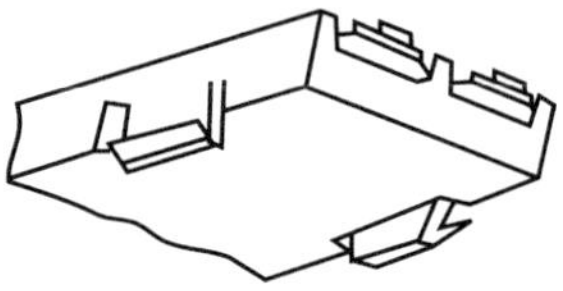

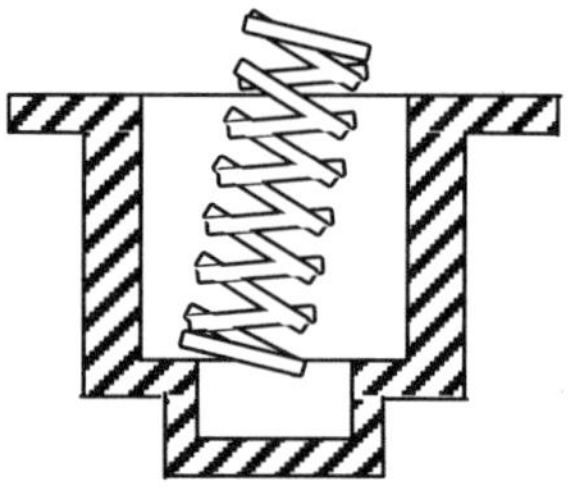

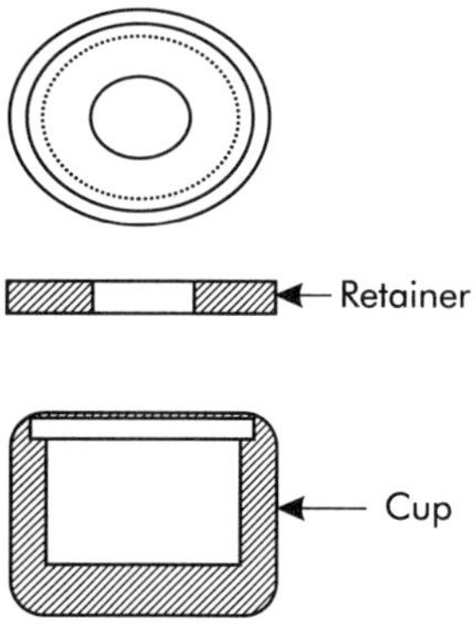

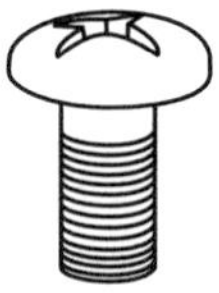

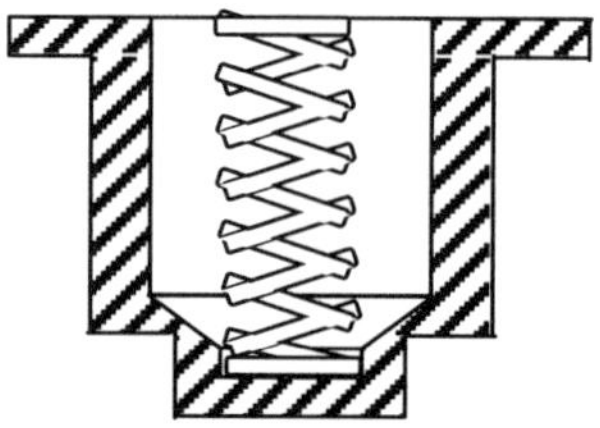

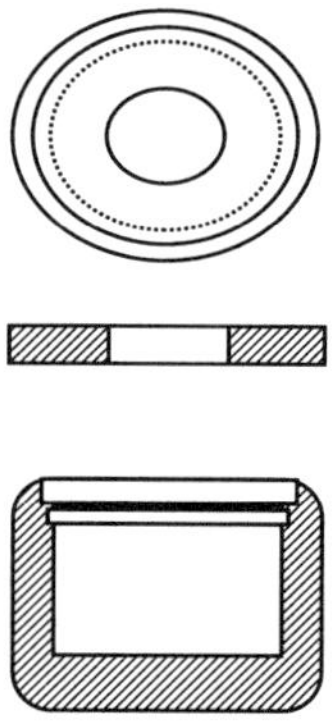

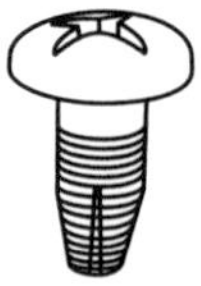

Table 3-6. (continued)

Guideline	Poor Design	Better Design
Use self-fastening parts	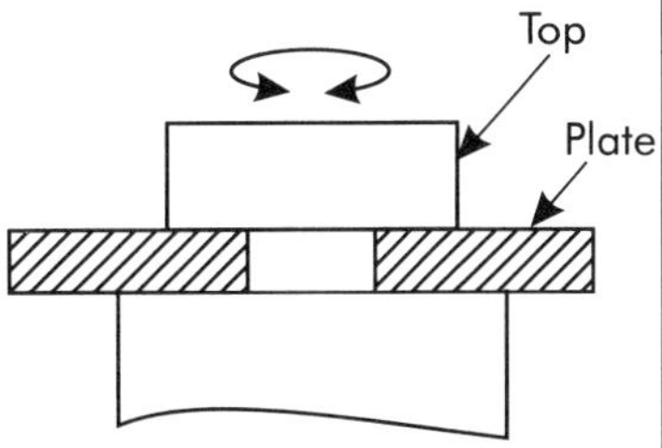 Top Plate Insert the plate and turn the tip.	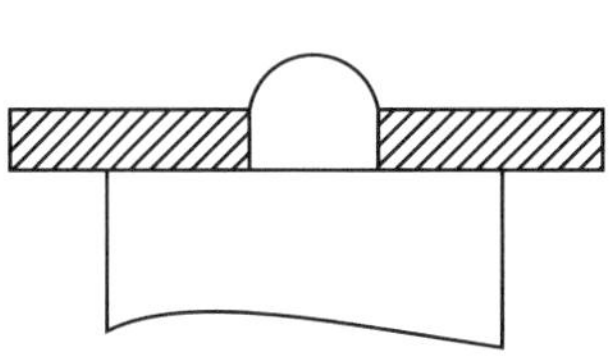 Press fit
Use self-locating parts	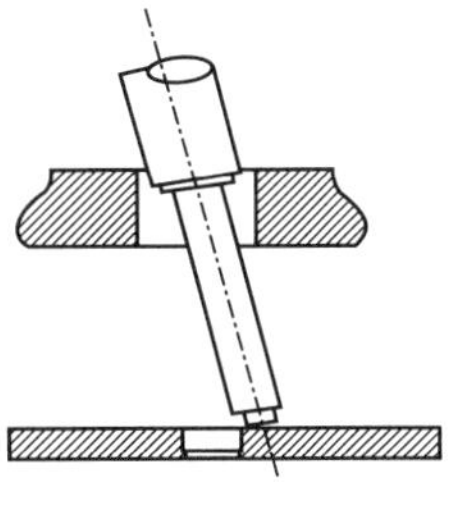 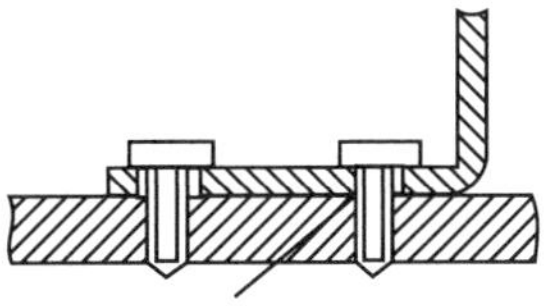 Enlarged hole for position adjustment 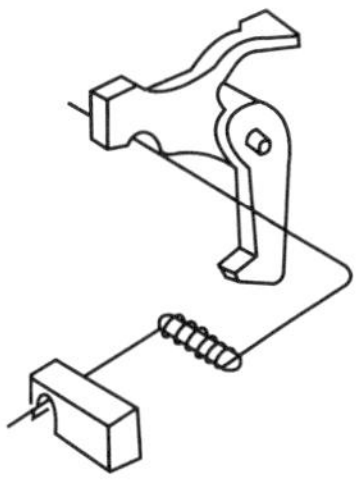	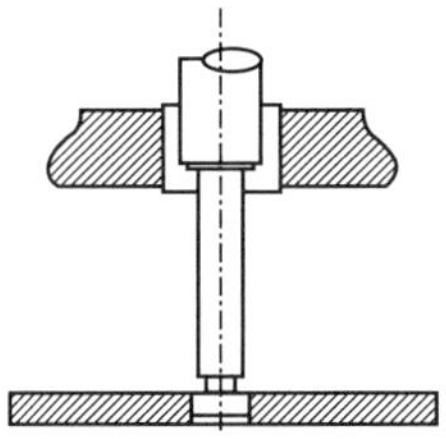 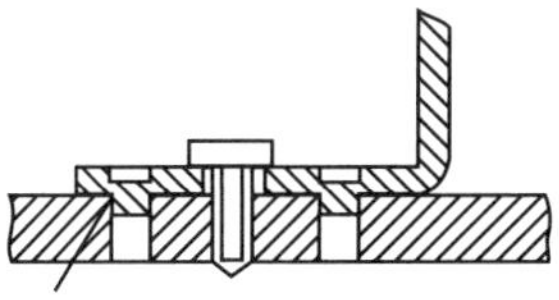 Position embossing pins into holes and no adjustment is needed. 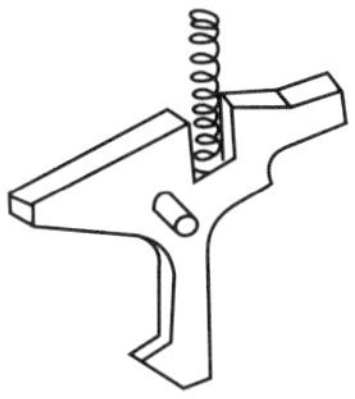
Shaft length exceeds free length of spring and thrust-washer thickness to prevent vibrating loose during assembly	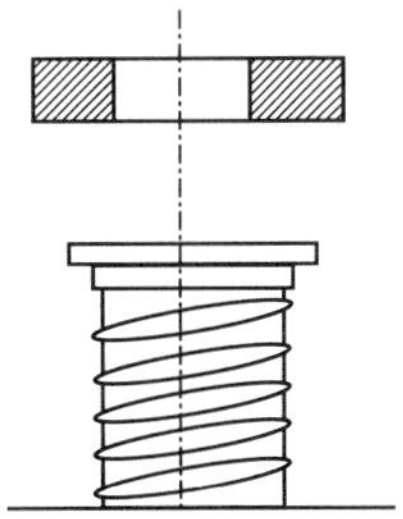	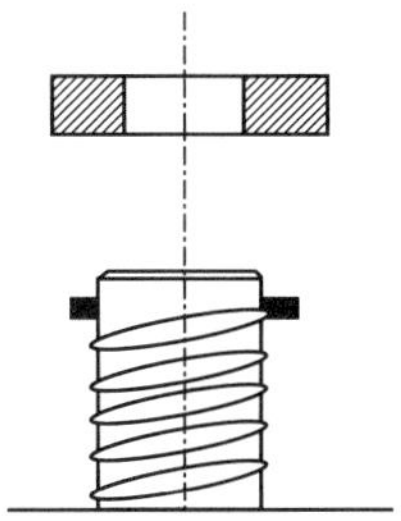

Table 3-6. (continued)

Guideline	Poor Design	Better Design
Minimize the number of parts	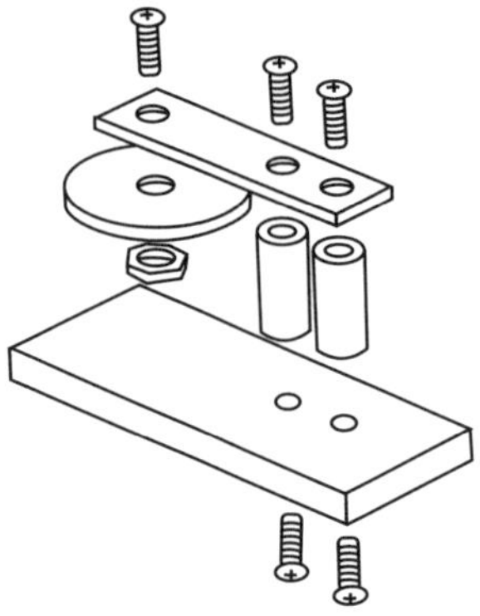	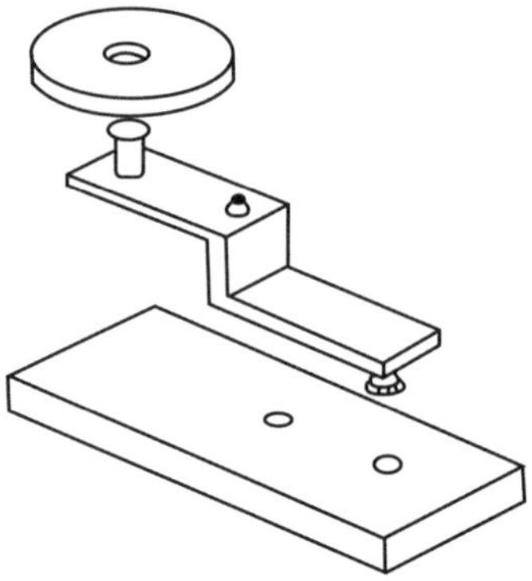
Slot compensates for stacking tolerances, $\pm\Delta X$, and design eliminates need for tight tolerance on frame and results in a quality assembly	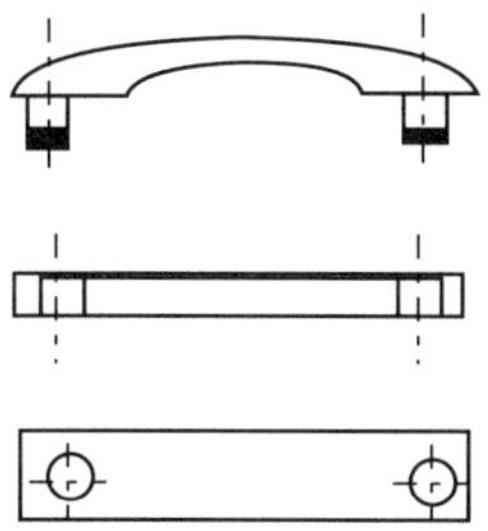	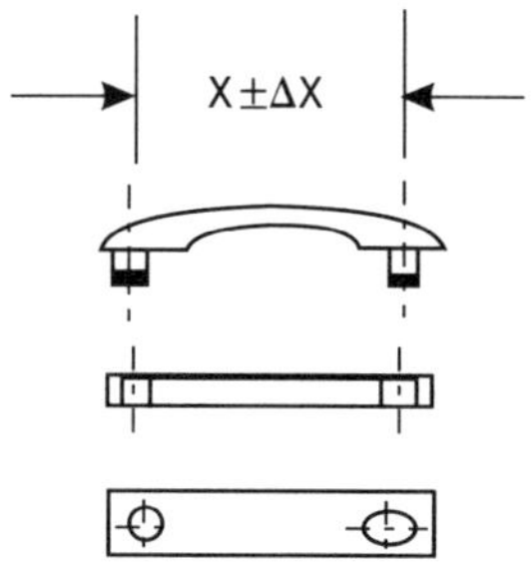 X±ΔX

of magnetic force or grease coating, are slippery, require careful handling, etc. Parts that nest or tangle are those that interlock when in bulk, but that can be separated by one simple manipulation of a single part. Examples of this concept include paper cups, closed-end helical springs, and circlips (ring washers). Parts that are slippery are those that easily slip from ringers or standard grasping tools because of their shape and/or surface condition. Parts that require careful handling are those that are fragile or delicate, have sharp corners or edges, or present other hazards to the operator. Parts that nest or tangle are those that interlock when in bulk and require both hands to apply a separation. Flexible parts are those that substantially deform during manipulation and also require use of two hands. Examples of such parts are paper or felt gaskets, rubber bands, and belts.

Part Location

A part is considered located if it will not require holding down or realignment for subsequent operations. A part is easy to align and position if the position is established by locating features on the part or on its mating part and if insertion is facilitated by well-designed chamfers (see Table 3-6). Resistance encountered during part insertion can be due to small clearances, jam-

Table 3-7. Design for ease of part handling

Guideline	Poor Design	Better Design
Avoid parts that tangle or nest	Springs with open loops will tangle	Closed-ended springs will not tangle
Avoid parts that interconnect	Parts that interconnect	Design barriers to prevent interconnection
Design parts to prevent nesting	Locking angle	Add ribs

ming, wedging, hang-up conditions, or insertion against a large force. For example, a press fit is an interference fit where a large force is required for assembly. The resistance encountered with self-tapping screws is similarly an example of insertion resistance. Assembly time can vary depending on whether the parts have clear access or obstructed access. Obstructed access causes a significant increase in assembly time. Restricted vision causes the operator to rely mainly on tactile sensors during assembly.

STEP-BY-STEP METHODOLOGIES

The Boothroyd-Dewhurst method starts by selecting an assembly method, followed by assembly analysis, and then the design improvement process. This method distinguishes between manual and automatic assemblies. Design improvement is focused on part-number reduction and shortening the associated process times. The Hitachi method is an empirical method that suggests three basic steps, including product design, assembly evaluation, and comparisons. The Lucas method is also an empirical method like the previous two methods. It consists of six steps that include product specification, design for assembly, functional analysis, handling analysis, fitting analysis, and redesign. The design-improvement process results in a reduction in the number of parts and improvement in the handling-and-fitting ratio.

Boothroyd-Dewhurst Method

The Boothroyd-Dewhurst method is an analytical procedure. It can be used to evaluate a new product design after engineering drawings are created or prototypes have been developed, re-evaluate an existing product, or evaluate the potential for automation of an existing product. The designer has to evaluate the geometry of each component in the product or its subassemblies and then determine the degree of difficulty of part handling and insertion. The result of the analysis is an estimated assembly cost and a direction for redesign to improve the product.

The main goal of the Boothroyd-Dewhurst method is to minimize product cost within constraints imposed by design features. The best way to achieve this goal is to reduce the number of components to be assembled and then to ensure that the remaining components are easy to install or assemble. In the early stages of design, the designer must evaluate assembly cost, which means he or she should be familiar with assembly processes. The designer should have a logical explanation for requiring parts that result in a longer assembly time. The designer should also be aware that combining two or more parts into one eliminates an assembly operation.

Determining Design Efficiency

The Boothroyd-Dewhurst method offers a means to judge design efficiency in terms of assembly. The whole procedure, shown in Figure 3-1 consists of three basic steps:

1. selecting the assembly method,
2. assembly analysis, and
3. design improvement.

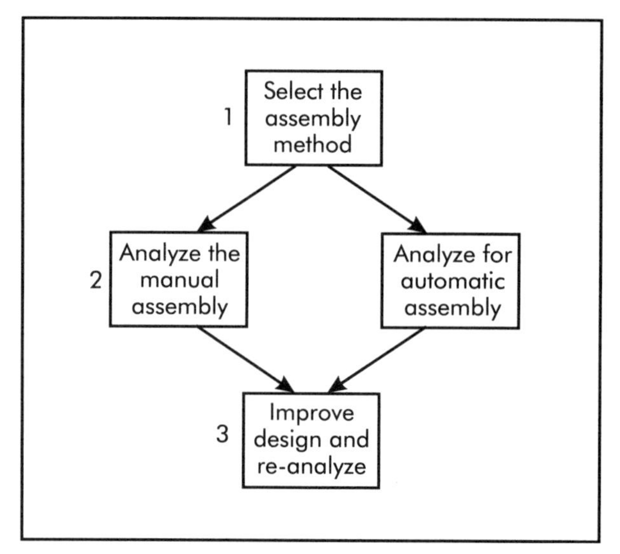

Figure 3-1. Stages in the Boothroyd-Dewhurst method.

Due to differences in the abilities of human operators and automated assembly lines, there is a significant difference between manual and automatic assemblies. The cost of assembly is related to both the product design and its assembly. The minimum product cost can be achieved when the appropriate assembly method is selected.

Design efficiency is determined by using the appropriate formula. Manual assembly design efficiency is found by:

$$E_m = \frac{3N_m}{T_m} \qquad (3\text{-}1)$$

where:

E_m = manual assembly design efficiency %
N_m = theoretical minimum number of parts
T_m = total assembly time

Automatic assembly design efficiency can be determined by:

$$E_f = \frac{0.09N_m}{C_A} \qquad (3\text{-}2)$$

where:

E_f = automatic assembly design efficiency %
N_m = theoretical minimum number of parts
C_A = total assembly cost

A redesign has to produce a better product with higher design efficiency. The most effective way of improving design efficiency is through reducing the number of parts. In the case of manual assembly, time reduction is another way of improving efficiency. In automatic assembly, the feeding and orienting efficiency needs to be improved.

Step-by-step procedure. Features of the design are examined in a systematic manner and design efficiency is calculated. The efficiency is then used to compare different designs. The technique involves two important steps for each part in the assembly:

1. the decision as to whether the part can be eliminated or combined with other parts in the assembly; and
2. an estimate of the time required to grasp, manipulate, and insert the part.

Phase One. The assembly is first taken apart. Each part is assigned a number. If the assembly contains subassemblies, these are first treated as parts. Subsequently, the parts are analyzed in the subassemblies.

Phase Two. The product is reassembled. The part with the highest identification number is assembled first to the work fixture, then the remaining parts are added one by one. The addition of parts could be done by using one or two hands, or by using a separate handling tool. Finally, an analysis of the assembly is done with the help of Table 3-8.

Enter the part identification number and part description. In Column 2, enter the number of times that identical part is used at only this level of assembly. For example, if there are six ½-in. (12-mm) machine screws used, there are six operations. It is assumed that all screws are inserted individually. Screws and washers are considered separate parts.

In Column 3 the symmetry of the handling part from the Table 3-9 is listed. It is important to always look at the part in relation to the *Z*-axis assembly direction to determine α and β before entering the two-digit handling process code.

Enter the handling time for this operation (found in Table 3-9) in Column 4.

Estimate the ease of assembly in Column 5, which can be determined by using Table 3-10 showing the estimated times for insertion. Carefully consider the attributes—hold down, align, and position-insertion resistance—to determine the insertion times. Enter the two-digit insertion-process code in Column 5.

Enter the insertion time (in seconds) in Column 6. The total assembly-operation time in seconds is calculated by adding handling and insertion in Columns 4 and 6, and multiplying this sum by the number of repeated operations in Column 2.

Total operation time is placed in Column 8. Total cost in Column 10 is obtained by multiplying the time in Column 8 by a labor and overhead rate.

In Column 9, an estimate of the theoretical minimum number of parts for the assembly is determined for each part by answering the following questions with practical feasibility in mind:

- During operation of the product, does the part move relative to all other parts already assembled? Only gross motion should be considered; small motions that can be accommodated by elastic hinges, for example, are not sufficient for a positive answer.
- Must the part be of a different material or be physically isolated from all other parts already assembled? Only fundamental reasons associated with material properties are accepted.
- Must the part be separate from all other parts already assembled? Necessary assembly or disassembly of separate parts for service or repair may be rendered impossible.

Table 3-8. Sample table for the calculation of design efficiency

	1	2	3	4	5	6	7	8	9	10
Part Identification Number	Part Description	Number of Identical Operations	$\alpha + \beta$ Symmetry (°)	Manual Handling Code	Manual Handling Time (seconds)	Manual Insertion Code	Manual Insertion Time (seconds)	Total Assembly Time (2) × [(5) + (7)] = T_m	Theoretical Minimum Number of Parts (N_m)	Costs

Table 3-9. Material handling—estimated times (seconds) (Boothroyd and Dewhurst 1987)

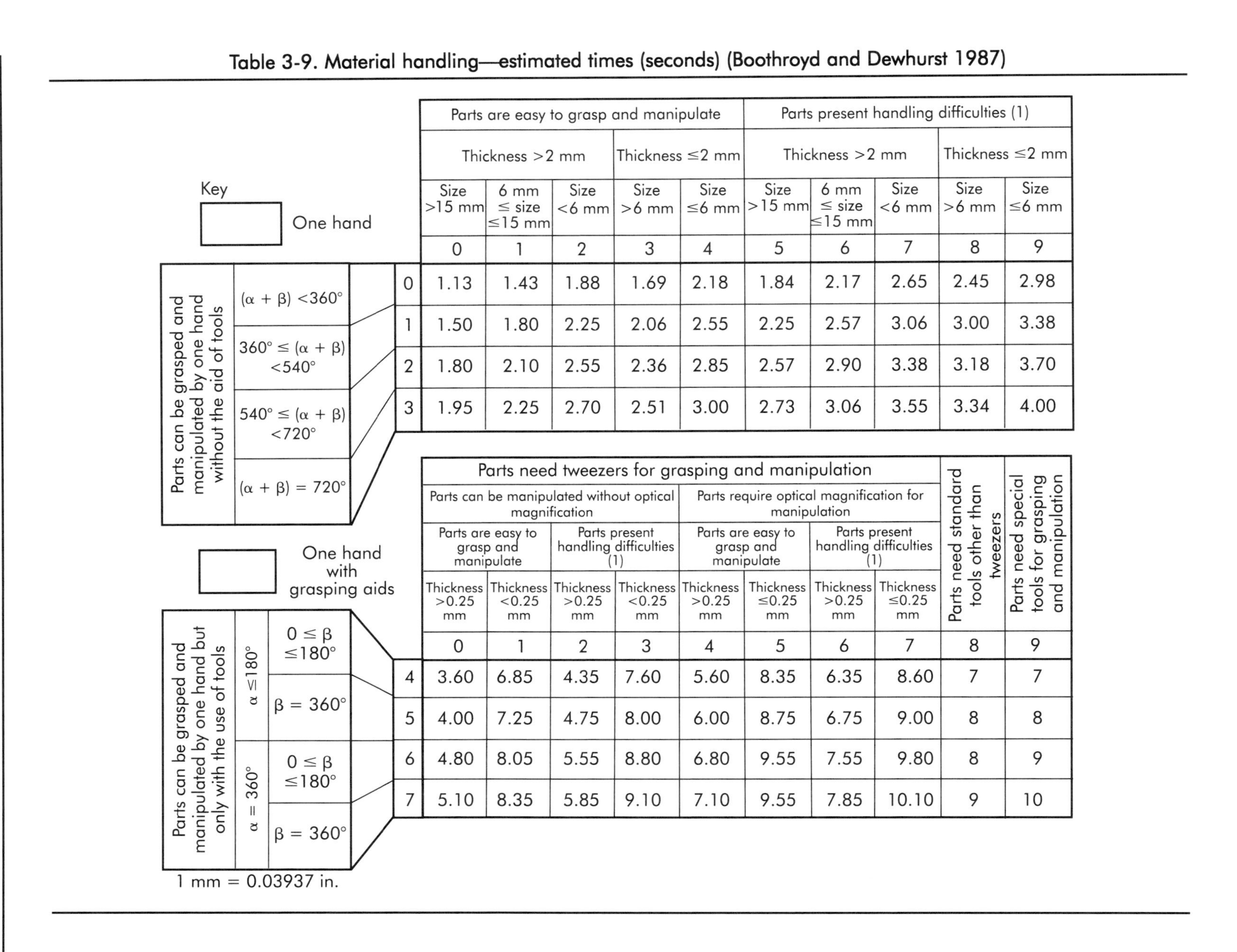

Key: One hand

Parts can be grasped and manipulated by one hand without the aid of tools			Parts are easy to grasp and manipulate					Parts present handling difficulties (1)				
			Thickness >2 mm			Thickness ≤2 mm		Thickness >2 mm			Thickness ≤2 mm	
			Size >15 mm	6 mm ≤ size ≤15 mm	Size <6 mm	Size >6 mm	Size ≤6 mm	Size >15 mm	6 mm ≤ size ≤15 mm	Size <6 mm	Size >6 mm	Size ≤6 mm
			0	1	2	3	4	5	6	7	8	9
	(α + β) <360°	0	1.13	1.43	1.88	1.69	2.18	1.84	2.17	2.65	2.45	2.98
	360° ≤ (α + β) <540°	1	1.50	1.80	2.25	2.06	2.55	2.25	2.57	3.06	3.00	3.38
	540° ≤ (α + β) <720°	2	1.80	2.10	2.55	2.36	2.85	2.57	2.90	3.38	3.18	3.70
	(α + β) = 720°	3	1.95	2.25	2.70	2.51	3.00	2.73	3.06	3.55	3.34	4.00

Key: One hand with grasping aids

Parts can be grasped and manipulated by one hand but only with the use of tools				Parts need tweezers for grasping and manipulation								Parts need standard tools other than tweezers	Parts need special tools for grasping and manipulation
				Parts can be manipulated without optical magnification				Parts require optical magnification for manipulation					
				Parts are easy to grasp and manipulate		Parts present handling difficulties (1)		Parts are easy to grasp and manipulate		Parts present handling difficulties (1)			
				Thickness >0.25 mm	Thickness <0.25 mm	Thickness >0.25 mm	Thickness <0.25 mm	Thickness >0.25 mm	Thickness ≤0.25 mm	Thickness >0.25 mm	Thickness ≤0.25 mm		
				0	1	2	3	4	5	6	7	8	9
	α ≤180°	0 ≤ β ≤180°	4	3.60	6.85	4.35	7.60	5.60	8.35	6.35	8.60	7	7
	α ≤180°	β = 360°	5	4.00	7.25	4.75	8.00	6.00	8.75	6.75	9.00	8	8
	α = 360°	0 ≤ β ≤180°	6	4.80	8.05	5.55	8.80	6.80	9.55	7.55	9.80	8	9
	α = 360°	β = 360°	7	5.10	8.35	5.85	9.10	7.10	9.55	7.85	10.10	9	10

1 mm = 0.03937 in.

Table 3-9. (continued)

Two hands for manipulation

		Parts present no additional handling difficulties					Parts present additional handling difficulties (for example, sticky, delicate, slippery, etc.)				
		$\alpha \leq 180°$			$\alpha = 360°$		$\alpha \leq 180°$			$\alpha = 360°$	
		Size >15 mm	6 mm ≤ size ≤15 mm	Size <6 mm	Size >6 mm	Size ≤6 mm	Size >15 mm	6 mm ≤ size ≤15 mm	Size <6 mm	Size >6 mm	Size ≤6 mm
		0	1	2	3	4	5	6	7	8	9
Parts nest or tangle severely or are flexible, but can be grasped and lifted by one hand (with the use of grasping tools if necessary)	8	4.10	4.50	5.10	5.60	6.75	5.00	5.25	5.85	6.35	7

Two hands are required for large size

		Parts can be handled by one person without mechanical assistance									Two persons or mechanical assistance are required for parts manipulation
		Parts do not nest or tangle severely and are not flexible								Parts nest or tangle severely or are flexible	
		Part weight <10 lb				Parts are heavy (>10 lb)					
		Parts are easy to grasp and manipulate		Parts create other handling difficulties		Parts are easy to grasp and manipulate		Parts create other handling difficulties			
		$\alpha \leq 180°$	$\alpha = 360°$	$\alpha \leq 180°$	$\alpha = 360°$	$\alpha \leq 180°$	$\alpha = 360°$	$\alpha \leq 180°$	$\alpha = 360°$		
		0	1	2	3	4	5	6	7	8	9
Two hands, two persons, or mechanical assistance are required for grasping and transporting parts	9	2	3	2	3	3	4	4	5	7	9

1 mm = 0.0397 in.
1 lb = 0.4536 kg

Table 3-10. Manual insertion—estimated times (seconds) (Boothroyd and Dewhurst 1987)

Key: Part added, but not secured

			After assembly, no holding down is required to maintain orientation and location				Holding down is required during subsequent processes to maintain orientation of location			
			Easy to align and position during assembly		Not easy to align and position during assembly		Easy to align and position during assembly		Not easy to align and position during assembly	
			No resistance to insertion	Resistance to insertion	No resistance to insertion	Resistance to insertion	No resistance to insertion	Resistance to insertion	No resistance to insertion	Resistance to insertion
			0	1	2	3	6	7	8	9
Addition of any part where neither the part itself nor any other part is secured immediately and finally	Part and associated tool (including hands) can easily reach the desired location	0	1.5	2.5	2.5	3.5	5.5	6.5	6.5	7.5
	Part and associated tool (including hands) cannot easily reach the desired location — Due to obstructed access or restricted vision	1	4.0	5.0	5.0	6.0	8.0	9.0	9.0	10.0
	Due to obstructed access or restricted vision	2	5.5	6.5	6.5	7.5	9.5	10.5	10.5	11.5

Table 3-10. (continued)

Part secured immediately		No screwing operation or plastic deformation immediately after insertion (snap/press fits, circlips, spire nuts, etc.)		Plastic deformation immediately after insertion						Screw tightening immediately after insertion	
				Plastic bending or torsion			Riveting or similar operation				
					Not easy to position or align during assembly			Not easy to position or align during assembly			
		Easy to align and position with no resistance to insertion	Not easy to align or position during assembly and/or resistance to insertion	Easy to align and position during assembly	No resistance to insertion	Resistance to insertion	Easy to align and position during assembly	No resistance to insertion	Resistance to insertion	Easy to align and position with no torsional resistance	Not easy to align or position and/or torsional resistance
		0	1	2	3	4	5	6	7	8	9
Addition of any part (1) where the part itself and/or other parts are being secured finally and immediately: Part and associated tool (including hands) can easily reach the desired location and the tool can be operated easily	3	2.0	5.0	4.0	5.0	6.0	7.0	8.0	9.0	6.0	8.0
Part and associated tool (including hands) cannot easily reach desired location or tool cannot be operated easily: Due to obstructed access or restricted vision	4	4.5	7.5	6.5	7.5	8.5	9.5	10.5	11.5	8.5	10.5
Due to obstructed access or restricted vision	5	6.0	9.0	8.0	9.0	10.0	11.0	12.0	13.0	10.0	12.0

Separate operation		Mechanical fastening processes (part[s]) already in place but not secured immediately after insertion)				Non-mechanical fastening processes (part[s]) already in place but not secured immediately after insertion)				Non-fastening processes	
		None or localized plastic deformation				Metallurgical processes					
							Additional material required				
		Bending or similar processes	Riveting or similar processes	Screw tightening or other processes	Bulk plastic deformation (large proportion of parts is plastically deformed during fastening)	No additional material required (for example, resistance, friction welding, etc.)	Soldering processes	Weld/braze processes	Chemical processes (for example, adhesive bonding, etc.)	Manipulation of parts or subassembly (for example, orienting, fitting, or adjustment of parts, etc.)	Other processes (for example, liquid insertion, etc.)
		0	1	2	3	4	5	6	7	8	9
Assembly processes where all solid parts are in place	9	4.0	7.0	5.0	3.5	7.0	8.0	12.0	12.0	9.0	12.0

If the answer to any of these questions is "yes," then a "1" is placed in Column 9, except where multiple identical operations are indicated in Column 2. If this is the case, the number of parts that must be listed separately is placed in Column 9.

The remaining parts are added one by one to the assembly. The manual assembly analysis worksheet is completed for each additional part.

When all of the rows of Table 3-8 are completed, the figures in Column 8 are all added to determine the total estimated manual assembly time. The figures in Column 10 are added to give the total manual assembly cost, and the figures in Column 9 are added to give the theoretical minimum number of parts for the complete assembly (or subassembly).

Redesign. The initial design for assembly analysis provides the designer with useful information for product redesign in two areas:

- Data contained in Column 9 of Table 3-8 indicates where it might be possible to reduce the number of parts.
- Columns 4 and 6 of Table 3-8 indicate those parts that are difficult to handle or insert.

When the number in column 9 is less than that in Column 2, there is a possibility of eliminating parts. A reduction in the part count is usually the most effective means of improving assemblability. In this way, design efficiency is greatly improved. Creative techniques should be applied to groups of parts in the assembly that can be combined. Alternate group parts should be sketched. The figures in Columns 4 and 6 indicate parts where there is the potential to decrease the handling time or insertion time. The design of parts that pose difficulty in handling and insertion should be reviewed.

- In the Boothroyd-Dewhurst design-for-assembly methodology, parts are added one at a time. However, for bench assembly and manual assembly lines, workers often handle two parts simultaneously. Under these circumstances, research has shown that assembly times can be reduced by one third. Thus, the design engineer can obtain a more accurate estimate of time by dividing the derived number by 1.5.
- In preparing a time-and-motion analysis used to derive manual handling time, it is often assumed that parts are randomly oriented in bins at the assembly station. However, many parts are available in proper orientation in trays or magazines. If the engineer is aware of how these functions affect the assembly time, he or she may incorporate this data into the analysis.

Electric Switch Case Study

Using the Boothroyd-Dewhurst method, an existing electric switch (see Figure 3-2) is analyzed and later redesigned. Changes in the number of parts as well as the design efficiency are calculated.

Existing design. The existing design for the switch contains a total of 14 separate parts and operations (see Table 3-11). The total assembly time is 183.11 seconds, and the corresponding design efficiency is calculated as 15%. This switch assembly is a good candidate for the application of design for assembly tools to obtain a more efficient design.

Proposed redesign. The goal of redesigning the switch is to minimize the number of parts, while maintaining the functionality of the original design. Basic guidelines for design for assembly analysis are applied for each part. The new design is detailed in Table 3-12 and Figure 3-3.

1. Switch base. The switch base was modified in this example to incorporate a snap fit into the switch cover. This eliminated the existing bent tabs used on the metal switch

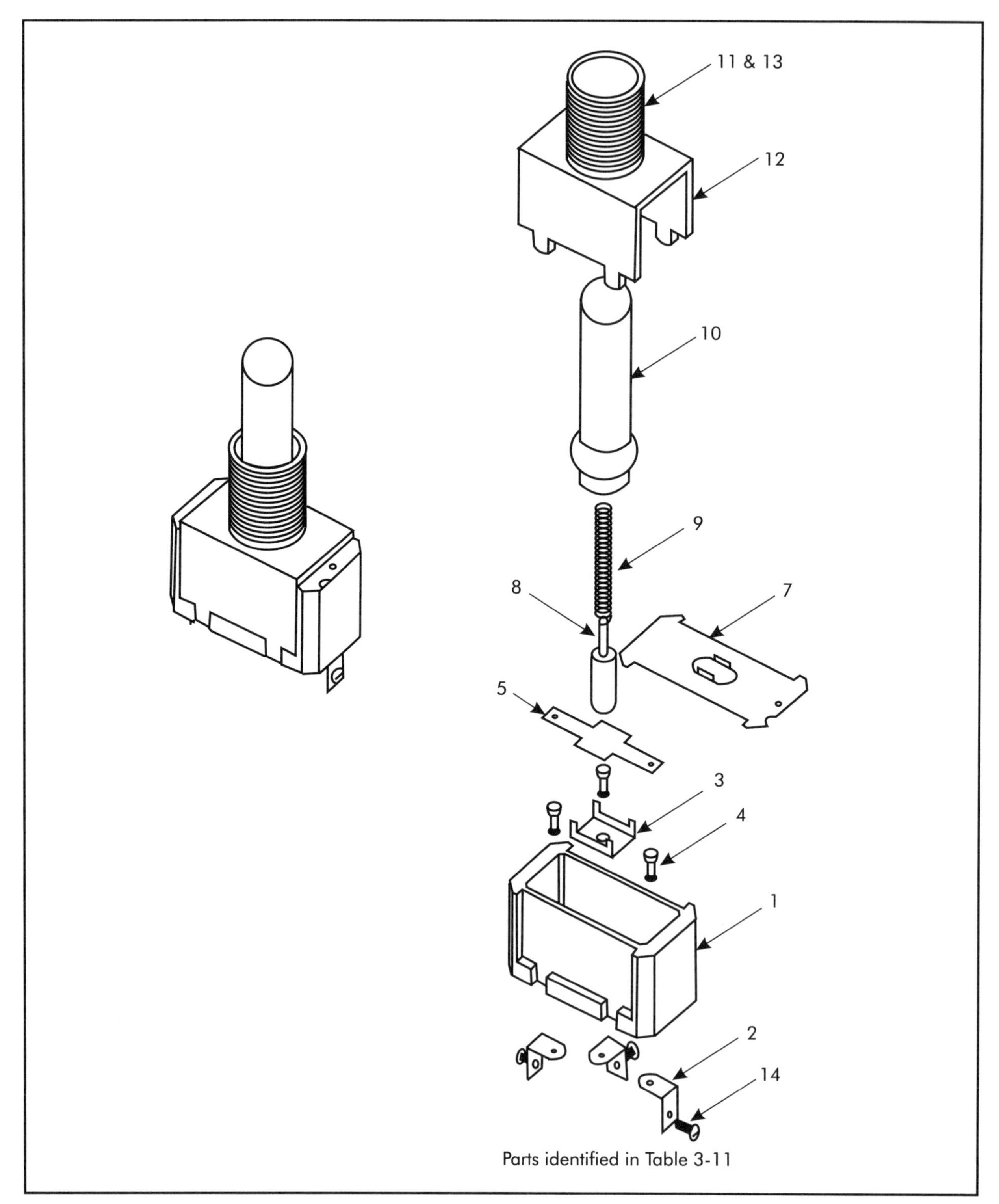

Figure 3-2. Single-pole double-throw switch.

Table 3-11. Components of the electric switch (original design)

Number	Part Name	Thickness in. (mm)	Size in. (mm)
1	Switch base	0.59 (15.0)	1.14 (29.0)
2	Terminals	0.35 (9.0)	0.32 (8.0)
3	Center terminal contact	0.24 (6.0)	0.32 (8.0)
4	Terminal rivets	0.28 (7.0)	0.16 (4.0)
5	Contact rocker	0.16 (4.0)	0.87 (22.0)
6	Add grease	—	—
7	Base cover	0.12 (3.0)	1.14 (29.0)
8	Switch plunger	0.16 (4.0)	0.63 (16.0)
9	Switch spring	0.12 (3.0)	0.79 (20.0)
10	Switch toggle	0.35 (9.0)	1.38 (35.0)
11	Mounting threads	0.47 (12.0)	0.55 (14.0)
12	Mounting cover	0.51 (13.0)	0.75 (19.0)
13	Mounting hardware	0.08 (2.0)	0.63 (16.0)
14	Terminal screws	0.28 (7.0)	0.28 (7.0)

Table 3-12. Components of the redesigned electric switch

Number	Part Name	Thickness in. (mm)	Size in. (mm)
1	Switch base	0.59 (15.0)	1.14 (29.0)
2	Wire-clinch terminals	0.20 (5.0)	0.47 (12.0)
3	Center terminal/rocker	0.59 (15.0)	0.87 (22.0)
4	Add grease	— —	
5	Plastic switch toggle	0.35 (9.0)	1.77 (45.0)
6	Switch cover	1.02 (26.0)	1.14 (29.0)
7	Mounting hardware	0.08 (2.0)	0.63 (16.0)

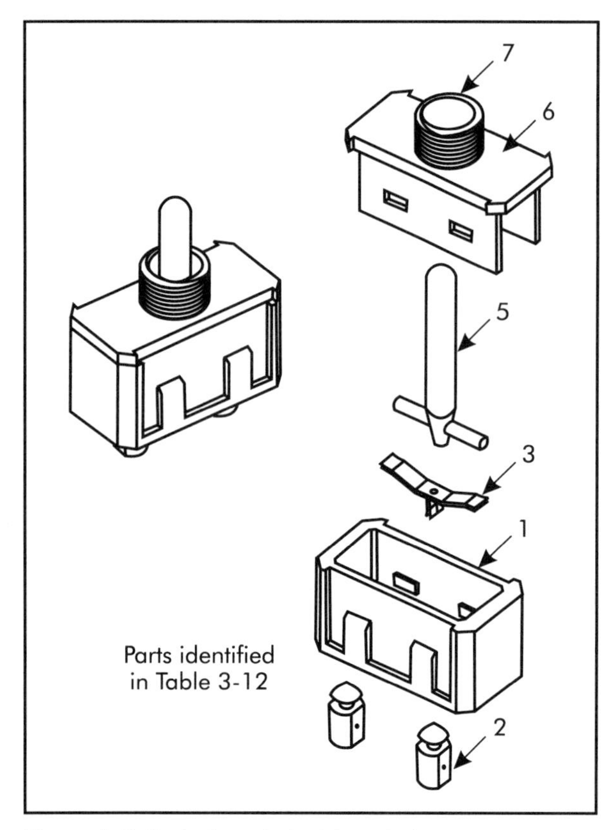

Figure 3-3. Redesigned electric switch.

cover to attach it to the plastic switch base. Other modifications were snap-fit sockets to hold two metal wire-contact terminals and the center terminal/rocker in place. Additional fabrication charges for this piece are not foreseen because a new plastic mold would have to be created.

2. Wire-clinch terminals. These parts replace the terminals, terminal rivets, and terminal screws of the original design. They perform the dual functions of holding the stranded wires and providing contact points for the terminal/rocker. Wires are held in place within terminals by a metal-locking spring action. The two wire-clinch terminals are formed from rolled brass sheets and they snap into the plastic switch base. Additional tooling and fabrication charges are incurred to create this specialized part.

3. Center terminal/rocker. This part replaces the center terminal contact, center terminal rivet, contact rocker, and switch spring of the original design. Like the wire-clinch terminals, this piece snaps into the plastic switch base. It is formed of brass and sheet metal and provides a flexible interface at the switch toggle. This part incurs extra tooling and fabrication charges.

4. Add grease. This separate operation remains the same as in the original design of the switch. Grease adds lubrication to the switch toggle/rocker interface and, thereby, increases the life of the switch during normal operation.

5. Plastic switch toggle. The plastic switch toggle was modified extensively from the original design. A molded plastic piece with snap-fit posts replaced the cast aluminum piece. The plastic design of the new toggle incorporates the original switch-plunger piece into the toggle itself. No extensive charges are foreseen in fabrication of the new part; a plastic-mold part simply replaces the casting process.

6. Switch cover. This part underwent an extensive redesign. It replaces the base cover, mounting threads, and mounting cover of the original design. This piece undergoes a complicated fabrication process. The overall shape is a metal casting, and several machining operations are performed to finish the piece. This makes the switch cover one of the most expensive parts in the new design. Design of the new switch cover allows for a snap fit at the switch-base interface, and posts on the toggle snap into the inner diameter of the threaded portion.

7. Mounting hardware. These parts are not changed from the original design. The switch assembly is redesigned to keep the same functionality as the original design. This includes the way that it is mounted to the electrical panel, chassis, etc.

The design changes lead to a faster and more efficient assembly of the single-pole, double-throw switch. The number of parts and operations is decreased from 14 to seven. This leads to an assembly time of 64.68 seconds—approximately 2.83 times faster than the original design. The assembly efficiency of the redesign is calculated as 42% (see Table 3-13). This is an increase from the original assembly efficiency of 15% (see Table 3-14). While these changes lead to a more efficient assembly and lower assembly time, the overall cost of production may not be reduced. This is due to modifications made to decrease the part count. New tooling and fabrication processes would have to be developed to create the required specialized combination parts. A design engineer would, therefore, have to calculate tooling and manufacturing charges to determine if using the redesign would compromise the production cost.

Electric Motor Case Study

This case study examines the mechanical components of an electric motor for ease of assembly. It does not consider the design of electromagnetic components. The original design efficiency index of 6.1% is increased to 11%. The number of components is reduced from 16 to 10. (See Figures 3-4 and 3-5 and Tables 3-15 and 3-16.)

In spite of redesign factors, the design efficiency index of the motor improves marginally. The size of the motor is a major contributing factor because it is heavy and big. The assembly process requires precision and delicate handling to fit the parts together correctly.

Hydraulic Shuttle Valve Case Study

The aerospace industry is one of the major users of shuttle valves to control the flow of various aerospace fluids. The shuttle valve as shown in Figure 3-6 is designed to isolate normal fluid flow from the emergency hydraulic system during normal operation. The pressure-actuated shuttle valve is completely self-contained and is a three-port valve. At each end of the valve is a supply port. In typical applications, one port provides normal flow in operation. When pressure is lost in the normal system and emergency pressure is applied, the poppet shuttles across to block the normal port. The third port is in the center of the shuttle valve. Fluid flows out of the center port through a series of windows cut at the outer diameter of the center of the valve.

Inside the valve body is a spring-loaded poppet, which normally closes off the emergency port and allows fluid flow from the normal supply port to the center discharge port. Each port is screened to prevent contamination from passing into the valve. The shuttle-valve construction must be of aerospace-grade stainless steel, except for the external nose seal, which is normally made from aerospace-grade aluminum.

During normal operation, normal and emergency ports are of equal pressure. An internal spring force drives the poppet against the emergency-valve seat, sealing off the emergency flow. With the poppet valve in this position, a free path is provided for fluid to flow from the normal supply port to the discharge port at the center.

In the event of a loss of normal operating-system pressure and flow, emergency-port pressure overcomes the poppet spring force and normal supply pressure. This forces the poppet to move against the normal port valve seat and moves the shuttle off the emergency port valve seat, allowing emergency flow to the discharge valve. The valve provides a means of automatic transfer of control in an aircraft landing system from normal hydraulic to redundant emergency hydraulic systems.

Initial assembly. The initial valve shown in Figure 3-6 had 11 individual parts. The valve was assembled with two subassemblies: the front and the rear body. Table 3-17

Table 3-13. Calculation of assembly time and design efficiency (redesign)

Part Identification Number	Item Type	Name	Repeat Count	Minimum Parts	Tool Acquisition Time (seconds)	Item Handling/ Acquisition Time (seconds)	Item Insertion Time (seconds)	Total Operating Time (seconds)	Total Operating Cost
1	Part	Switch base	1	1	0	1.95	1.5	3.45	0.03
2	Part	Wire-clinch terminal	2	2	0	1.80	5.0	13.60	0.11
3	Part	Center terminal	1	1	0	1.80	5.0	6.80	0.06
4	Operation	Apply grease to area	1	—	3.0	—	—	7.00	0.06
5	Part	Plastic switch toggle	1	1	0	1.80	2.6	4.40	0.04
6	Part	Switch cover	1	1	0	1.80	1.8	3.60	0.03
7	Part	Mounting hardware	1	0	2.9	1.69	7.5	12.09	0.10

Design efficiency = 42%

Table 3-14. Calculation of assembly time and design efficiency (original)

Part Identification Number	Item Type	Name	Repeat Count	Minimum Parts	Tool Acquisition Time (seconds)	Item Handling/ Acquisition Time (seconds)	Item Insertion Time (seconds)	Total Operating Time (seconds)	Total Operating Cost
1	Part	Switch base	1	1	0	2.73	1.5	4.23	0.04
2	Part	Terminals	3	3	0	5.10	7.4	37.50	0.31
3	Part	Terminal center contact	1	0	0	4.80	7.4	12.20	0.10
4	Part	Terminal rivets	3	0	2.9	4.80	11.2	50.90	0.42
5	Part	Contact rocker	1	1	0	4.35	7.4	11.75	0.10
6	Operation	Apply grease to area	1	—	3.0	—	—	7.00	0.06
7	Part	Base cover	1	0	0	2.73	5.2	7.93	0.07
8	Part	Switch plunger	1	1	0	2.06	3.0	5.06	0.04
9	Part	Switch spring	1	0	0	5.60	6.5	12.10	0.10
10	Part	Switch toggle	1	1	0	1.50	2.6	4.10	0.03
11	Part	Mounting cover	1	0	0	1.80	2.6	4.40	0.04
12	Part	Mounting thread	1	0	0	1.80	5.2	7.00	0.06
13	Part	Mounting washer	1	0	2.9	2.06	7.5	12.46	0.10
14	Part	Terminal screws	1	0	2.9	1.80	9.2	13.90	0.12

Design efficiency = 15%

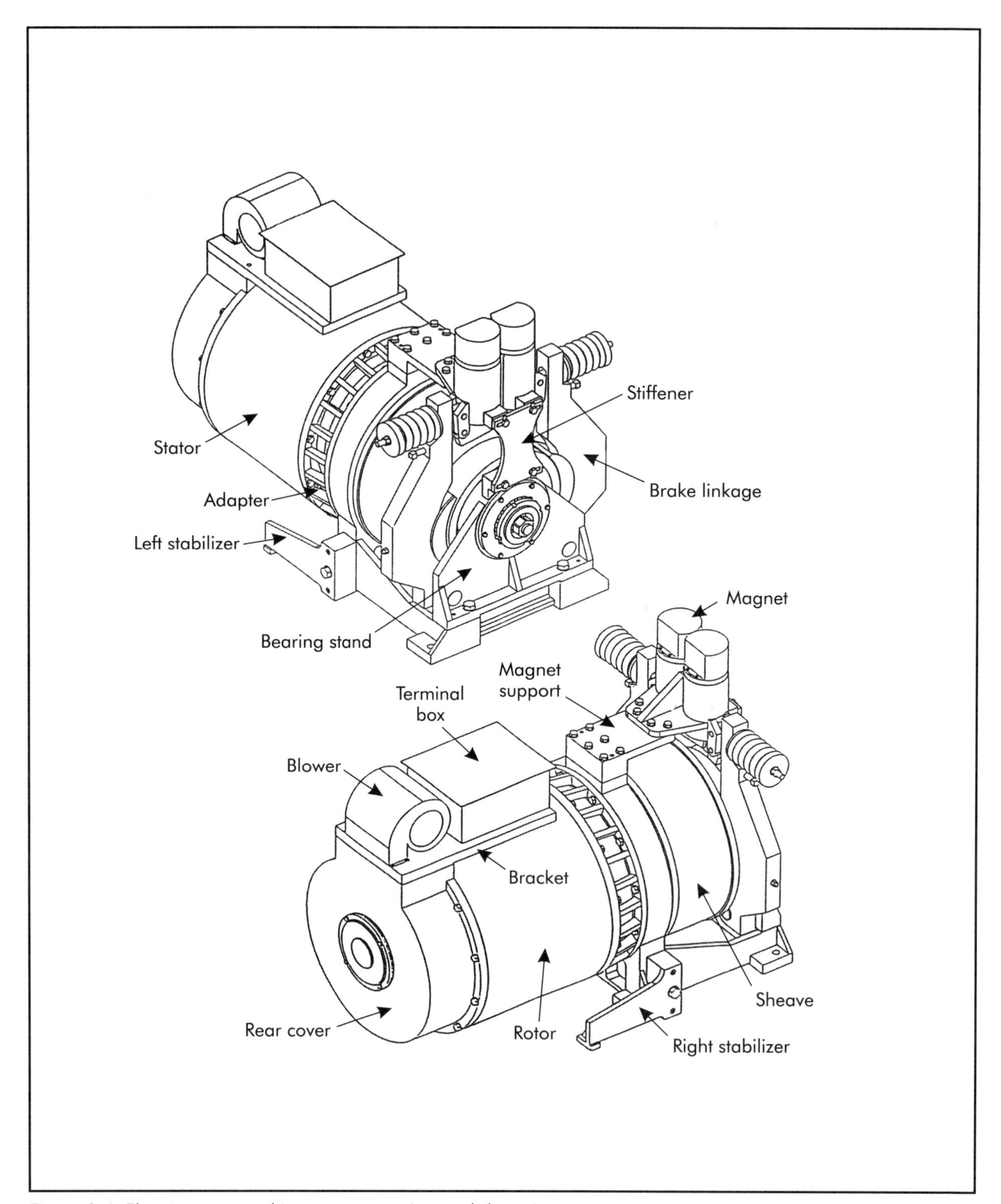

Figure 3-4. Electric motor and its components (original design).

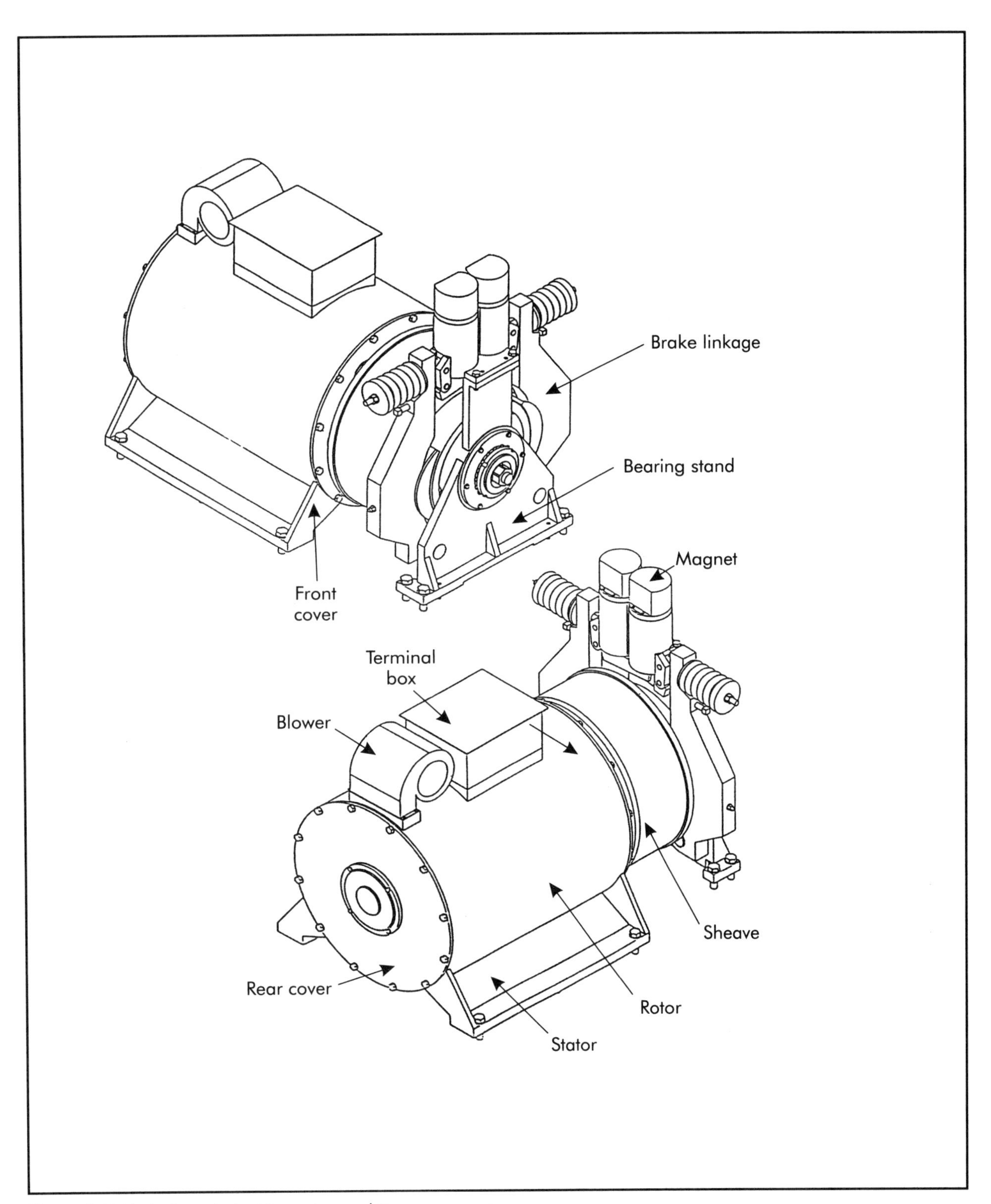

Figure 3-5. Redesigned electric motor and its components.

Table 3-15. DFM worksheet for the original motor

	1	2	3	4	5	6	7	8	9	10
Part Identification Number	Part Description	Number of Identical Operations	Manual Handling Code	Manual Handling Time (seconds)	Manual Insertion Code	Manual Insertion Time (seconds)	Total Assembly Time (2) × [(4) + (7)] = T_m	Theoretical Minimum Number of Parts (N_m)	Part Costs	Total Part Cost (10) × (2) = T_p
1	Terminal box	1	95	4.0	38	6.0	20.0	1		0
2	Blower	1	91	3.0	38	6.0	18.0	1		0
3	Bracket	1	95	4.0	49	10.5	29.0	0		0
4	Stiffener	1	70	5.1	39	8.0	26.2	0		0
5	Brake linkages	1	99	9.0	92	5.0	28.0	1		0
6	Magnet	1	99	9.0	38	6.0	30.0	1		0
7	Magnet support	1	99	9.0	38	6.0	30.0	0		0
8	Bearing stand	1	99	9.0	37	8.0	34.0	1		0
9	Sheave	1	99	9.0	92	5.0	28.0	1		0
10	Rotor	1	99	9.0	92	5.0	28.0	1		0
11	Adapter	1	99	9.0	49	10.5	39.0	0		0
12	Stator	1	99	9.0	49	10.5	39.0	1		0
13	Rear cover	1	99	9.0	39	8.0	34.0	1		0
14	Right stabilizer	2	95	4.0	38	6.0	20.0	0		0
15	Left stabilizer	2	95	4.0	38	6.0	20.0	0		0
16	Frame	1	99	9.0	00	1.5	21.0	0		0
							444.2	9		0

Design efficiency = 6.1%

Table 3-16. DFM worksheet for the redesigned motor

	1	2	3	4	5	6	7	8	9
Part Identification Number	Part Description	Number of Identical Operations	Manual Handling Code	Manual Handling Time (seconds)	Manual Insertion Code	Manual Insertion Time (seconds)	Total Assembly Time (2) × [(5) + (7)] $= T_m$	Theoretical Minimum Number of Parts (N_M)	Part Costs
1	Terminal box	95	4	38	6.0	20	8.0	1	
2	Blower	91	3	38	6.0	18	7.2	1	
3	Brake linkage	99	9	92	5.0	28	11.2	1	
4	Magnet	99	9	38	6.0	30	12.0	1	
5	Bearing stand	99	9	39	8.0	34	13.6	1	
6	Sheave	99	9	92	5.0	28	11.2	1	
7	Front cover	99	9	39	8.0	34	13.6	1	
8	Rotor	99	9	92	5.0	28	11.2	1	
9	Rear cover	99	9	39	8.0	34	13.6	1	
10	Stator	99	9	00	1.5	21	8.4	1	
						275		10	
Design efficiency = 11%									

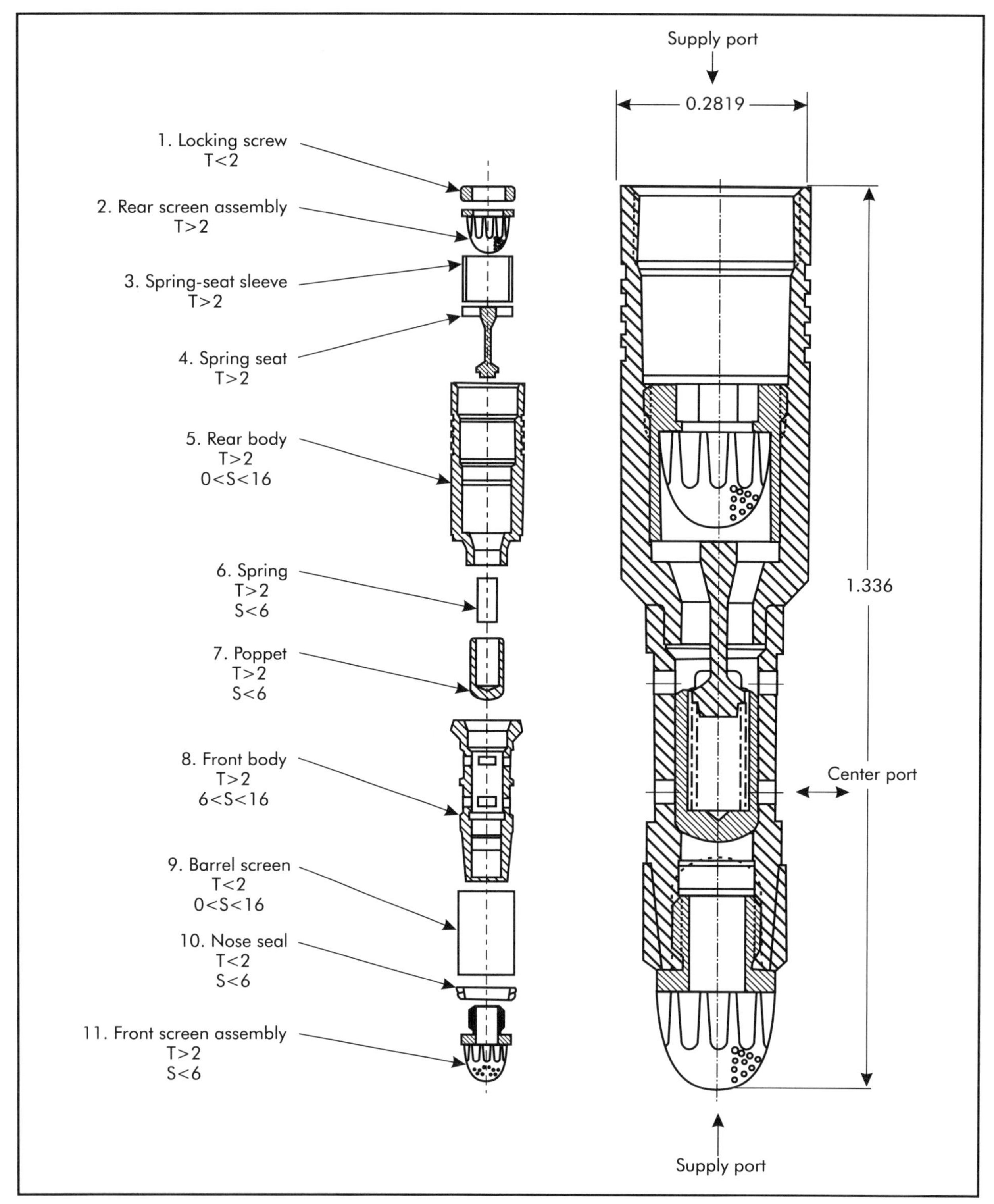

Figure 3-6. Schematic of original shuttle valve.

Table 3-17. DFM worksheet for original valve

	1	2	3	4	5	6	7	8	9	10
Part Identification Number	Part Description	Number of Identical Operations	$\alpha + \beta$ Symmetry (°)	Manual Handling Code	Manual Handling Time (seconds)	Manual Insertion Code	Manual Insertion Time (seconds)	Total Assembly Time (2) × [(5) + (7)] = T_m	Theoretical Minimum Number of Parts (N_m)	Costs
1	Locking screw	1		82	5.10	49	10.5	15.60	0	
2	Rear screen	1		12	2.25	02	2.5	4.75	1	
3	Spring-seat sleeve	1		02	1.88	00	1.5	3.38	0	
4	Spring seat	1		89	7.00	23	7.5	14.50	0	
5	Rear body	1		11	1.80	96	12.0	13.80	1	
6	Spring	1		02	1.88	01	2.5	4.38	1	
7	Poppet	1		12	2.25	00	1.5	3.75	1	
8	Front body	1		11	1.80	00	1.5	3.30	1	
9	Barrel screen	1		02	1.88	96	12.0	13.88	1	
10	Nose seal	1		16	4.80	01	2.5	7.30	1	
11	Front screen	1		12	2.25	49	10.5	12.75	1	
		11						97.39	8	

Design efficiency 24.6%

provides complete data on the components, their handling time, the time for insertion, and the parameters needed to calculate the efficiency index.

DFA analysis of the shuttle valve indicates that there are candidates for elimination. Results of the analysis are as follows:

- design efficiency is 24.6%;
- total assembly time is 97 seconds;
- total number of parts is 11; and
- theoretical minimum number of parts is 8.

About 40% of the 97-second assembly time can be reduced. The three components that took the longest to assemble in the example were the locking screw (15.60 seconds), the barrel screen (13.88 seconds), and the spring seat mounted in the rear body (14.50 seconds).

Redesign. The redesign is focused on eliminating unnecessary parts and operations and on combining parts (see Table 3-18). The most significant component redesigns are the poppet valve and return spring. A separate spring seat mounted in the rear body is seated against the poppet. An additional spacer supports the spring seat over the bleed screen. The assembly is then attached to the rear body through a locking screw that retains the internal screen, spacer, seat, and spring components. By combining parts, the redesign eliminates the spacer and seat by creating a spring-seat surface on the nose of the rear body, where it is swaged into the front body (see Figure 3-7).

A screen brazed to the rear body eliminates the locking screw. The poppet is redesigned to be of a stepped diameter, where the smaller outside diameter (OD) moves within the rear body, and the larger OD moves within the front body. The step surface is utilized as a seat on which the spring force is applied to the poppet.

As shown in Table 3-19, the analysis profile indicates that for the redesigned part, the assembly of necessary parts takes 60% of the assembly time. Standard operations take the remaining amount of time.

The redesigned product shows a significant reduction in assembly time. Design efficiency is increased as a result of the redesign. The new version has two sets of windows that allow flow to discharge from the valve; one set is for the normal flow path and the second set is for the emergency flow path. Although these two ports discharge into the same downstream annulus, they are internally isolated by the poppet valve, except for the minimal lapped leakage.

Redesign also causes additional operations. In the redesigned valve, the rear body is swaged inside the front body. The valve moves inside the rear body to close off the normal discharge flow path in an emergency. This requires that a discharge window be cut into both the rear and front bodies, adding a machining operation. On the assembly of the two bodies, the windows must line up perfectly to eliminate blocking the discharge port. This adds additional assembly movement because the alpha and beta angles in the redesign are different and the components require additional manipulation. During the assembly process, additional features create a perfect alignment in the rear and front body.

Mechanical Press Case Study

Presses provide a means of compressing and shaping components by exerting high pressure. Mechanical presses use various drive systems. In the screw press, a screw spindle is rotated on a fixed nut, whereby a longitudinal force is transmitted through the spindle to the workpiece. On larger presses, the upper end of the screw spindle has a large flywheel that, when rotating, contains a large reserve of stored energy.

The assembly layout of a mechanical press is shown in Figure 3-8. The present design

Table 3-18. DFM worksheet for redesigned valve

	1	2	3	4	5	6	7	8	9	10
Part Identification Number	Part Description	Number of Identical Operations	$\alpha + \beta$ Symmetry (°)	Manual Handling Code	Manual Handling Time (seconds)	Manual Insertion Code	Manual Insertion Time (seconds)	Total Assembly Time (2) × [(5) + (7)] = T_m	Theoretical Minimum Number of Parts (N_m)	Costs
1	Rear screen	1		12	2.25	90	4.0	6.25	1	
2	Rear body	1		11	1.80	96	4.0	5.80	1	
3	Spring	1		02	1.88	01	2.5	4.38	1	
4	Poppet	1		12	2.25	00	1.5	3.75	1	
5	Front body	1		11	1.80	00	1.5	3.30	1	
6	Barrel screen	1		02	1.88	96	12.0	13.88	1	
7	Nose seal	1		16	4.80	01	2.5	7.30	1	
8	Front screen	1		12	2.25	90	4.0	6.25	1	
		8						50.91	8	

Design efficiency = 47.1%

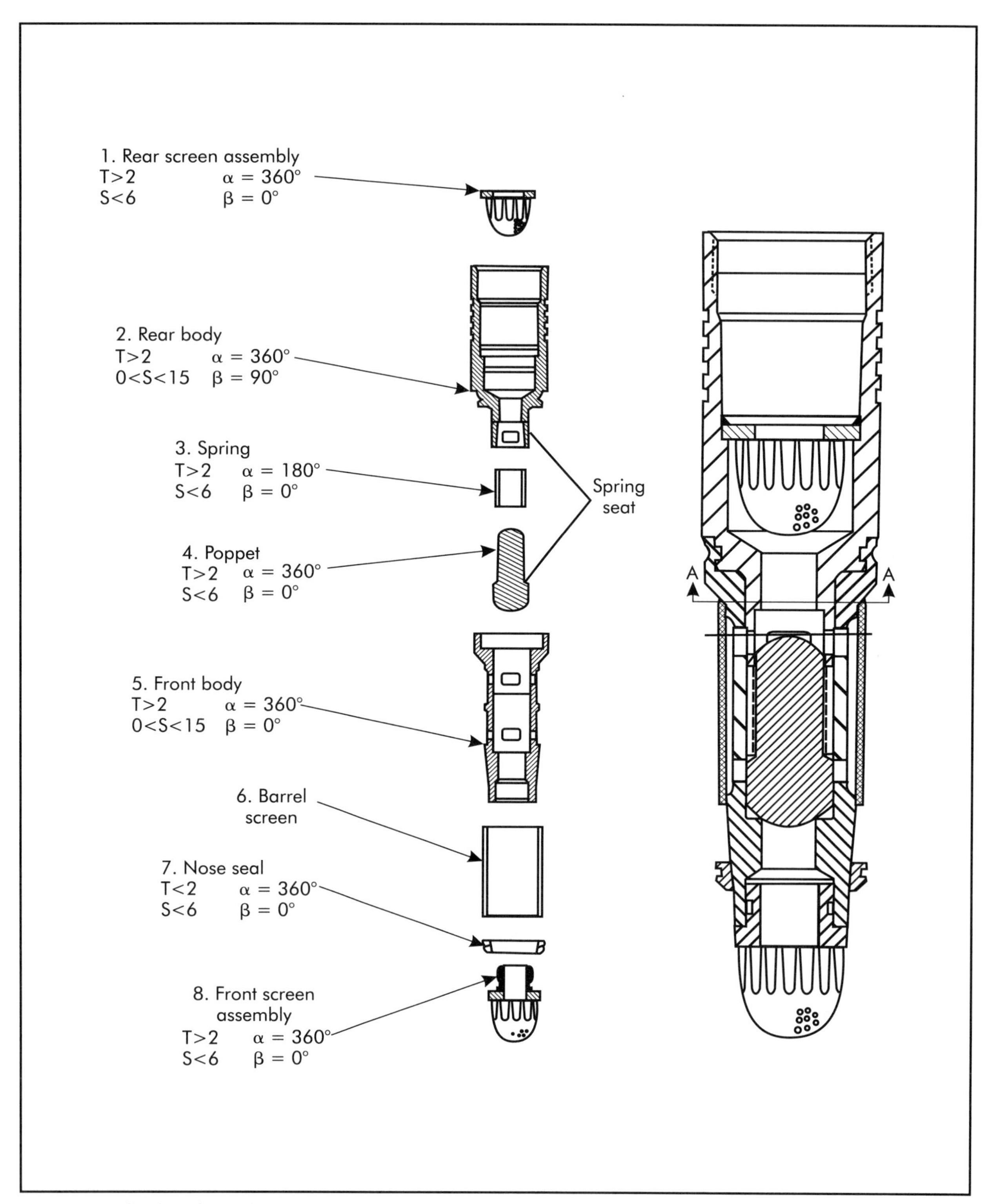

Figure 3-7. Shuttle valve redesign.

Table 3-19. Analysis profile of hydraulic shuttle valve

DFA Metric	Original	Redesign	Improvement (%)
Design efficiency index (%)	24.6	47.1	91.5
Total assembly time (seconds)	93.3	50.9	47.7
Total number of parts	11	8	27.3
Theoretical minimum number of parts	8	8	
Necessary parts (%)	72.6	100	27.3

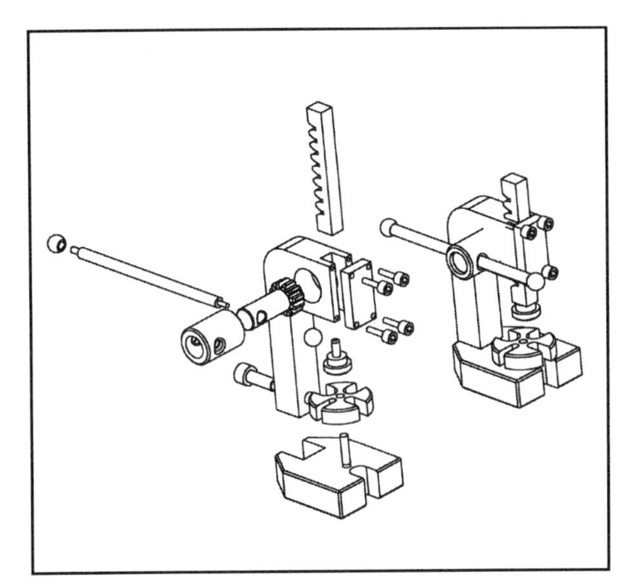

Figure 3-8. Assembly layout of the original mechanical press.

of the press uses a rack-and-pinion combination and a long lever arm. It has 13 components and 17 total parts—such as a base, column, table, etc. The design uses six components that require screws. Table 3-20 shows the press components and their dimensions. Press assembly involves securing a series of components to the base, connecting the column to the base, and connecting the handle to the column. Table 3-21 shows the completed analysis of the tabulated list of operations, its related handling, insertion times, and assembly times. Each of the assembly operations is divided into handling and insertion; the corresponding code for each process is also given.

Step-by-step assembly operation. The assembly operation starts by placing the base on an assembly table and attaching the remaining parts to the base. As an example, consider the following process of handling the column (or Part No. 2) and attaching it to the base of the press.

The insertion axis for a column is vertical along the direction of beta symmetry. The alpha symmetry for the column is 360° and the beta symmetry for the column is 360°. Thus, the total angle of symmetry is 720°. Table 3-21 provides the database for handling time for the column. The column is handled and manipulated with one hand without the aid of tools. For a total angle of alpha and beta of 720°, the first digit of the handling code is 3. The column presents no handling difficulties and can be easily separated from the bulk. Its thickness is greater than 0.08 in. (2 mm) and its size is greater than 0.59 in. (15 mm). Therefore, the second digit is 0, giving it a handling code of 30. A handling time of 1.95 seconds corresponds to the handling code of 30.

The column is not secured on insertion. It gets fastened in the next operation. Since there is no restriction to vision, the first digit of the insertion code is 0. Holding down is necessary while subsequent operations are carried out if the column is not easy to align. It has to be aligned to screw holes with no

Table 3-20. Components of the original mechanical press

Part Identification Number	Item	Quantity	Length in. (mm)	Width in. (mm)	Thickness in. (mm)	Diameter in. (mm)
1	Base	1	3.35 (85)	2.36 (60)	0.98 (25)	
2	Column	1	5.91 (150)	2.95 (75)	0.98 (25)	
3	Machine screw	1	0.98 (25)			0.24 (6)
4	Table	1			1.97 (50)	0.47 (12)
5	Table pin	1	1.97 (50)			0.24 (6)
6	Sleeve	1	1.58 (40)			1.06 (27)
7	Handle	1	4.72 (120)			1.57 (40)
8	Ball end	2				0.59 (15)
9	Gear	1			1.97 (50)	1.06 (27)
10	Rack	1	4.33 (110)	0.47 (12)	0.55 (14)	
11	Cover plate	1	1.77 (45)	0.98 (25)	0.02 (0.4)	
12	Cap screws	4	0.59 (15)			0.24 (6)
13	Rack pad	1			0.98 (25)	0.04 (0.90)

resistance to insertion. Therefore, the second digit of the insertion code is 9. An insertion of 7.50 seconds corresponds to the insertion code of 09.

The total operation time is the sum of the handling and insertion times multiplied by the number of items. For the part under consideration, the total operation time is 9.45 seconds. The cost of an assembly depends on the hourly cost of operators. The rate includes the overhead cost of an organization, which varies from region to region.

As discussed previously, the identification of the theoretical minimum number of parts is a way to identify whether the specific part is a candidate for elimination under the following conditions:

1. During the operation of the product, the specific part under discussion does not move relative to the other parts already assembled.
2. The column is a different material or is physically isolated from other parts already assembled.
3. The column need not be separate from other parts to facilitate assembly and disassembly.

Since the answer to each of these questions is "No," a 0 is placed in Column 9.

The remaining parts are added one by one to the assembly and the manual assembly analysis table is completed for each additional part. The total number of parts is 13 and there are two fastening operations (items 4 and 14). The operations do not have handling time associated with them and are considered only as a separate insertion. Total assembly time is 87.62 seconds. The

Table 3-21. Compilation of assembly time for the original mechanical press

	1	2	3	4	5	6	7	8	9	10
Part Identification Number	Part Description	Number of Items	$\alpha + \beta$ Symmetry (°)	Manual Handling Code	Manual Handling Time (seconds)	Manual Insertion Code	Manual Insertion Time (seconds)	Total Assembly Time (2) × [(5) + (7)] = T_m	Theoretical Minimum Number of Parts (N_m)	Costs
1	Base	1	540	91	3.00	00	1.5	4.50	1	
2	Column	1	720	30	1.95	09	7.5	9.45	0	
3	Machine screw	1	360	18	3.00	02	2.5	5.50	0	
4	Fastening	Operation	—	—	—	92	5.0	5.00	—	
5	Table	1	180	00	1.13	02	2.5	3.63	1	
6	Table pin	1	180	00	1.13	00	1.5	2.63	0	
7	Sleeve	1	540	20	1.80	31	5.0	6.80	1	
8	Handle	1	180	00	1.13	00	1.5	2.63	1	
9	Ball end	1	180	01	1.43	00	1.5	2.93	0	
10	Gear	1	540	20	1.80	00	1.5	3.30	1	
11	Rack	1	720	30	1.95	09	7.5	9.45	1	
12	Cover plate	1	540	20	1.80	08	6.5	8.30	1	
13	Cap screws	4	360	10	1.50	00	1.5	12.00	0	
14	Fastening	Operation	—	—	—	92	5.0	5.00	—	
15	Rack pad	1	360	10	1.50	31	5.0	6.50	1	
	Total	13						87.62	8	
Design efficiency = 27%										

theoretical minimum number of parts is eight; these are essential parts and cannot be combined or eliminated. The assembly design efficiency index for manual assembly is obtained by using Equation 3-1:

$$E_m = \frac{3N_m}{T_m}$$

where:

E_m = manual assembly design efficiency %
N_m = theoretical minimum number of parts
T_m = total assembly time

$$E_m = \frac{3(8)}{87.62}$$

$$E_m = 0.27\%$$

Some components used in the design can be eliminated because they lack a functional purpose. In redesign, the rack-and-pinion combination is still being used, but there are only eight components. The column and cover plate are combined into a single machinable part. The redesigned subassembly rotary set includes a sleeve, gear, handle, and ball end.

Figure 3-9 and Table 3-22 provide design details and assembly data for the redesigned press. Table 3-23 outlines the improvements gained by the redesign.

In redesign, the rotary set subassembly consists of a sleeve, gear, handle, and ball end.

Hitachi Method

The *Hitachi method*, also known as the *assembly-evaluation method*, is an alternative method used to assess the manufacturability of a product. Manufacturability to a large extent depends on design, material costs, processing costs, and other indirect costs (Miyakawa and Ohashi 1986).

Some features of Hitachi methodology are:

- comparison of concept designs and consideration of the advantages of each;
- ranking of concept designs and comparison to competitors' products;
- ranking of product in terms of assemblability;
- facilitation of design improvements on product;
- identification of key points that need improvement;
- estimation of effects of improvement; and
- assembly cost estimate.

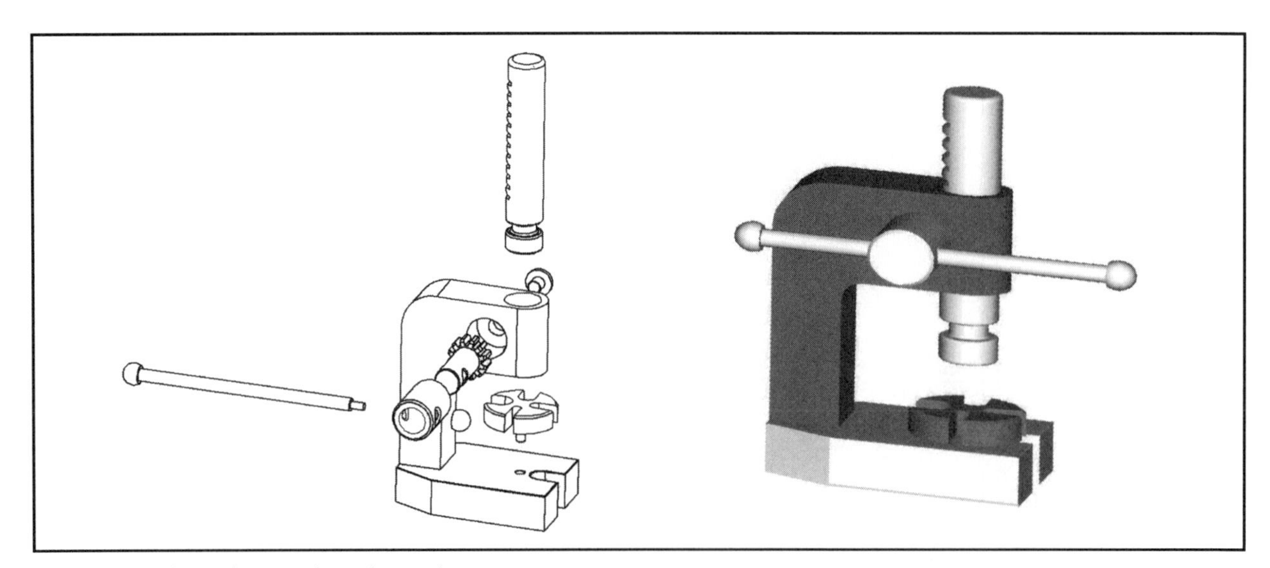

Figure 3-9. The redesigned mechanical press.

Table 3-22. Assembly time for the redesigned mechanical press

Part Identification Number	Part Description	Number of Items	Manual Handling Time (seconds)	Manual Insertion Time (seconds)	Total Assembly Time (2) × [(5) + (7)] = T_m	Theoretical Minimum Number of Parts (N_m)
1	Body	1	3	1.5	4.5	1
2	Table	1	2.5	1.5	4.1	1
3	Apply grease	1	3		7.0	
4	Rack	1	3	2.6	5.6	1
5	Rotor assembly	1	4.8	3	7.8	4
6	Align screw	1	2.55	7.5	12.9	1
	Total				41.9	8
Design efficiency = 60%						

Table 3-23. Analysis profile of mechanical press

DFA Metric	Original	Redesign	Improvement (%)
1. Design efficiency index	27	60	122
2. Total assembly time (seconds)	87.62	41.9	51
3. Total number of parts	13	8	38
4. Theoretical minimum number of parts	8	8	—
5. Necessary parts %	61	100	39

Figure 3-10 shows the steps used in the Hitachi method.

Assembly evaluation is generally carried out with completed product design drawings. However, evaluation of a conceptual design is also part of process improvement. Design improvement is performed based on data obtained by reviewing the evaluation results. Design after improvement is again subjected to an assemblability evaluation process with the purpose of evaluating the effects of improvements. It is important to point out that assemblability evaluation does not distinguish between manual and automatic assemblies. This is because of the belief that there is a strong correlation between the degree of difficulty of manual and automatic assembly.

The assemblability evaluation procedure consists of the steps shown in Figure 3-11.

1. Preparation begins with collecting drawings, conceptual and completed samples, etc. Evaluation results are more accurate if more precise drawings and data are available.
2. The attachment sequence is determined, and the names and numbers of

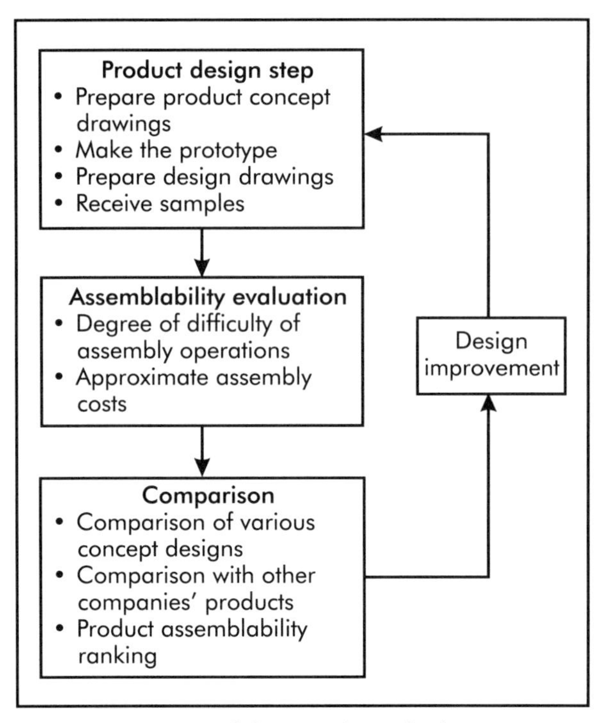

Figure 3-10. Steps of the Hitachi method.

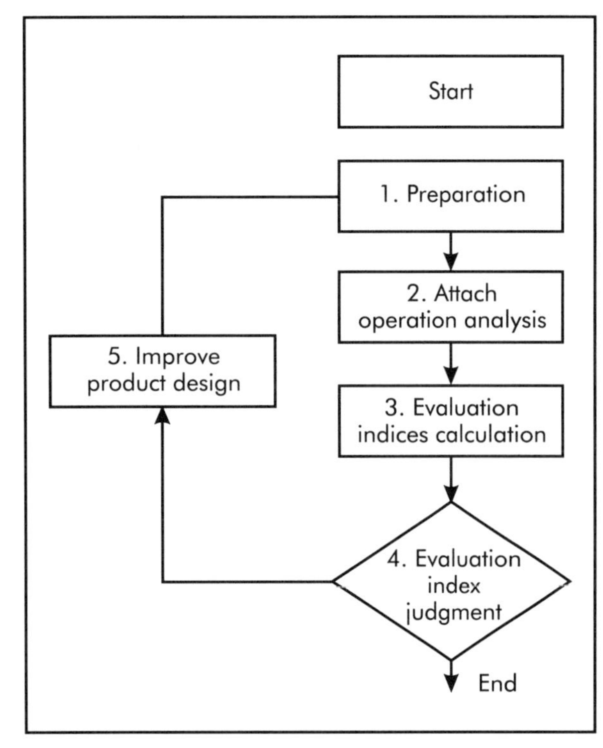

Figure 3-11. Assemblability evaluation procedure.

parts with corresponding symbols are filled into evaluation forms in the same order as the assembly sequence. The evaluation forms are then used to determine the attaching method.

3. Simple calculations are used to determine the evaluation indices (parts and product assemblability evaluation scores).
4. The assemblability evaluation indices are compared to target values. Lower assembly costs are obtained when the product assemblability evaluation score is higher than the target value. Opportunities for assembly automation increase with higher scores. If the product assemblability evaluation score is lower than the target value, then product improvement is needed.
5. The product design improvement process consists of: the identification of weak points and causes of a low score; the creation of alternative designs that eliminate weak points; and the evaluation of the effects of improvement by comparing the assembly evaluation scores of the improved and basic product.

The effects of the assemblability evaluation are summarized:

- The amount of assembly labor is an easy identifier of weak points in the design. It enables rapid improvement, resulting in advanced assemblability and a reduction in manual assembly labor.
- When the evaluation and improvement of a product can be reached at an early stage of design, simpler production and assembly operations result.
- A reduction in the total design period occurs when the time for design improvement is significantly shortened because of evaluation taking place in the early stages of design. The designer

should perform a cost evaluation for assemblability, which will contribute greatly to a reduced design process.
- Simplification of parts production and assembly operations, as well as automation, improves product and process reliability.

Lucas Design Method

The Lucas design method enables a designer to identify nonfunctional and difficult-to-fit design elements, thus indicating areas that will benefit from further scrutiny before the design is finalized (El Wakil 1998). The technique highlights non-essential elements that result not only in part-number reduction, but also advantages such as lower inventories, assembly times, and production control costs. The whole process can be divided into the following steps:

1. product design specification,
2. functional analysis,
3. handling analysis,
4. fitting analysis, and
5. redesign.

Product Design Specification

The product design specification (PDS) is a crucial document for the purpose of analysis. It contains all of the requirements, including customer and business data, which the product must satisfy to be successful. A well-researched PDS provides solutions for frequently conflicting requirements of customer need and component functionality. The PDS is considered a reference point for an emerging design. Every component must be present for a definite purpose and the purpose must be outlined in the specification.

A major factor in the Lucas method is the determination of whether the product is unique or has a relation to other products from the company, indicating there are similarities and opportunities for rationalization and standardization of parts and procedures. If the organization can establish a product family where identical components are used across a range of products, the Lucas methodology becomes very efficient. Standardization enables a single assembly system to be used across a range of products. The product family enables the creation of product groups with a high enough demand to justify automated production and assembly. Product groups keep assembly system designs from becoming obsolete as long as new products are within the product design profile. The objectives are:

- Use standard parts for the product and a range of products to maximize tooling and utilization and to minimize variety.
- Assemble from the same direction and in the same sequence, eliminating the need for duplicate tooling.
- Use common handling and feeding features on larger components—again, to minimize the degree of handling-tool dedication.

The Lucas methodology provides the benefits of keeping handling and tooling technology simple and low cost, while still maintaining an acceptable level of versatility within the system. The benefit of standardization is that a system can be rapidly changed over between batches of different products.

Evaluation Procedure

The evaluation procedure consists of three steps and is illustrated in Figure 3-12.

1. functional analysis;
2. handling analysis (manual or automatic); and
3. fitting analysis (manual or automatic).

Functional analysis. Functional analysis is the first element of the design iteration procedure and it continually repeats

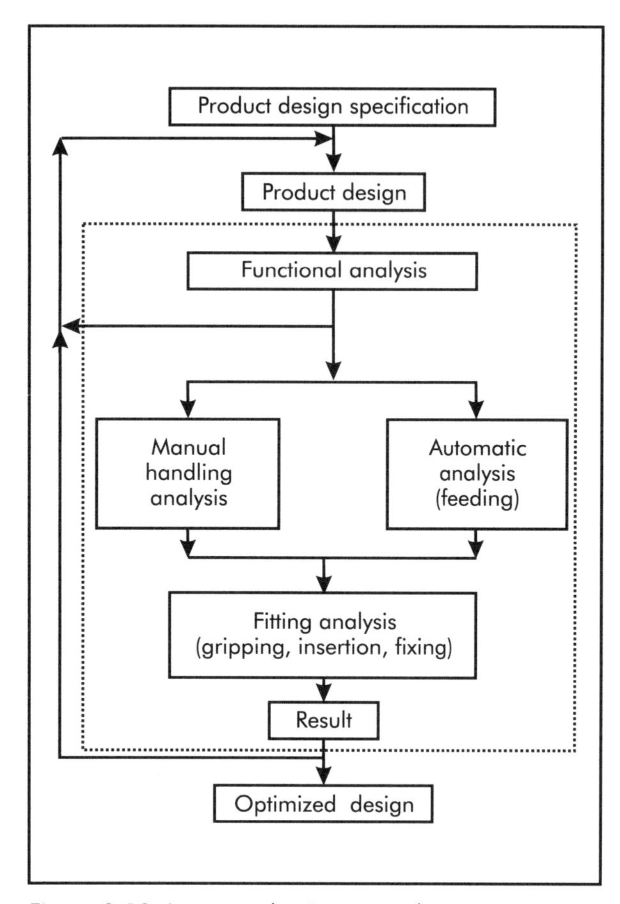

Figure 3-12. Lucas evaluation procedure.

itself until a satisfactory level of design efficiency is reached. Using this technique, the design team evaluates alternatives and selects the best choice. Functional analysis may be undertaken in the early design stage. Every component has to be itemized by name and number in a logical sequence for assembly. Functional analysis is carried out in the following steps:

1. Determine the functional requirements of the product.
2. Decide whether the product can be considered as a whole or as a series of functional sub-sections. (The product should be considered as a whole to avoid duplication of parts in adjacent subsections.)
3. Divide the components into two categories:
   A. Components that carry out functions vital to performance of the product such as drive shafts.
   B. Components like fasteners and locators that are not critical to the product's function.
4. Categorize the mating components in a logical progression until every component has been considered.

Design efficiency (E) is used to functionally assess product design using the following formula:

$$E = \frac{A_N}{C_t} \times 100\% \tag{3-3}$$

where:

A_N = number of A components
C_t = total number of components

The Lucas method suggests that design efficiency be at a level of at least 60% to produce a quality new product. While performing a functional analysis of an existing product, it is important to assess what the design is, but not what it should be. If there is any doubt about a component's category, the component is classified as "B." The goal of this analysis is to determine the components needed for function of the product (under classification A) and to highlight theoretically nonessential parts (under classification B). There is little point in simplifying the assembly if the manufacture of the redesigned component does not contribute to the savings (see Figure 3-13).

Handling analysis. As shown in Figure 3-14, handling analysis helps the designer assess whether a product design with a satisfactory level of functionality is acceptable from an assembly point of view. In the case of a manual assembly, the manual-handling index has to be calculated. In the case of an

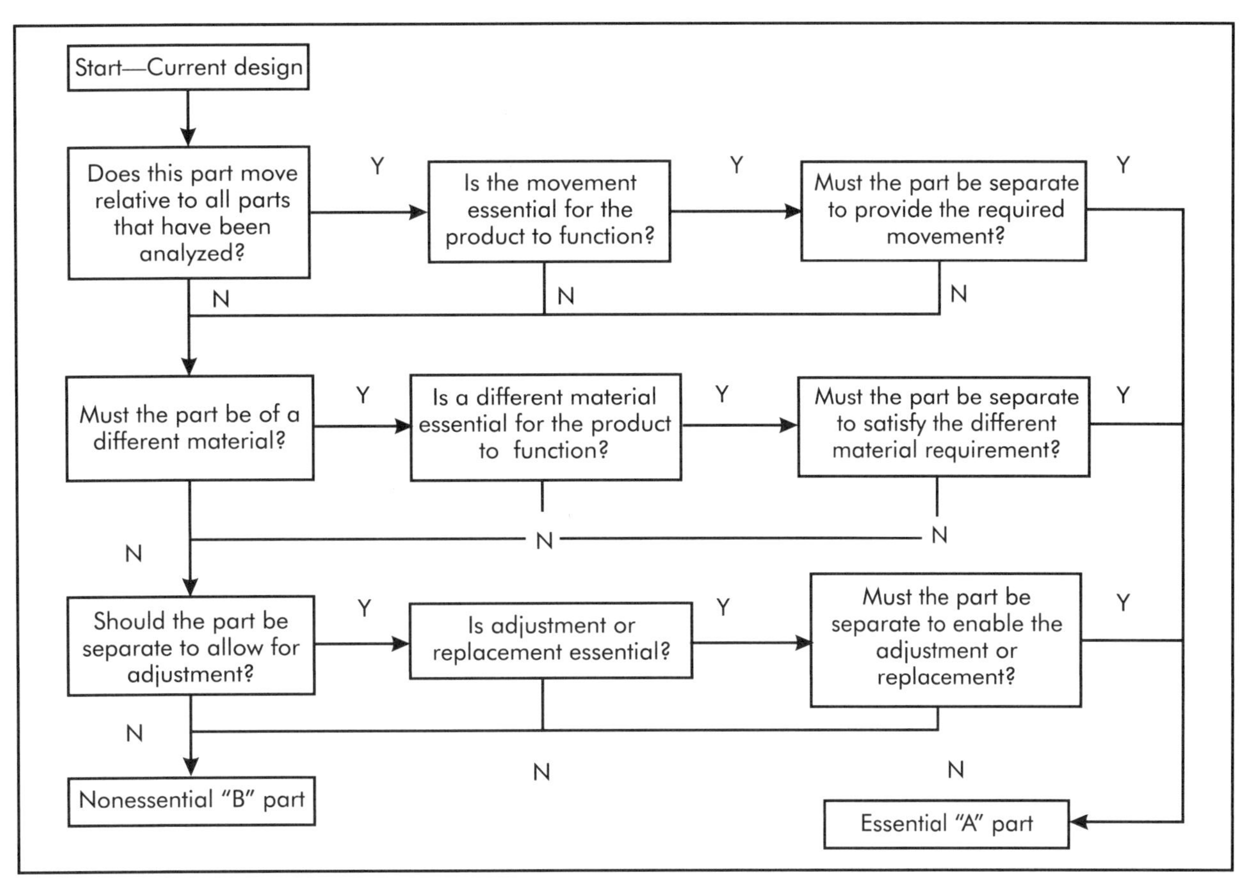

Figure 3-13. Functional analysis of redesigned components.

automated assembly, the feeding index has to be determined. If the index does not meet expectations, a redesign may result in an index reduction. The next important step in the analysis is to determine how the components and subassemblies are going to be assembled. There are two modes to be considered: manual handling and automated feeding.

Manual handling. The manual handling analysis is based on:

- size and weight of the part,
- handling difficulties, and
- orientation of the part.

For a manual assembly, a less complex process is used than for automated assemblies.

Automated feeding. For an automated assembly, calculation of the relative handling cost is the same as the calculation of the feeding ratio. A useful measure of overall effectiveness of product design from the feeding point of view is the feeding ratio. Automatically assembled parts are subject to a three-step analysis for obtaining the automated feeding ratio:

1. Determine whether the components are best transported in a retained orientation, or in the form of bulk supply that is reoriented at the input point.
2. Assess the general physical properties of those components that will not be transported with a retained orientation.
3. Examine the suitability of a detailed design of those components proposed for automatic feeding.

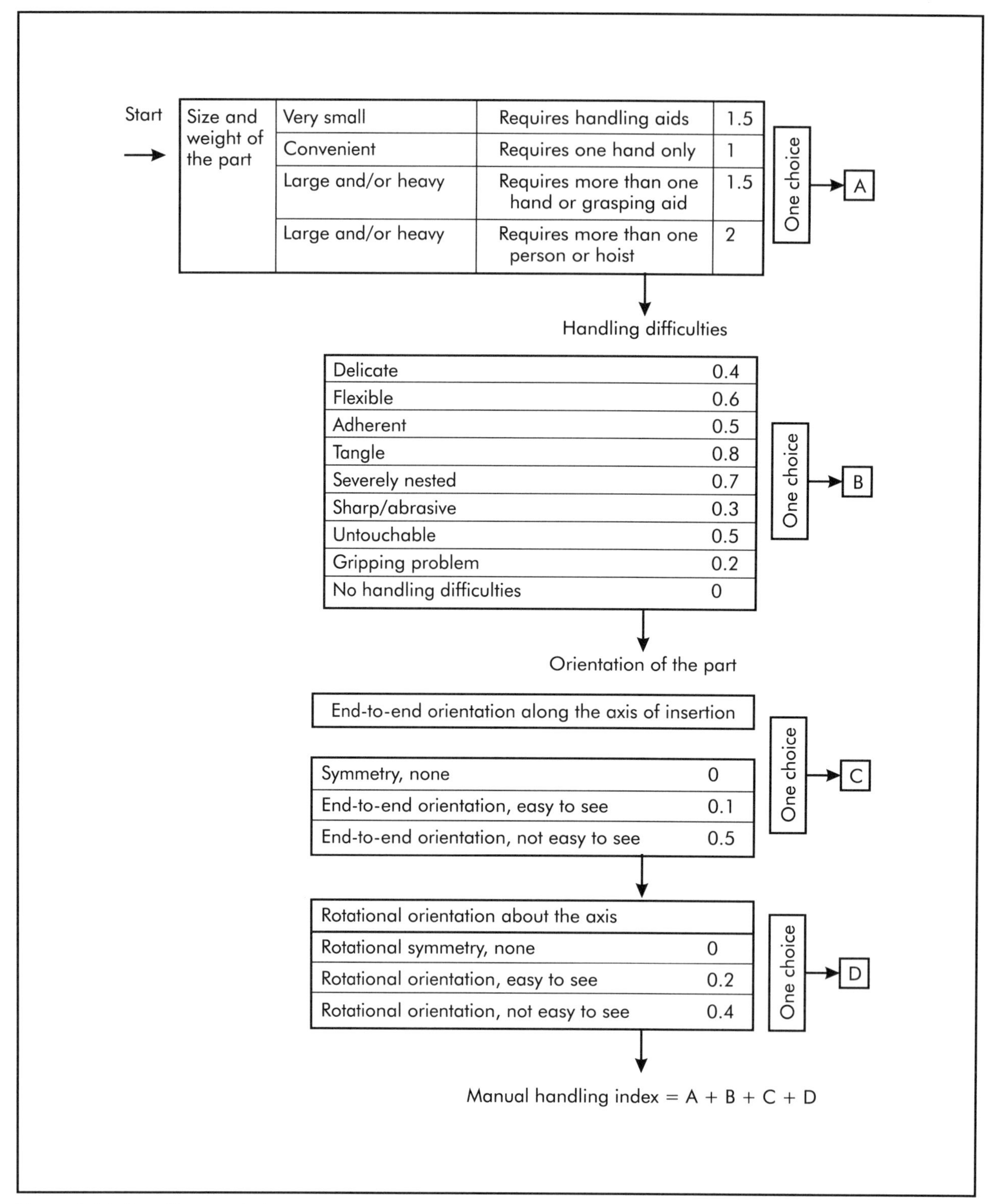

Figure 3-14. Handling analysis.

Feeding indices. For the calculation of feeding indices, the feeding ratio is expressed as:

$$\text{Feeding ratio} = \frac{F_c}{A_N} \quad (3\text{-}4)$$

where:

F_c = total relative feeding cost
A_N = number of A parts

The total relative feeding cost can be obtained by summing all the individual feeding indices A, B, and C. As shown in Figure 3-15, Stage A indices provide information on parts not suitable for mechanical orientation. Stage B provides information on parts that can be mechanically oriented and those with end-to-end orientations. Stage C provides information on parts with rotational orientations. Large numbers of experiments have indicated that the feeding ratio for an acceptable design is generally less than 2.5.

Fitting analysis. Fitting analysis follows handling analysis and indicates what types of gripping, insertion, and fixing operations are required. Each operation is rated and the whole assembly task produces a fitting ratio. The analysis is primarily intended for automated assembly, but if manual assembly is required, then manual ratings are used. Fitting analysis assesses the relative ease or difficulty of carrying out each task required to assemble the complete product from its constituent parts. This varies depending on whether a process is carried out manually or by automated methods and is reflected in respective costs. Individual index values, experimentally determined as 1.5 or greater, indicate the presence of a fitting problem.

A gripping assessment examines how each part is held for transportation, from point of presentation within the automatic assembly system, to the stage where insertion is completed. This applies primarily to automated assembly applications. A fitting assessment examines the overall effectiveness of a product design from a fitting point of view. In non-assembly situations, the procedure looks at an individual non-assembly process and identifies a relative cost that contributes to the total assembly cost. Examples of such processes would be tightening pre-placed screws, welding, and adhesive bonding of pre-placed parts.

The fitting ratio is calculated as:

$$F_r = \frac{G_c + I_i + N_i}{A_N} \quad (3\text{-}5)$$

where:

F_r = fitting ratio
G_c = gripping cost index
I_i = insertion and fixing cost index
N_i = non-assembly cost index
A_N = number of A parts

In the analysis, an appropriate surface is one that enables a component to be carried at the required gripping force. A surface is said to be available when it is possible for the component to be assembled satisfactorily without the gripper obstructing the insertion process. Part characteristics (center of mass, gripping area, etc.) may be such that it is difficult to hold a part securely enough during transport accelerations and decelerations.

A measure of the effectiveness of a product design is its fitting ratio. For a good design, the fitting ratio is less than 2.5, although the aim is to minimize this factor. The fitting ratio is based on the gripping, insertion, and non-assembly costs (see Figures 3-16, 3-17, and 3-18). After analyzing a product, certain aspects of its design, part feeding, and assembly are re-examined for more evaluation. Attention should to be paid to the efficiency of the feeding components. All B-category parts should be eliminated or combined with A-category components. During the concept design stage, emphasis should be placed on increasing the design efficiency rating by looking at the suitability

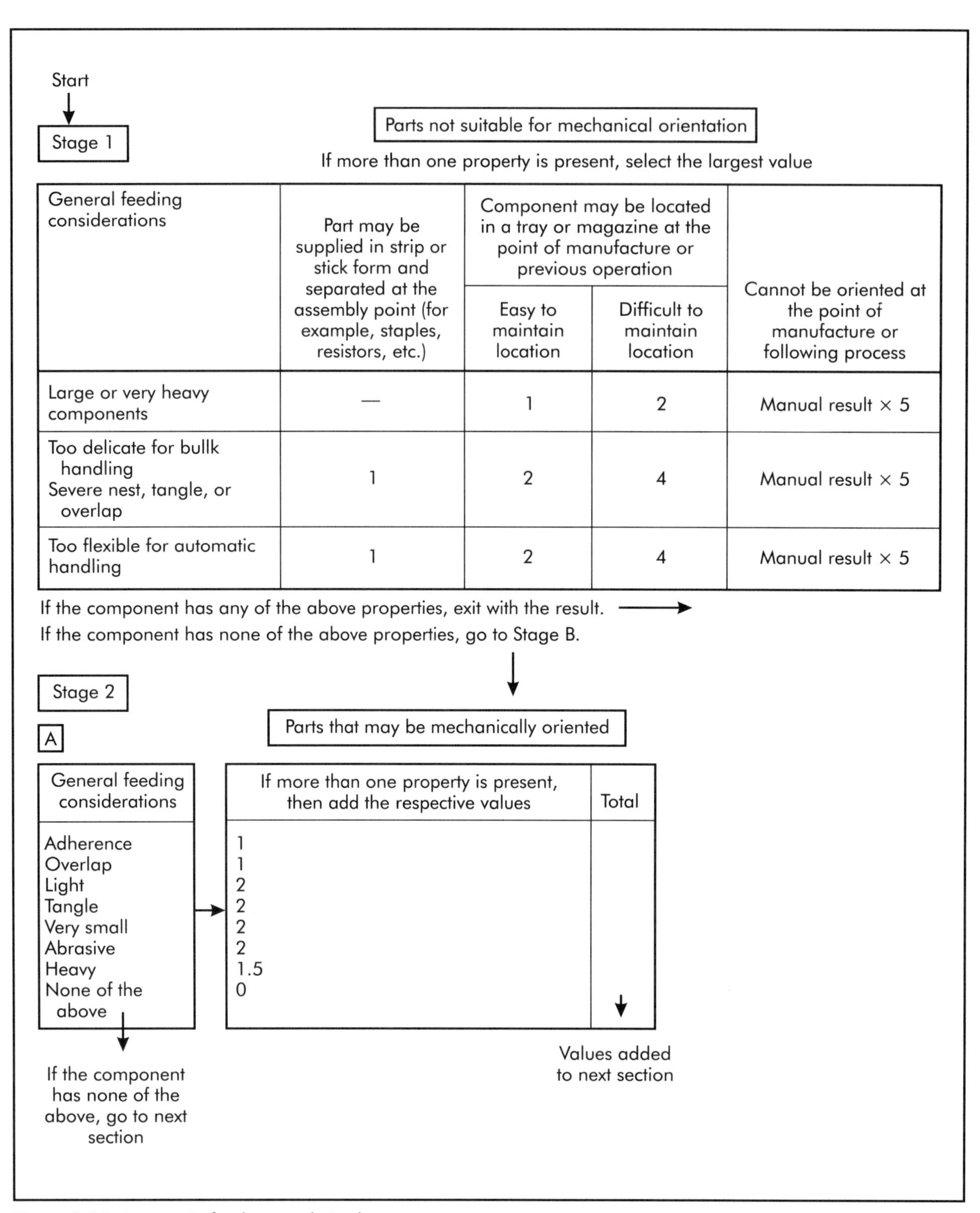

Start

Stage 1

Parts not suitable for mechanical orientation

If more than one property is present, select the largest value

General feeding considerations	Part may be supplied in strip or stick form and separated at the assembly point (for example, staples, resistors, etc.)	Component may be located in a tray or magazine at the point of manufacture or previous operation		Cannot be oriented at the point of manufacture or following process
		Easy to maintain location	Difficult to maintain location	
Large or very heavy components	—	1	2	Manual result × 5
Too delicate for bullk handling Severe nest, tangle, or overlap	1	2	4	Manual result × 5
Too flexible for automatic handling	1	2	4	Manual result × 5

If the component has any of the above properties, exit with the result.

If the component has none of the above properties, go to Stage B.

Stage 2

A

Parts that may be mechanically oriented

General feeding considerations	If more than one property is present, then add the respective values	Total
Adherence	1	
Overlap	1	
Light	2	
Tangle	2	
Very small	2	
Abrasive	2	
Heavy	1.5	
None of the above	0	

If the component has none of the above, go to next section

Values added to next section

Figure 3-15. Automatic feeding analysis chart.

B | End-to-end orientation

Directional orientation of the major axis	End-to-end orientation not required	Can be fed in a slot supported by its head	External features	Internal features	Non-geometric features too small for mechanical orientation
L/I>1.5	MT 0.5	MT 0.7	MT 1.6	MT 1.6	LT 3
L/I<1.5	MT 0.5	MT 0.7	MT 1.0	Manual handling × 5	LT 3

C | Rotational side-to-side orientation

		Rotational/side-to-side orientation not required	Features seen in end view/silhouette	Features not seen in end view/silhouette	Non-geometric features or features too small for mechanical orientation
Poor stability at major axis	L/I>1.5	MT 0.5	MT 0.8	MT 1.5	Manual handling × 5 (A and B ignored)
	L/I<1.5	MT 0.5	MT 0.9	MT 0.8	Manual handling × 5 (A and B ignored)
Rotational stability at major axis	L/I>1.5	MT 0.5	MT 0.8	MT 1.2	LT 3 (ignore B)
	L/I<1.5	MT 0.5	MT 0.9	MT 1.2	LT 3 (ignore B)

Feeding index = A + B + C

MT = mechanical tooling (1)
LT = laser tooling (2)
M = manual orientation (3)
L = longest dimension
I = intermediate dimension

Figure 3-15. (continued)

Properties	Component has appropriate gripping surface				Component has no suitable gripping surface
	Surface is available during the insertion process		Surface is not available during the insertion process		
	Component is easy to grip securely during transport	Component is not easy to grip securely during transport	Component is easy to grip securely during transport	Component is not easy to grip securely during transport	
Index	0	0.5	1	1.5	2.5

Figure 3-16. Gripping difficulty index.

of the component part design for handling and feeding. The result of a redesign should be consistent with the gripping provisions if automation is to be applied. In all events, the product redesign stage must consider the task of actually assembling parts into their final position.

Electric Motor Case Study

Using the Lucas design method, the functional, handling, and fitting analyses are carried out on an electric motor. Charts for functional, handling, and fitting analyses are drawn. The function analyses charts are used to identify essential parts (A) and non-essential parts (B). Tables 3-24a and b list 16 components, of which nine are classified as "A" parts. The design efficiency in this functional analysis is identified as 56%.

The manual handling index and fitting ratios are calculated for the original design and improved design of the electric motor (see Tables 3-24 to 3-28).

Handling analysis. As outlined in Figure 3-14, the handling analysis considerations include the size and weight of the part (A), handling difficulty (B), and orientation of the part (C and D). The handling index is the sum of A, B, C, and D. Table 3-24 shows the individual as well as cumulative handling difficulties for all 16 parts of the electric motor.

Fitting ratio analysis. In Figure 3-17, the fitting process, the process and volume, access to the process, and aligning and insertion difficulties are identified. Table 3-25 shows individual, as well as cumulative values of the insertion and fixing index for all 16 parts of the electric motor.

The fitting ratio is determined by calculating the sum of the gripping cost, insertion, fixing cost, and non-assembly cost indices. In this case, the gripping-cost index is zero because the component has an appropriate gripping surface available during the insertion process (see Table 3-28).

Using Equation 3-5 and substituting the indice values from Table 3-25:

$$F_r = \frac{0 + 45.9 + 72.5}{9} = 13.15$$

Table 3-29 outlines the improvement realized from redesign of the electric motor. Although design efficiency has improved substantially, the manual handling and fitting ratios have not improved proportionally.

COMPARISON OF DFM METHODS

Different DFM techniques provide systematic and disciplined ways of raising the importance of manufacturing and assembly in the mind of the designer. The aim is to concentrate early in the design stage on cre-

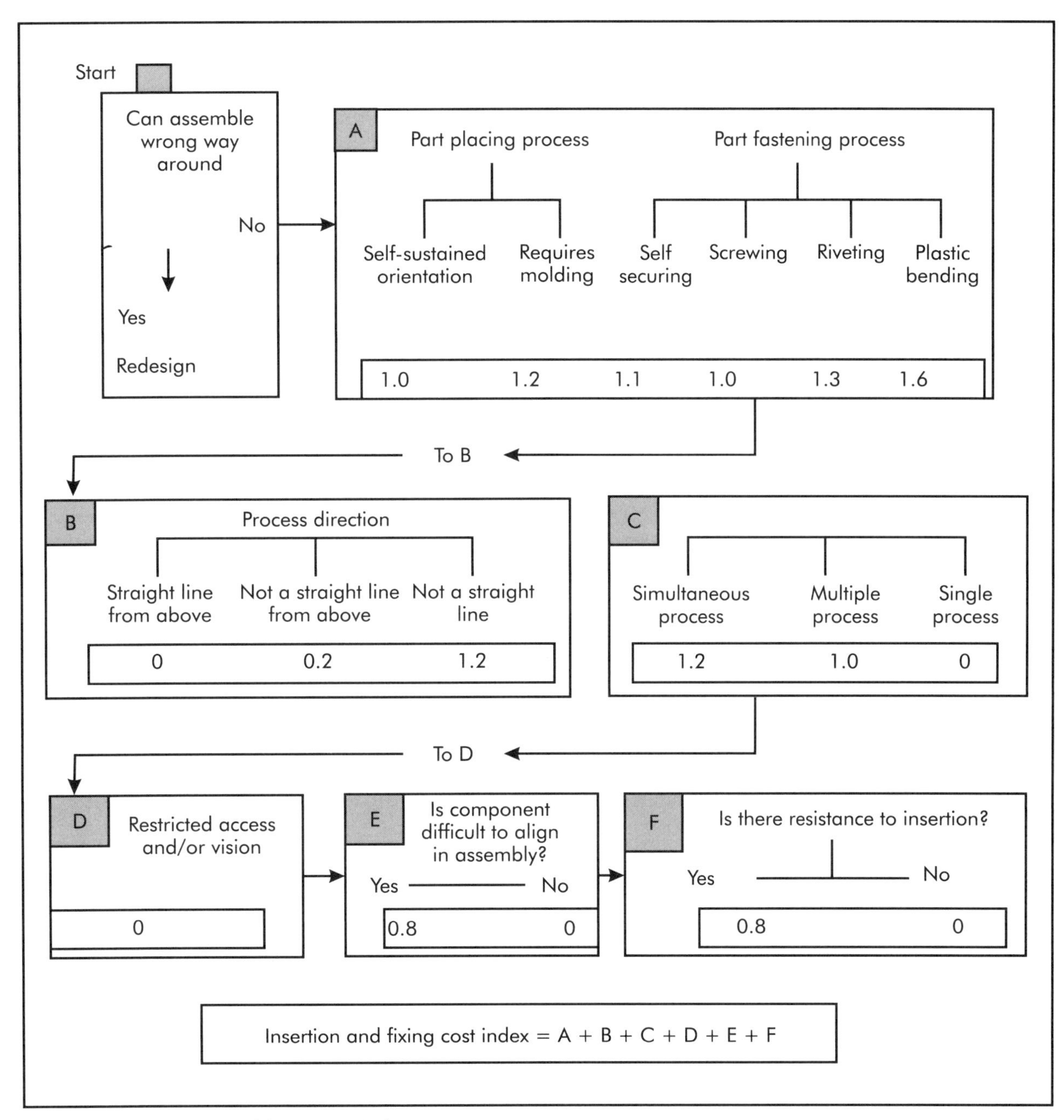

Figure 3-17. Insertion and fixing cost index.

ating products that are easy to manufacture and assemble, before much effort and cost is expended in pursuing another design, which might be unnecessarily expensive. DFM methods provide the basis from which to develop an integrative prospective of design, manufacture, and assembly.

The methods covered in this book, except for the axiomatic method discussed in Chapter 2, apply mainly to mechanism-based

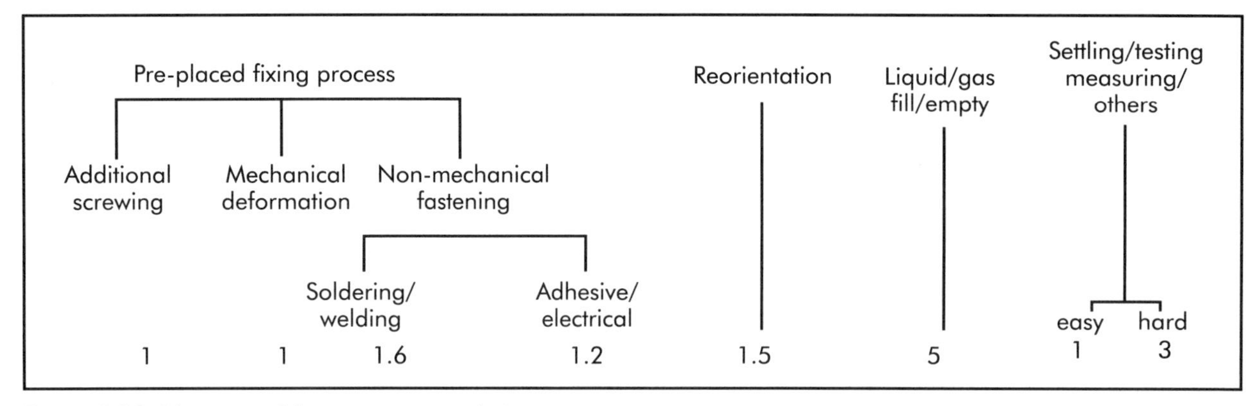

Figure 3-18. Non-assembly processes cost index.

assemblies of sizes that can be conveniently assembled on a desktop. Typical products are tape/video recorders, car alternators, and water pumps. The methods are not appropriate for large products such as complete cars. This is because there is a lack of data for these large products.

The Hitachi, Boothroyd-Dewhurst, and Lucas methods are supported by computer software systems that provide step-by-step instructions. An advantage of the software is that it aids the evaluation procedure by prompting the user with help screens in context, and by conveniently documenting the analysis. The user can quickly assess the effect of a proposed design change by editing a current analysis. The Boothroyd-Dewhurst and Lucas methods distinguish manual from automatic assemblies, while the Hitachi method does not. Based on a comparison analysis, it becomes clear that the axiomatic technique cannot really be compared with other methods. This is because the axiomatic technique gives a conceptual understanding of a product design, while the other three methods give detailed assembly evaluations.

In the conceptual stage, the axiomatic method is superior. However, the axiomatic approach has two major weaknesses when manufacturing is considered in the early stages of product design. It does not provide any means of making judgments between centrally important tradeoffs posed by possible alternative choices of materials and processes.

In the analysis stage, the Boothroyd-Dewhurst method has some strong points. This is because manufacturing guidelines are invariably intended to make individual processing steps as efficient as possible. The Boothroyd-Dewhurst method focuses attention mostly on the handling and insertion of parts, with detailed consideration given to automation.

An evaluation of product designs for automated assembly has real value to anticipate difficulties that would otherwise occur during product installation. The design efficiency reflects the scope for parts reduction and for improving the handling and insertion (manual) processes. Software for the Boothroyd-Dewhurst method provides an extensive range of analysis output and report formats for the user. It also facilitates the freedom to present results in a number of ways.

The most complete calculation of design efficiency is provided by the Boothroyd-Dewhurst method. It determines design efficiency that takes into account parts reduction and handling and insertion improvement. The Hitachi method calculates design efficiency based on the insertion pro-

Table 3-24a. Function analysis of the electric motor

Part Identification Number	Part Description	Function	Rating
1	Terminal box	Connection to power source	A
2	Blower	Cools the motor	A
3	Bracket	Holds blower and terminal box	B
4	Stiffener	Stiffens magnet	B
5	Brake linkages	Transmits braking force	A
6	Magnet	Creates braking	A
7	Magnet support	Supports	B
8	Bearing stand	Forces from rotor	A
9	Sheave	Provides motion and traction	A
10	Rotor	Enables torque creation	A
11	Adapter	Grips between stator and frame	B
12	Stator	Magnet circuit creation	A
13	Rear cover	Support for end bearing	A
14	Right stabilizer	Stability of motor	B
15	Left stabilizer	Stability of motor	B
16	Frame	Grip to stator and stand	B
Design efficiency = Number of A parts/Number of total parts = 9/16 = 56%			

cess only, while the Lucas method is focused on a reduction of the number of parts. The axiomatic method does not deal explicitly with design efficiency.

On the other hand, the Hitachi technique gives a process overview of assembly, sequence, and insertion operations. There is no explicit criterion for a minimum parts count. The Hitachi method does not offer direct analysis for parts feeding and orientation. For this reason, design for automated assembly is not an option, the argument being that an assessment of a product design for automated assembling is sensitive to part configuration and is difficult to assess precisely at the early design stages. These aspects should be dealt with at later design stages.

The Lucas method is based on a symbolic logic programming paradigm. The advantage of this is that it is easier to encode and derive the design for assembly rules embodied in the method and, at the same time, provide the user with generalized suggestions

Table 3-24b. Handling analysis of the electric motor

Part Identification Number	Part Description	Size and Weight	Handling Difficulty	Orientation	Manual Handling Index
1	Terminal box	1.5	0.4	0.4	2.3
2	Blower	1.5	0.4	0.1	2.0
3	Bracket	1.5	0	0.1	1.6
4	Stiffener	1.5	0	0.1	1.6
5	Brake linkages	2.0	0.2	0.1	2.3
6	Magnet	2.0	0.4	0.1	2.5
7	Magnet support	2.0	0	0.1	2.1
8	Bearing stand	2.0	0.2	0.1	2.3
9	Sheave	2.0	0.2	0.1	2.3
10	Rotor	2.0	0.4	0.1	2.5
11	Adapter	2.0	0	0.5	2.5
12	Stator	2.0	0.4	0.7	3.1
13	Rear cover	2.0	0	0.3	2.3
14	Right stabilizer	1.5	0	0.1	1.6
15	Left stabilizer	1.5	0	0.1	1.6
16	Frame	2.0	0	0.1	2.1
					34.7
Manual handling index = 34.7					

for possible design changes as the evaluation proceeds. The Lucas method adopts aspects of both the Hitachi and Boothroyd-Dewhurst methods by dealing with handling and insertion with some consideration of automation. It gives a good overview of the assembly process. The design efficiency of the Lucas method is based solely on the scope for reducing the number of parts in a product design and is not as comprehensive as the Boothroyd-Dewhurst method.

All methods are useful to create products with superior quality levels, shorter marketing times, and low costs.

Table 3-30 gives an evaluation of the various DFM methods.

The Boothroyd-Dewhurst method shows an excellent feasibility score compared to other methods. It is straightforward and relatively well documented with step-by-step instructions. The Lucas method has similar advantages, while the Hitachi method does

Table 3-25. Fitting ratio analysis for an electric motor

Part Identification Number	Part Description	A	B	C	D	E	F	Insertion and Fixing Index	Non-assembly Index
1	Terminal box	2	0	0	0	0.7	0	2.7	5.5
2	Blower	2	0	0	0	0	0	2.0	5.5
3	Bracket	2	0	0	1.5	0.7	0	4.2	5.5
4	Stiffener	2	0.1	0	0	0.7	0	2.8	5.5
5	Brake linkages	2	0.1	0.7	0	0	0	2.8	5.5
6	Magnet	2	0	0	0	0	0	2.0	4.0
7	Magnet support	2	0	0	0	0	0	2.0	4.0
8	Bearing stand	1	0.1	0	1.5	0.7	0	3.3	5.5
9	Sheave	2	0.1	0	1.5	0.7	0	3.3	1.5
10	Rotor	1	0.1	0	1.5	0.7	0	4.3	5.5
11	Adapter	2	0.1	0	0	0.7	0	2.8	4.0
12	Stator	2	0.1	0	1.5	0.7	0	4.3	4.0
13	Rear cover	2	0.1	0	0	0.7	0	2.8	5.5
14	Right stabilizer	2	0.1	0	0	0.7	0	2.8	5.5
15	Left stabilizer	2	0.1	0	0	0.7	0	2.8	5.5
16	Frame	1	0	0	0	0	0	1.0	0
								45.9	72.5

Insertion and fixing index = 45.9
Non-assembly index = 72.5
Gripping index = 0
Fitting ratio = 13.15

Table 3-26. Function analysis of redesigned electric motor

Part Identification Number	Part Description	Function	Rating
1	Terminal box	Connects to power source	A
2	Blower	Cools the motor	A
3	Brake linkages	Transmits braking force	A
4	Magnet	Creates braking	A
5	Bearing stand	Forces from rotor	A
6	Sheave	Provides motion and traction	A
7	Rotor	Enables torque creation	A
8	Front cover	Takes rotor force	A
9	Rear cover	Supports end bearing	A
10	Stator	Magnet circuit creation	A

Design efficiency = Number of A parts/Number of total parts
= 10/10 =100%

Table 3-27. Handling analysis of redesigned electric motor

Part Identification Number	Part Description	Size and Weight	Handling Difficulty	Orientation	Manual Handling Index
1	Terminal box	1.5	0.4	0.4	2.3
2	Blower	1.5	0.4	0.1	2.0
3	Brake linkages	2.0	0.2	0.1	2.3
4	Magnet	2.0	0.4	0.1	2.5
5	Bearing stand	2.0	0.2	0.1	2.3
6	Sheave	2.0	0.2	0.1	2.3
7	Rotor	2.0	0.4	0.1	2.5
8	Front cover	2.0	0	0.5	2.5
9	Rear cover	2.0	0	0.3	2.3
10	Stator	2.0	0.4	0.7	3.1
					24.1

Manual handling index = 24.1

Table 3-28. Fitting ratio analysis for redesigned electric motor

Part Identification Number	Part Description	A	B	C	D	E	F	Insertion and Fixing Index	Non-assembly Index
1	Terminal box	2	0	0	0	0.7	0	2.7	5.5
2	Blower	2	0	0	0	0	0	2.0	5.5
3	Brake linkages	2	0.1	0.7	0	0	0	2.8	5.5
4	Magnet	2	0	0	0	0	0	2.0	4.0
5	Bearing stand	1	0.1	0	1.5	0.7	0	3.3	5.5
6	Sheave	2	0.1	0	1.5	0.7	0	3.3	1.5
7	Rotor	1	0.1	0	1.5	0.7	0	4.3	5.5
8	Front cover	2	0.1	0	0	0.7	0	2.8	5.5
9	Rear cover	2	0.1	0	0	0.7	0	2.8	5.5
10	Stator	2	0.1	0	1.5	0.7	0	4.3	4.0
								30.3	48.0

Insertion and fixing index = 30.3
Non-assembly index = 48.0
Gripping index = 0
Fitting ratio = 7.83

Table 3-29. Analysis profile for electric motor

DFA metric	Original	Redesign	Improvement %
Number of subassemblies	16	10	37.5
Design efficiency	56	100	78
Manual handling ratio	34.7	23.9	31
Fitting ratio	13.15	7.83	39

not follow the sequential steps of the other two techniques and is not well represented in technical literature. The axiomatic method is based on a fixed set of axioms and looks at a design from a conceptual, rather than a design for assembly point of view. The dependency of functional requirements and design parameters is analyzed fundamentally, giving the designer a solid base to start with. The axiomatic method offers the right methodology to manage a design process in the conceptual stage. The other three methods do not provide such features.

REFERENCES

Boothroyd, Geoffrey. and Dewhurst, Peter. 1987. *Product Design for Assembly Handbook*. Wakefield, RI: Boothroyd-Dewhurst, Inc.

Table 3-30. Evaluation table of DFM methods (Zlatko 1995)

	Hitachi	Boothroyd-Dewhurst	Lucas
Method's feasibility from the practical viewpoint	Medium	Excellent	High
Effectiveness in design-efficiency calculation	Medium	Excellent	Medium
Effectiveness in dealing with design as a part of design improvement	N/A	N/A	N/A
Use of computer software	High	High	High
Diversity of application	Low	High	Medium

El Wakil, Sherif, 1998. *Processes and Design for Manufacturing*. Second edition. Boston, MA: PWS Publishing Company ITP.

Hartley, John R. 1992. *Concurrent Engineering*. Portland, OR: Productivity Press.

Miyakawa, S. and Ohashi, T. 1986. "The Hitachi Assemblability Evaluation Method." Proceedings of the First International Conference on Product Design for Assembly. Newport, RI: April.

Zlatko, Strbuncelj. 1995. "A Study in the Comparative Analysis of DFM Methods." Masters thesis in Mechanical Engineering. University of Hartford, CT.

Chapter 4

Manufacturing, Disassembly, and Life Cycle

SYSTEMATIC PROCESS SELECTION

It is extremely important to make good manufacturing decisions early in the design process. Such decisions can influence the cost of the product and the selection of the appropriate process. A systematic procedure can be established for process selection that considers all processes and eliminates those that cannot satisfy the design requirements. Process-selection charts with capable databases are useful in ranking processes based on their costs.

In the early stages of product design, computer-aided design tools help with product-modeling geometry and selection criteria for manufacturing and assembly. Close interaction between designers and those involved in manufacturing is essential for making the right manufacturing decisions. Product designers should be aware of the manufacturing consequences of their decisions, since minor changes in design during the early stages can often prevent major manufacturing problems later. For example, a product might be required to perform a completely new function or satisfy a need not previously filled. The identification of the most economical manufacturing process is one of the major prerequisites for making a competitive product. Similarly, a product modification may require the use of a different manufacturing process to satisfy new requirements.

Typically, a product consists of assemblies and components. An air conditioner consists of a set of assemblies that includes a compressor, fan, cooling unit, frame to mount, etc. The process selection in this instance generally takes place at the component level. Each component is considered individually. The manufacturing process capable of economically making each component is then identified.

The design process starts with identifying the market need and it proceeds through the stages of conceptual design, refinement, and detail. The output of these stages leads to a set of specifications or constraints, which dictate how the product should be made. In the conceptual stage of design, little information is available and few constraints are specified—so all possible manufacturing processes can be considered. As the design process progresses, more information on the product becomes available. This information is used to recommend the best processes that can be used to make the product. As a design reaches its final stages and becomes detailed enough to allow a cost evaluation to be performed, a single process can be selected.

MANUFACTURING PROCESS SELECTION

There are a large number of available processes. The selection of a particular process depends on many factors. A product can be made by more than one method and it is important that the designer be familiar with conventional as well as special processes. Capabilities and data on various manufacturing processes can be acquired from databases and handbooks that provide descriptions of various processes, principles, equipment, process parameters, process capabilities, and application examples. However, data about different processes is not stored in a standardized format to facilitate process comparison. A comparison of process capabilities can provide only qualitative guidance toward the selection of a particular process. To develop a systematic method for selecting a manufacturing process, information should be structured in a way to make it easy to compare the capabilities of various processes.

How is a manufacturing process selected? The processes are selected based on certain product attributes such as size, shape, finish, strength, volume, and cost. Other influencing parameters are material availability, useful life, physical loading, chemical environment, disposal, and recycling. The process of manufacturing and the shape are closely related to the material chosen. Table 4-1 shows useful parameters that should be observed when selecting a material (Hundal 1997).

The material selection is based on functional consideration—choosing the right material with suitable properties and strength. Additional considerations include cost and fabrication ability.

Joseph Datsko lists 11 design rules that should be considered for ease of production (Datsko 1997):

1. Select a material based on ease of fabrication, function, and original cost.
2. Use the simplest configuration and specify standard sizes whenever possible.
3. Use a configuration requiring the least number of separate operations.
4. Use configurations attainable with efficient manufacturing.
5. Design the fabrication process to achieve the desired strength distribution in the finished part.
6. Provide clamping, locating, and measuring surfaces.
7. Specify the tolerances and surface finish while considering the functional requirements and processes (see Figure 4-1).
8. Determine the specific function of each part.
9. Determine the specific feature of a part that enables it to perform its function.
10. Prepare a systematic process sheet listing processes needed for a component.
11. Evaluate a preliminary design by considering any changes in the design that can simplify the fabrication process.

Table 4-2 provides a matching of manufacturing process attributes to those of materials under consideration. Each individual process is characterized by a set of attributes. Features are referred to as process attributes. They are also the factors that influence the process selection decision.

The principle behind the systematic selection procedure to identify possible processes is to match desired attributes dictated by the design to the process attributes. As shown in Table 4-3, characteristics are broken down into sets of features such as surface finish, dimensional accuracy, complexity, production rate, size, and relative cost.

Table 4-1. Parameters for manufacturing processes (Hundall 1997)

Mechanical Parameters	Other Parameters	Suitability for Processing
Density	Thermal conductivity	Machining
Strength	Magnetic properties	Casting
Elasticity	Melting point	Forging
Toughness	Specific heat	Drawing
Hardness		Joining
Wear resistance		
Fatigue		

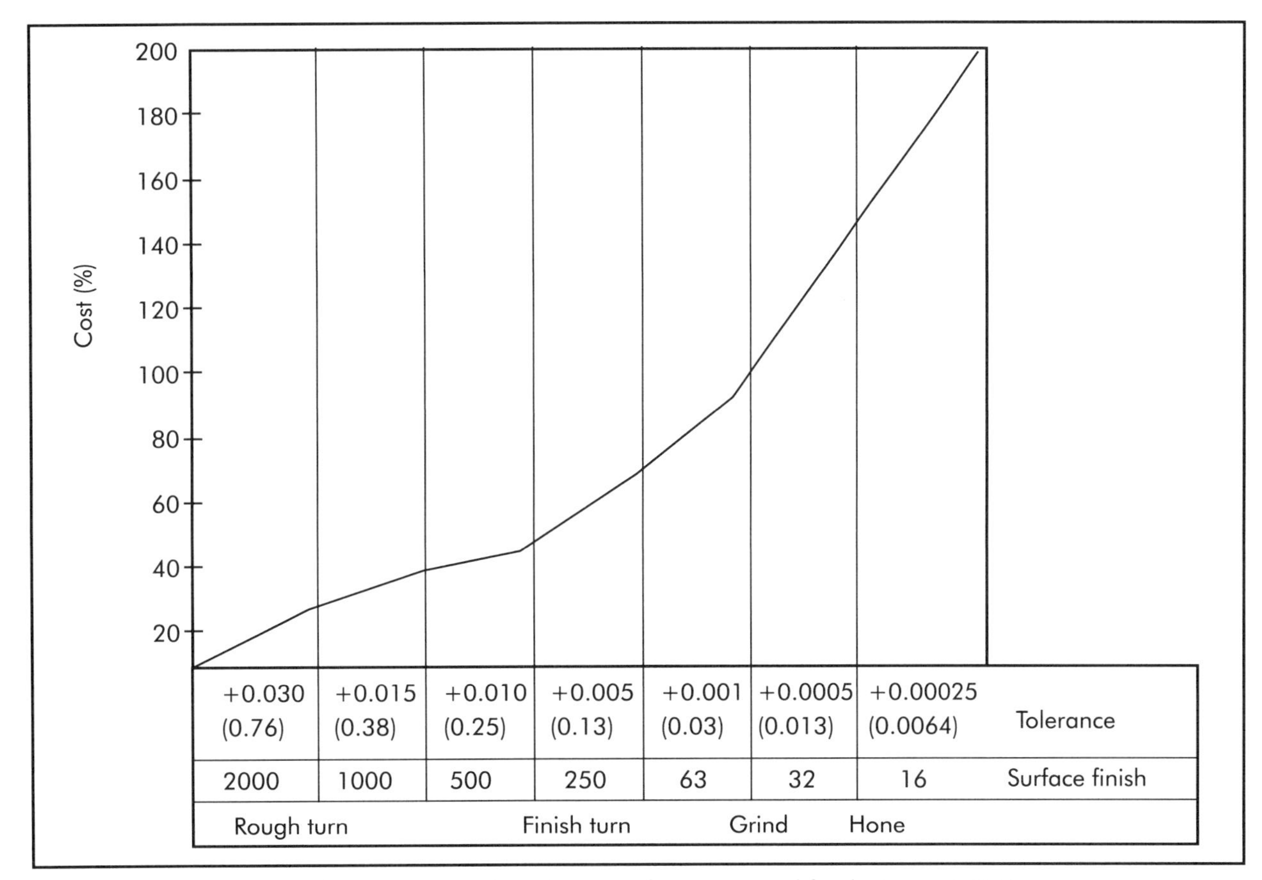

Figure 4-1. General cost relationship of various degrees of accuracy and finish (Tanner 1991).

Manufacturing Processes and Attributes

The process-selection procedure consists of an initial screening phase based on the process class, type, and attributes. The attributes are matched with the design attributes and the design is broken down into components. Then the ranking phase follows, which ranks the processes based on cost.

Figure 4-2 shows the breakdown of processes into classes such as deformation,

Table 4-2. Suitability of materials and manufacturing processes (Ashby 1992)

	Manufacturing Process													
Material	Casting (Die)	Casting (Centrifugal)	Injection Molding	Sand Casting	Investment Casting	Milling	Grinding	Electrical Discharge Machining (EDM)	Forging	Rolling	Extrusion	Powder Metallurgy	Sheet Metal Working	Blow Molding
Low-carbon steel	—	E	—	E	E	G	E	E	G	G	G	E	G	—
High-carbon steel	—	E	—	E	E	G	E	E	G	G	G	E	G	—
Low-alloy steel	—	E	—	E	E	G	E	E	G	G	G	E	G	—
Stainless steel	—	G	—	E	E	—	—	E	G	G	G	E	G	—
Malleable iron	—	E	—	E	E	G	E	E	S	S	S	E	G	—
Alloy cast iron	—	E	—	E	E	G	E	E	S	S	S	E	G	—
Zinc alloys	E	—	—	G	S	—	S	E	S	S	G	E	E	—
Aluminum alloys	E	E	—	E	E	E	G	E	E	E	E	E	E	—
Titanium alloys	—	—	—	—	S	—	S	E	G	S	S	E	—	—
Copper alloys	G	E	—	E	G	E	G	E	E	E	E	E	E	—
Nickel alloys	—	E	—	E	G	—	S	E	S	G	G	E	G	—
Tungsten alloys	—	—	—	—	G	—	S	E	S	—	—	E	—	—
ABS	—	—	—	—	—	G	G	—	—	—	E	—	—	G
Nylons	—	—	E	—	—	G	G	—	—	—	G	—	—	G
Polystyrene	—	—	E	—	—	G	G	—	—	—	E	—	—	G
PVC	—	—	—	—	—	G	G	—	—	—	E	—	—	G
Polyurethane	—	—	—	—	—	G	G	—	—	—	G	—	—	G
Polyethylene	—	—	E	—	—	G	G	—	—	—	E	—	—	E
Acrylics	—	—	—	—	—	G	G	—	—	—	S	—	—	—
Epoxies	—	—	E	—	—	G	G	—	—	—	S	—	—	—
Silicones	—	—	—	—	—	—	—	—	—	—	S	—	—	—
Polyester	—	—	—	—	—	G	G	—	—	—	S	—	—	—
Rubbers	—	—	E	—	—	—	—	—	—	—	S	—	—	—

E = Excellent—material is most suitable for the process.
G = Good—material is a good candidate for the process.
S = Seldom used—material is seldom used in the process.
— = Unsuitable—material is not used or is unsuitable for the process.

Table 4-3. Process attributes (Ashby 1992)

Process / Attributes	Surface Roughness	Dimensional Accuracy	Complexity	Production Rate	Production Run	Relative Cost	Size (Projected Area)
Pressure die casting	L	H	H	H/M	H	H	M/L
Centrifugal casting	M	M	M	L	M/L	H/M	H/M/L
Compression molding	L	H	M	H/M	H/M	H/M	H/M/L
Injection molding	L	H	H	H/M	H/M	H/M/L	M/L
Sand casting		M	M	L	H/M/L	H/M/L	H/M/L
Shell-mold casting	L	H	H	H/M	H/M	H/M	M/L
Investment casting	L	H	H	L	H/M/L	H/M	M/L
Machining	L	H	H	H/M/L	H/M/L	H/M/L	H/M/L
Grinding	L	H	M	L	M/L	H/M	M/L
EDM	L	H	H	L	L	H	M/L
Sheet metalworking	L	H	H	H/M	H/M	H/M/L	L
Forging	M	M	M	H/M	H/M	H/M	H/M/L
Rolling	L	M	H	H	H	H/M	H/M
Extrusion	L	H	H	H/M	H/M	H/M	M/L
Powder metallurgy	L	H	H	H/M	H	H/M	L
Units	in. (mm)	in. (mm)		parts/hour	parts		ft^2 (m^2)
High (H)	>0.0025 (0.064)	<0.005 (0.13)	High	>100	>5,000	High	>5.38 (0.5)
Medium (M)	>0.00006 (0.016)	>0.005 (0.13)	Medium	>10	>100	Medium	>0.215 (0.02)
	<0.00025 (0.0064)	<0.05 (1.3)		<100	<5,000		<5.38 (0.5)
Low (L)	<0.00006 (0.0016)	>0.05 (1.3)	Low	<10	<100	Low	<0.215 (0.02)

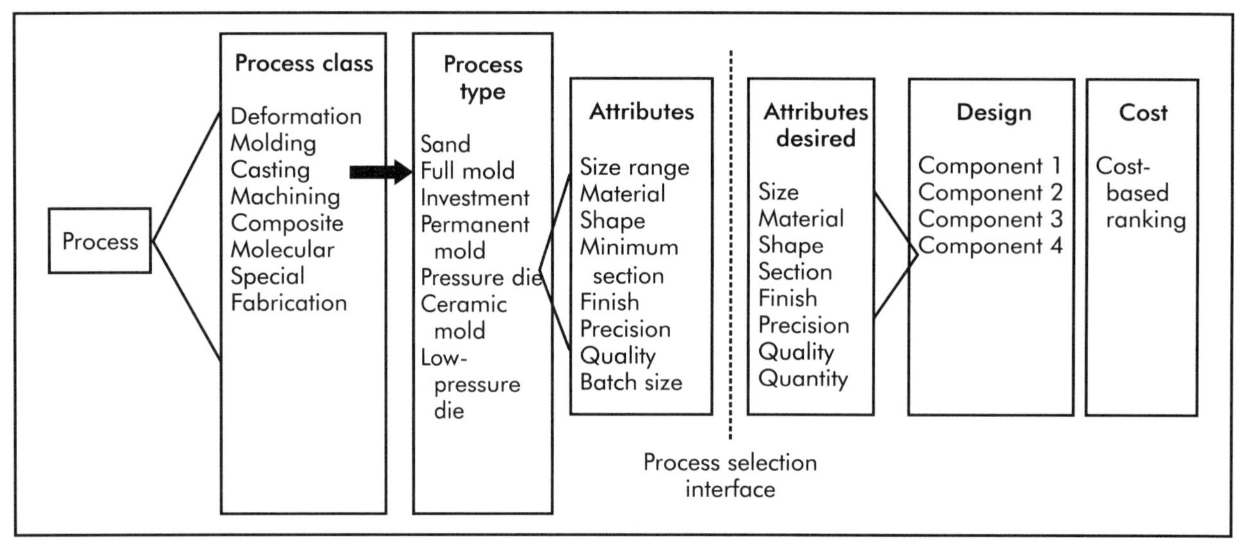

Figure 4-2. Matching process attributes with design needs.

machining, and molding. This breakdown is further subdivided into a set of processes, such as different casting methods. Each individual process has its own attributes, which are matched with the design need.

The successive application of some tables based on part size, shape, and complexity can narrow the material choice to a short list of viable processes. The presence of additional features such as holes, threads, undercuts, bosses, and re-entrant shapes—items that cause manufacturing difficulties or that require additional operations to produce—add to the complexity of the component. The tables have obvious limitations, but they do provide an initial, at-a-glance, graphical comparison of the capabilities of various manufacturing processes, and thus, can be used as a quick reference for designers.

The costs of manufacturing are a major criterion. Certain factors that influence the final cost—material cost, batch size, production rate—can be built into models. It is very difficult to build in other costs, such as experience, idle plant costs, and the cost of holding stock. The output of the screening stage is a short list of possible processes that satisfy design requirements. These processes are again ranked by cost. A procedure for cost-based ranking of successful processes is very useful.

Table 4-4 gives the approximate relative costs to achieve different grades of surface finish with different cutting operations. Rough-machining operations involve standard unit costs; other processes are comparable to rough machining.

An increase in the cost of attaining a greater degree of accuracy and finer surface finish is illustrated by the curve rising to the right in Figure 4-3. However, when a manufactured component consists of an assembly of parts, the cost of assembly and fabrication will usually be reduced if more accurate parts are used. This is reflected in the curve falling to the right in Figure 4-3.

A combination of two effects leads to an optimum where the overall cost of manufacturing and assembly is the least amount. The shape of this curve is similar to the relationship observed in the calculation of optimum cost versus machining speed in machining economics. A similar type of optimum relationship occurs in reliability cal-

Table 4-4. Surface finish and cost comparison for various cutting operations

Surface Finish	Root Mean Square (RMS) µin. (µm)	Relative Cost
Very rough (machined)	1,969 (50)	1
Rough	1,024 (26)	3
Semi-rough	512 (13)	6
Medium	256 (6.5)	9
Semi-fine	126 (3.2)	13
Fine	63 (1.6)	18
Coarse (ground)	32 (0.8)	20
Medium	16 (0.4)	30
Fine	8 (0.2)	35
Super fine (lapped)	4 (0.1)	40

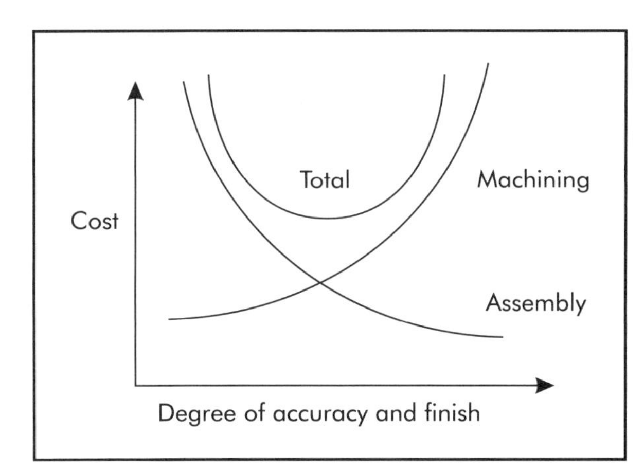

Figure 4-3. Cost versus accuracy and finish.

culations. Costs associated with design and manufacturing must increase with the increased reliability of the product. After-delivery costs, such as warranties, should fall with improved reliability.

GENERAL GUIDELINES FOR EFFICIENT MANUFACTURING

The product developer who selects the manufacturing process should make a design not only to serve the function, but also to facilitate the process of fabrication. The material selected largely dictates the manufacturing processes used. Certain manufacturing processes allow for the creation of complex shapes. For example, metal casting and plastic molding are bulk deformation processes. These processes lend themselves easily to any changes in shape. The design of a part should be such that it takes full advantage of the particular manufacturing process. There are certain parameters that the designer should keep in mind (see Table 4-5).

This section examines the current state of the art in design for disassembly and looks at existing methodologies applicable to product design. A new methodology using a combination of tables outlining damage, tool, reuse, and access-area ratings is also explained. General guidelines for efficient manufacturing are outlined in Table 4-6.

DESIGN FOR RECONDITIONING

Design for reconditioning is based on the quality of the parts. Depending on the quality, the decision is made whether to reuse, remanufacture, recycle, or dispose of the product. A savings on the investment in raw materials, labor, and energy is one consideration. At the same time, the quality of the product waste is looked at to decide whether the product can fulfill a secondary role (for example, can car tires be used as boat bumpers?). Design for reconditioning addresses the reuse of products by undertaking the following procedures:

- complete disassembly,
- cleaning,
- testing,
- reusing the good parts,
- replacing unusable parts,
- reassembling, and
- final testing.

Table 4-5. Parameters of machining characteristics

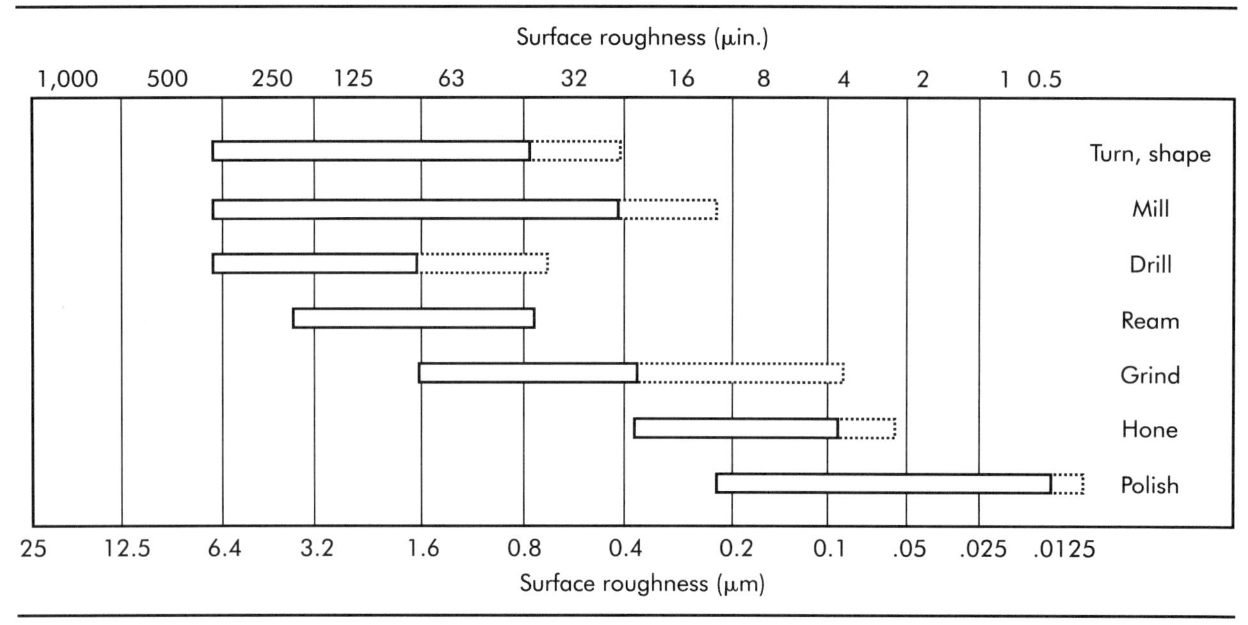

DESIGN FOR MAINTENANCE

Design for maintenance is a proactive step incorporating easier maintenance capability into a product during its initial design. It is a preventive measure that has become increasingly important as systems continue to become more complex. Design for maintenance involves monitoring and assessing the actual condition of a system and maintaining and recovering the intended condition. The following advantages are realized with design for maintenance:

- the prevention of damage and increased reliability;
- avoidance of simple errors during assembly and reassembly;
- confirmation of results; and
- simplification of inspection procedures.

There are two strategies for dismantling a product:

1. Remove the most valuable parts first and stop the dismantling when a marginal return on the operation is reached.
2. Maximize each dismantling operation's yield by having it release many parts at once.

DESIGN FOR RECYCLABILITY

A proactive step, design for recyclability incorporates recyclability characteristics into a product during its initial design. It can reduce recycling costs at the end of the product's life (or during the maintenance of the product).

Automobiles are one of the most highly recycled products. About 75% of the weight of the materials is recovered and recycled, including most of the metal components. Three primary operations in automobile recycling are: dismantling the automobile, shredding and separating the iron component, and separating the nonferrous component. The automotive industry has used design for disassembly (DFD) principles for some time, allowing the replacement of a wide range of rebuilt automotive components.

Table 4-6. General guidelines for efficient manufacturing

Standardization	1. Use standard components as much as possible. 2. Pre-shape the workpiece, if appropriate, by casting, forging, or welding. 3. Utilize standard pre-shaped workpieces, if possible. 4. Employ standard machined features, if possible.
Raw material	1. Choose raw material that will result in minimum component cost. 2. Use raw material in standard forms supplied.
Machining	1. If possible, design a component that can be produced in one machine. 2. If possible, design a component so that machining is unnecessary on the nonfunctional parts. 3. Design the component so that the workpiece, when gripped on the workholding device, is sufficiently rigid to withstand machining forces. 4. Make sure that, when features are machined, the part, tool, and tool-holder do not interfere with each other. 5. Ensure that auxiliary holes are parallel or normal to the workpiece axis or reference surface, and that they are related by a drilling pattern. 6. Make sure end blind holes are conical, and in the case of a tapped blind hole, thread should not continue to the bottom of the hole. 7. Avoid bent holes and dogleg holes.
Cylindrical components	1. Try to ensure that diameters of cylindrical surfaces increase from the exposed face of the workpiece. 2. Ensure that the diameters of internal features decrease from the exposed face of the workpiece. 3. Avoid internal features for long components. 4. Avoid components with large or very small length/depth ratios. 5. For internal corners on components, specify radii equal to the tool radius.
Non-rotational parts	1. Provide a base for workholding and reference. 2. Ensure that exposed surfaces of the components consist of a series of mutually perpendicular plane surfaces parallel and normal to the base. 2. Avoid cylindrical bores in long components. 3. Avoid extremely long and thin components. 4. If possible, restrict plane surface machining (slots and grooves) to one surface of the component.
Assembly	1. During assembly, ensure that internal corners do not interfere with corresponding external corners on the mating component. 2. Specify the widest tolerances and roughest surface that will give acceptable performance for operating surfaces.
Kinematics	Base the initial design on kinematics principles, and modify as necessary to meet the requirements of load and wear.

DESIGN FOR DISASSEMBLY

Design for disassembly (DFD) is the proactive step that incorporates disassembly characteristics into a product's initial design. DFD can reduce production costs, maintenance costs, and recycling costs. Table 4-7 shows the DFD guidelines for materials and the rationale for various processes.

Most products are not designed for easy dismantling or disassembly. As manufacturers become responsible for their products at the end of the products' operational lives, the dismantling of products has emerged as a serious component of manufacturing. In situations involving integrated design principles, certain assembly procedures or joining techniques can make it very difficult to disassemble a product and separate materials into noncontaminated groups. The strategy is to include design for disassembly (DFD) guidelines and analysis within existing product design processes. The intent is to design and produce a product that is easier to work with during assembly, maintenance, disassembly, and recycling. The inclusion of procedures for design for disassembly, recyclability, and remanufacture will save resources by prolonging the useful life of the product.

Design for disassembly is closely related to design for the environment and design for maintainability. When employing the procedures of DFD, the conceptual product-design stage through the life-cycle phases of development, production, distribution, use, and disposal should be considered. This will mandate the establishment of certain procedures for disposal, recycling, and occupational health reasons. These requirements are not only for new products, but also for other types of products, especially ones that are currently in the marketplace.

Planning

While planning for the implementation of DFD, the designer should consider the requirements of shape, size, geometry, tooling, and the nature of handling and manipulation. Some considerations are:

- the size and shape of part details;
- the required expertise for disassembly, adjustments, and reassembly;
- special tooling requirements;
- handling and manipulation (how heavy is the part or tool?);
- cleaning operations;
- product fragility (can it be easily damaged?);
- can details be reused in other components?;
- liquids (volume, disposal, or hazardous); and
- is a sterile room required?

Recommended Guidelines

It is critical that the designer weigh many factors and options before he or she goes forward with a specific design. The designer must understand the basic purpose of the part, where it is to be installed, and if and when it can be reused. Additional factors have to be understood such as: if maintenance is a factor, how long it will be in service, and how often it will be used. General guidelines are established with a focus on the design and reuse aspects of the materials, fasteners, and snap-on connections. The rules are categorized under classification of materials, fasteners, and product architecture. Table 4-8 lists guidelines for fasteners and other parts and their rationale. It also presents guidelines for the product architecture and their influence on assembly and disassembly, some of which are interdependent.

Recycling Practices in the Automotive Industry

One of the aims of design for disassembly is to carry out the maximum amount of disassembly with the savings from recycling

Table 4-7. Guidelines for efficient manufacturing (Bralla 1986; Pahl and Beitz 1996; Magrab 1997)

Material Guideline	Rationale for Various Processes
Minimize the number of different types of materials.	Simplify the recycling process.
Make subassemblies and connected parts from the same or compatible material.	Reduce the need for disassembly and sorting; improve assembly time.
Use materials that can be recycled.	Minimize waste; increase the end-of-life value of the product.
Use recycled materials.	Stimulate the market for recycling.
Ensure standard, easy identification for all materials.	Maintain the maximum value of recovered material.
Hazardous parts should be clearly marked and easily removable.	Rapidly eliminate parts of negative value.

Milling Guideline	Poor	Better
Provide flat surfaces.	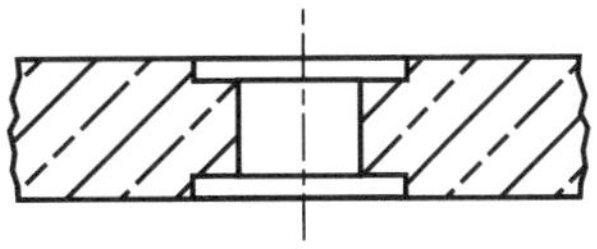	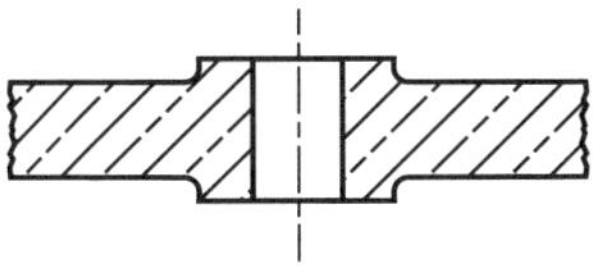
Avoid undercuts.	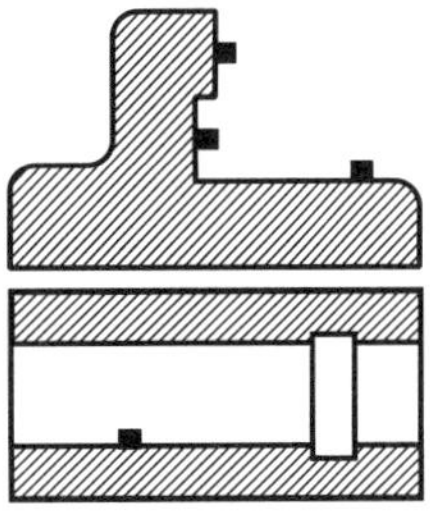	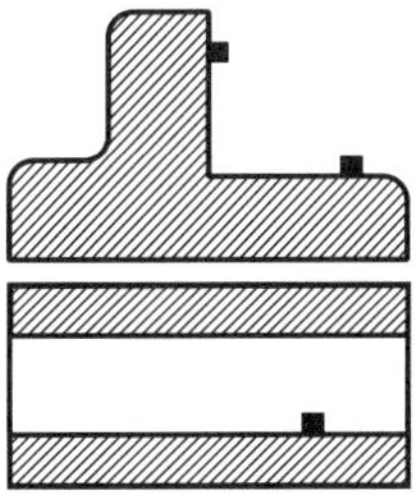
Make changes that reduce machining time.	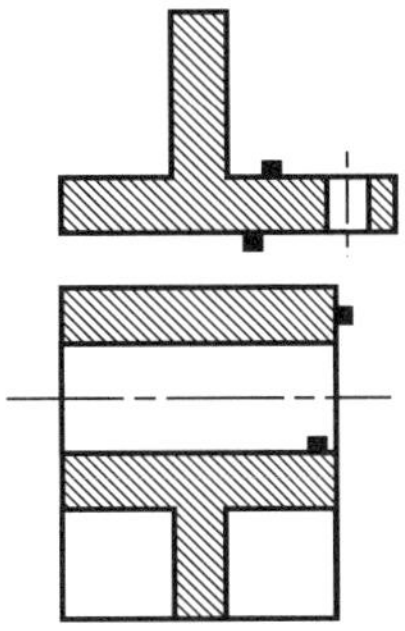	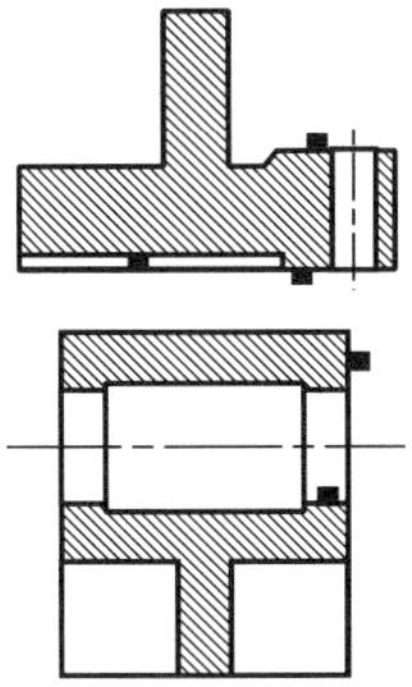

Table 4-7. (continued)

Milling Guideline	Poor	Better
Make sure that parts and tools do not deflect.	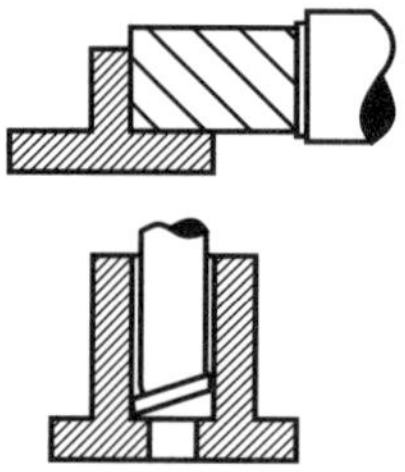	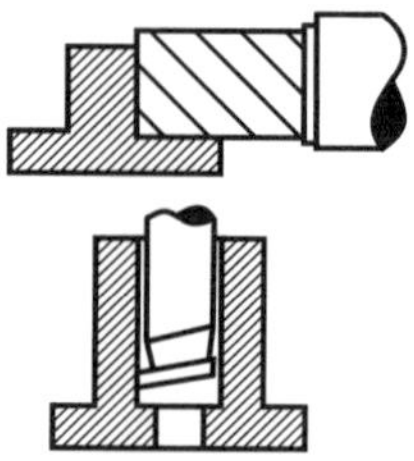

Grinding Guideline	Poor	Better
Mimimize surface to be ground and reduce part weight.	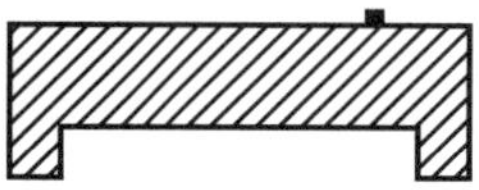	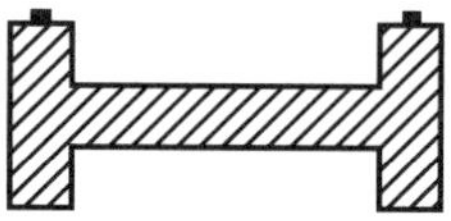
Aim for unimpeded grinding.	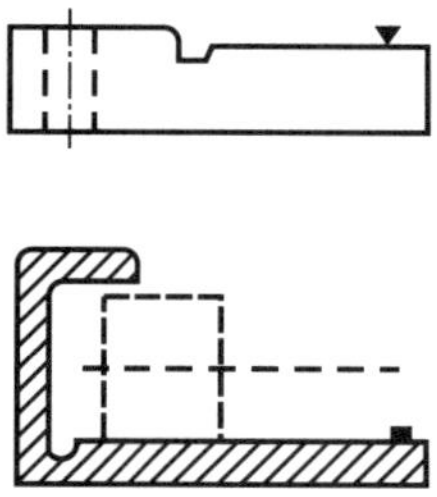	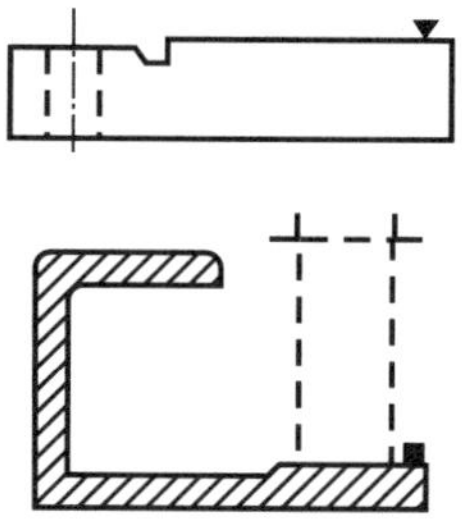
Avoid edge limitations for grinding wheels.	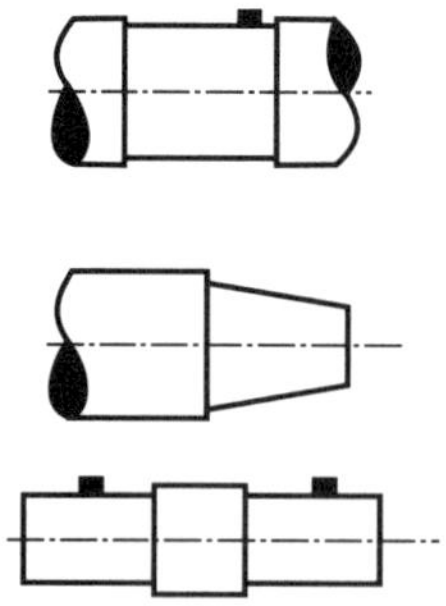	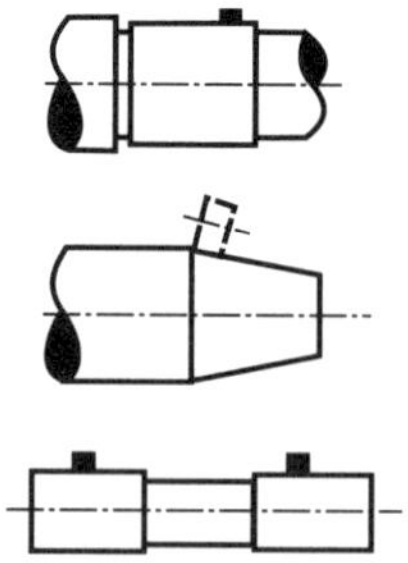

Table 4-7. (continued)

Grinding Guideline	Poor	Better
Provide runouts for grinding wheels.	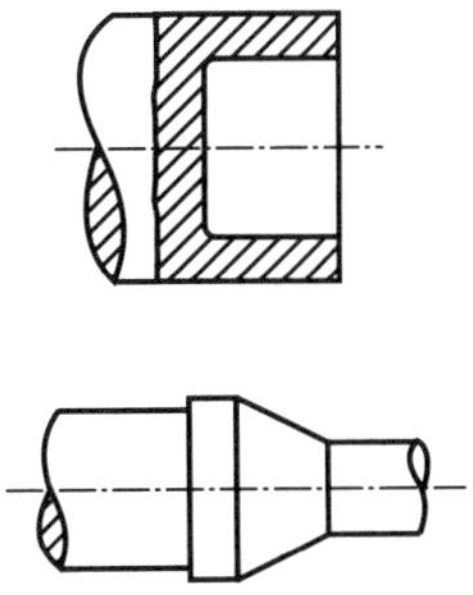	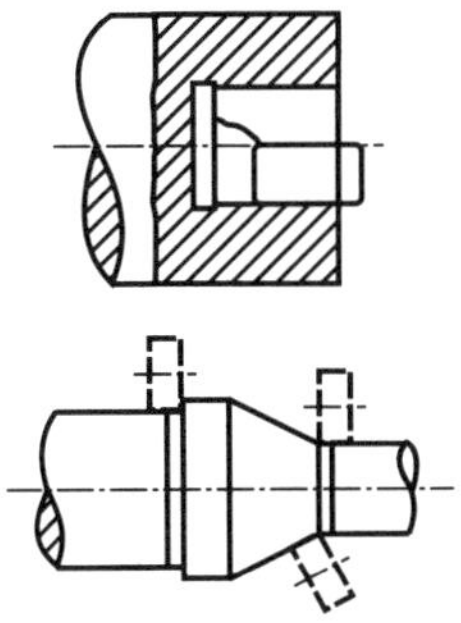

Sheet Metalworking Guideline	Poor	Better
Improve material utilization through part redesign.	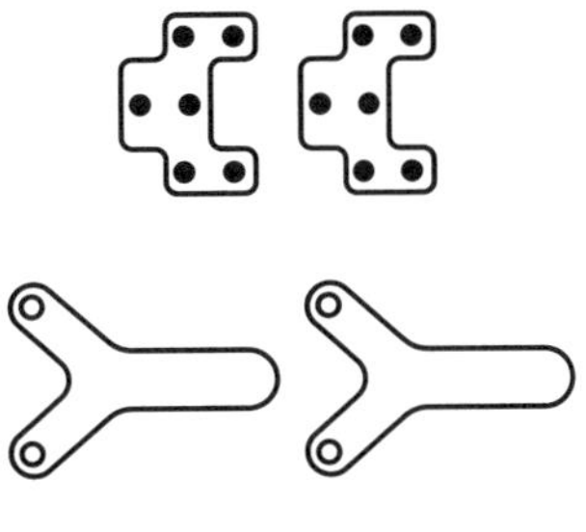	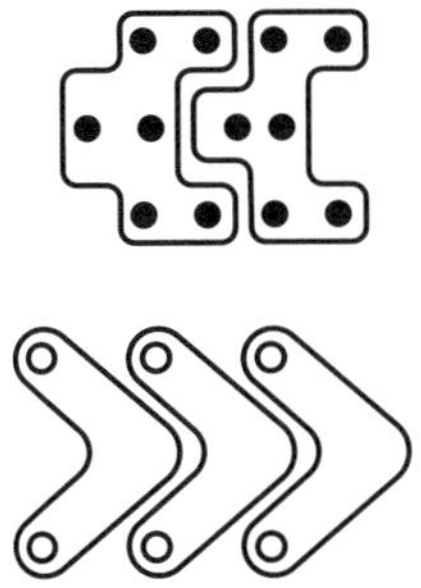
Improve material utilization by combining parts.	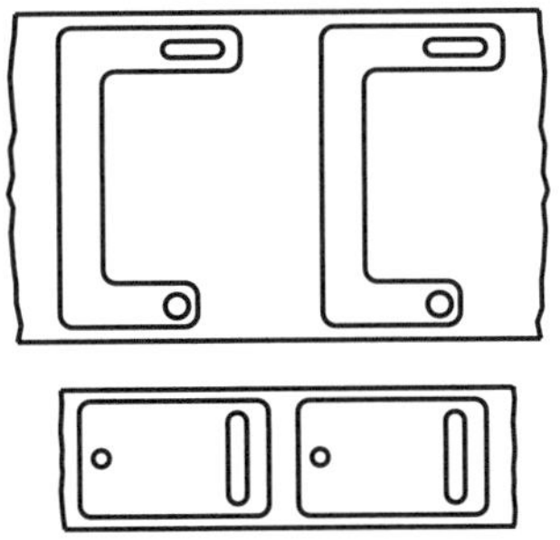	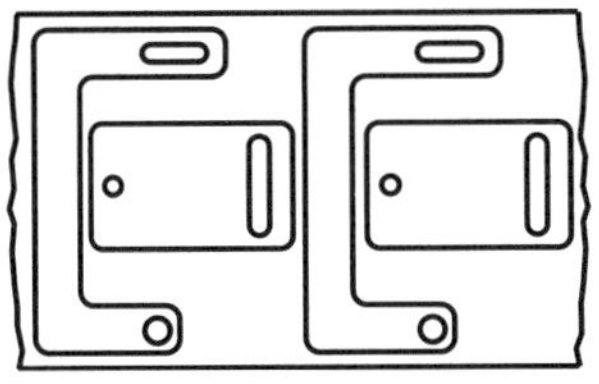
Use width of stock to improve material utilization.	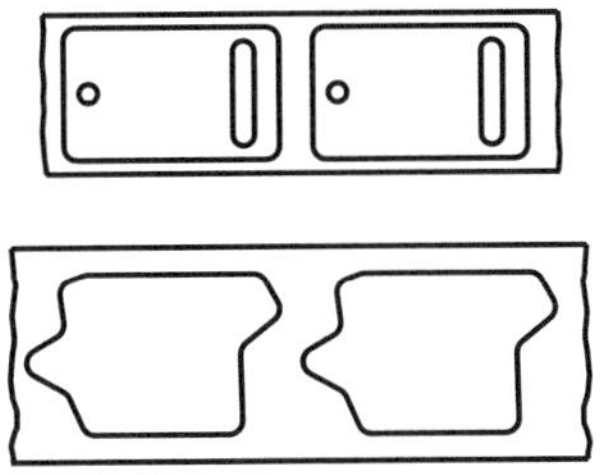	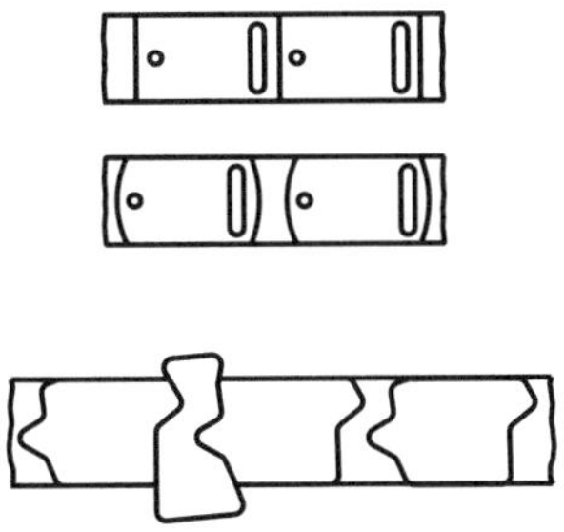

Table 4-7. (continued)

Forging Guideline	Poor	Better
Avoid non-planar parting lines.		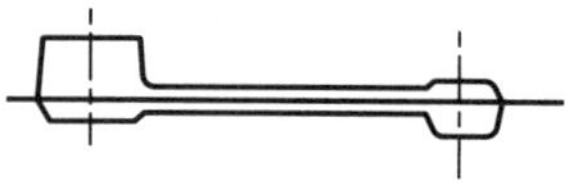
Provide tapers.	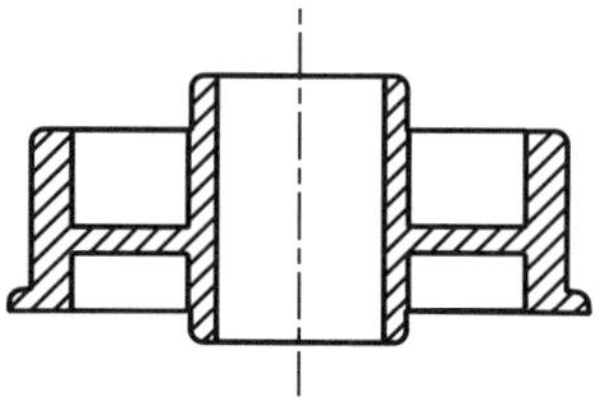	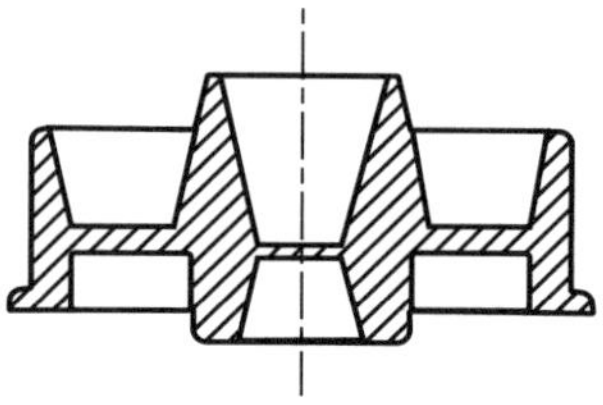
Locate parting line so that metal will flow parallel to the parting line.	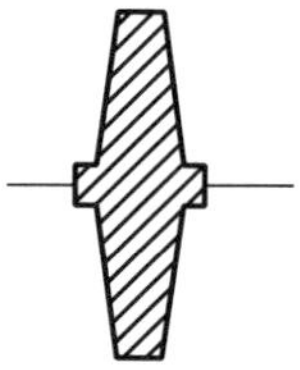	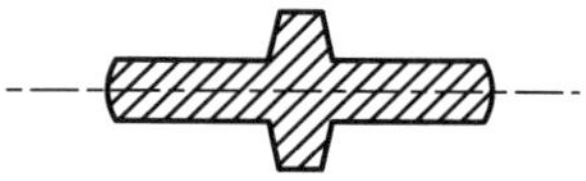
Design for parting lines at about half height.	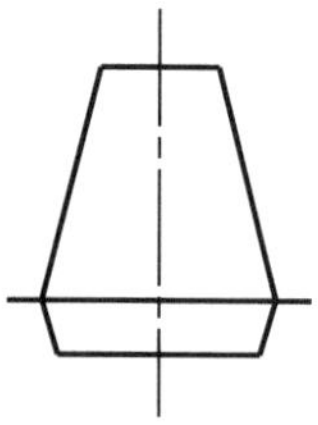	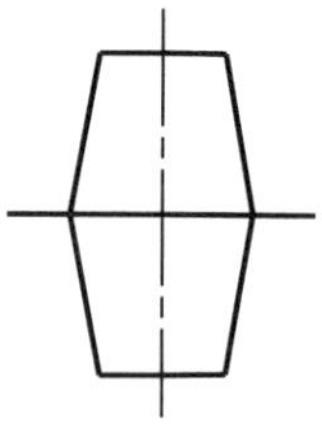
Avoid sharp changes in cross sections.	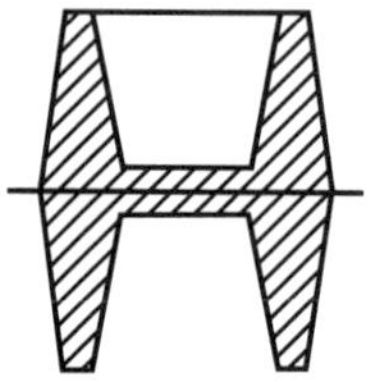	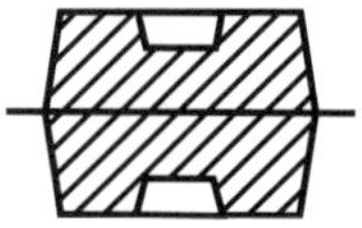

Table 4-7. (continued)

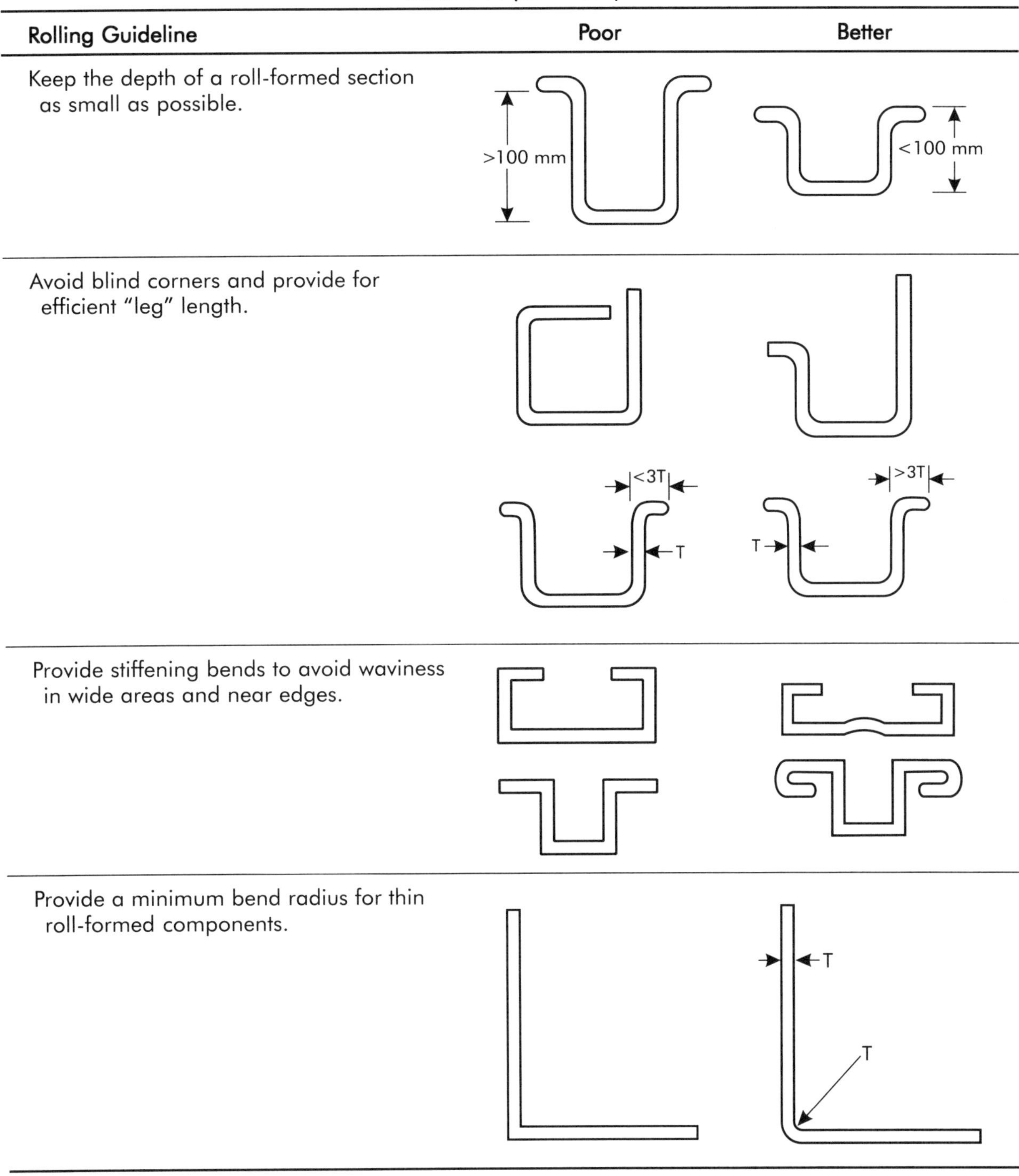

Rolling Guideline	Poor	Better
Keep the depth of a roll-formed section as small as possible.		
Avoid blind corners and provide for efficient "leg" length.		
Provide stiffening bends to avoid waviness in wide areas and near edges.		
Provide a minimum bend radius for thin roll-formed components.		

Table 4-7. (continued)

Extrusion Guideline	Poor	Better
Impact-extruded parts should be symmetrical.	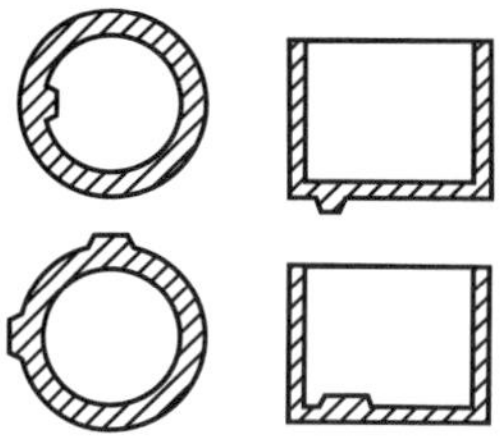	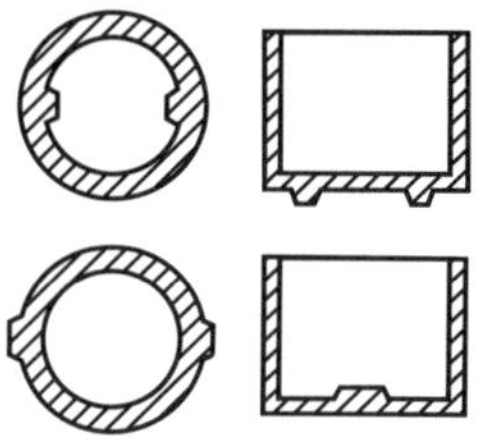
Avoid sharp changes in cross-section, sharp edges, and fillets.	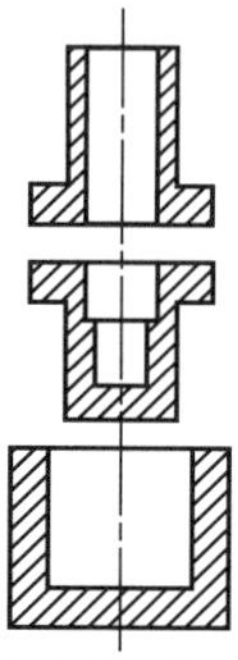	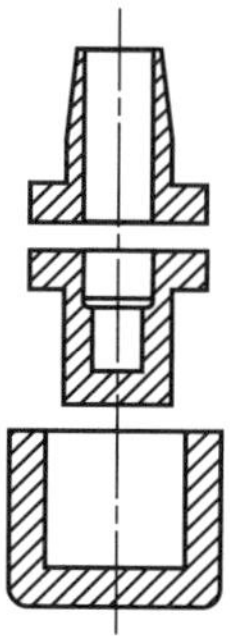
Avoid tapers and almost-equal diameters.	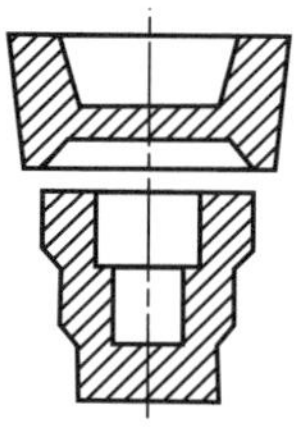	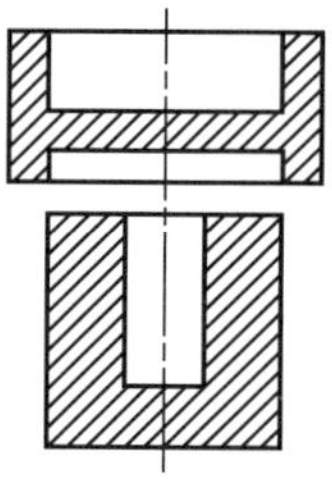
Avoid hollows and maintain uniform wall profiles.	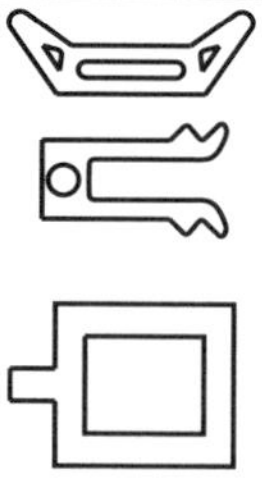	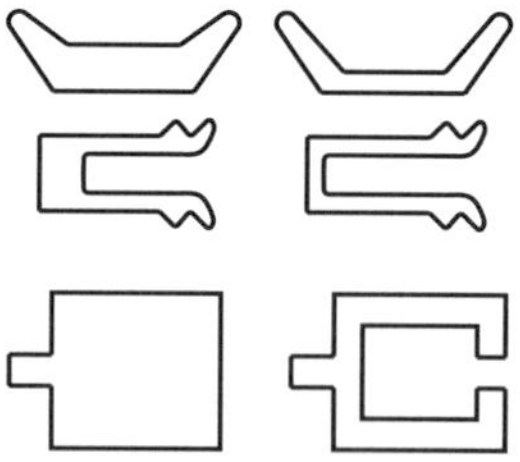
Provide rotationally symmetrical parts without material protrusions.	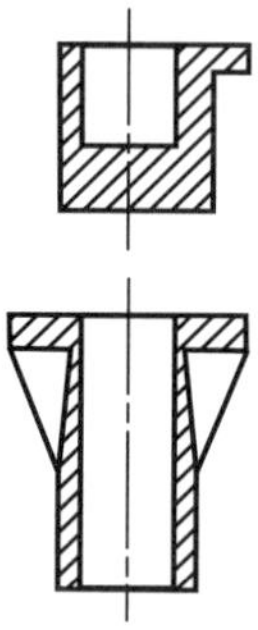	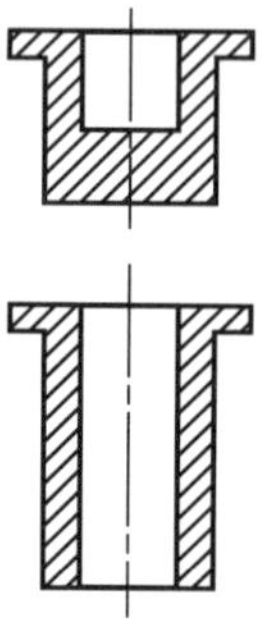

Table 4-7. (continued)

Powder Metallurgy Guideline	Poor	Better
Observe the recommended mimimum dimensions.	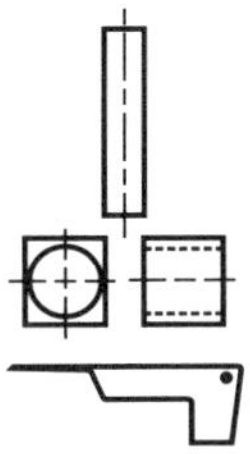	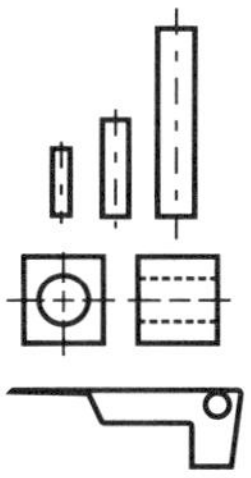
Avoid sharp corners and sharp re-entry corners.	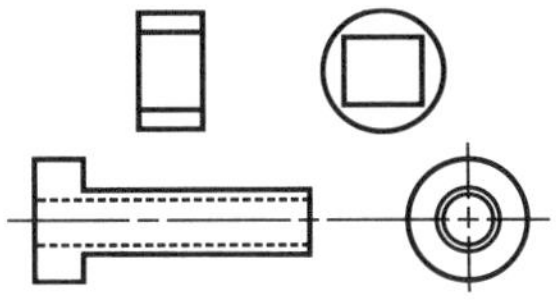	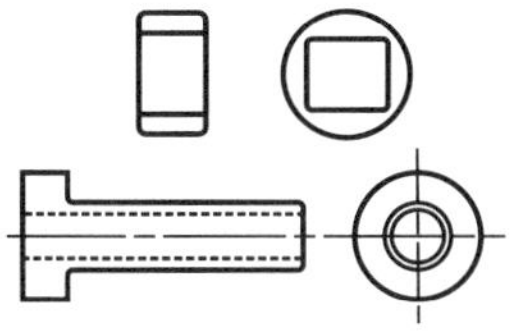
Avoid undercuts.	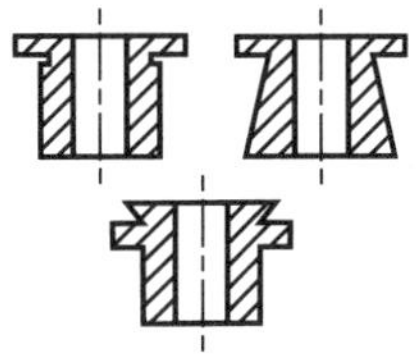	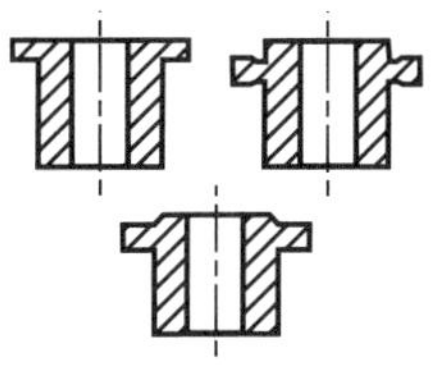
Avoid blind holes where the blind end is opposite the flange.	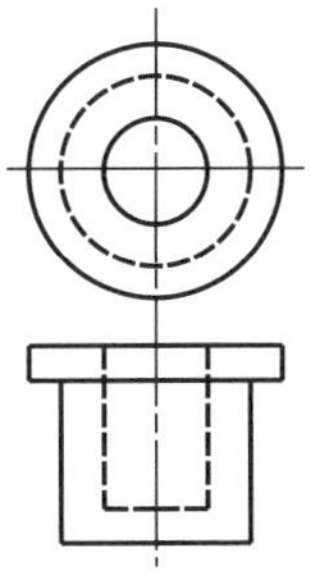	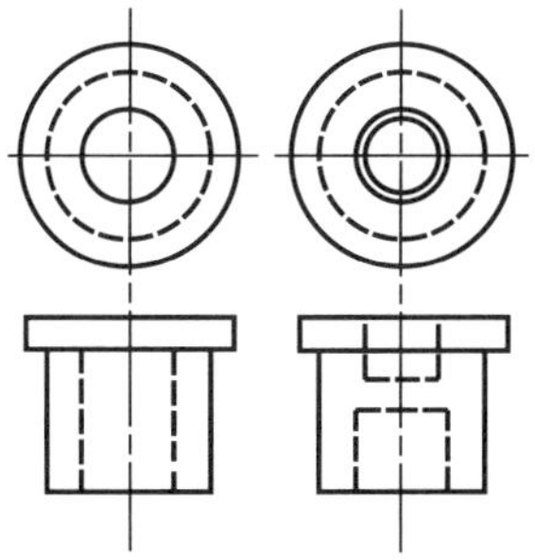
Smaller radii are preferred.	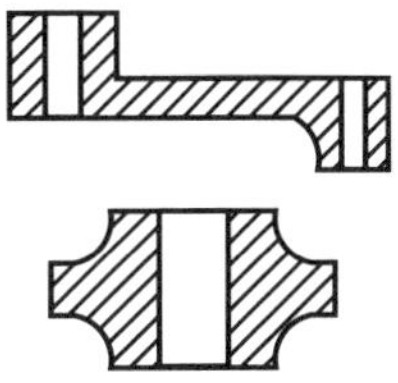	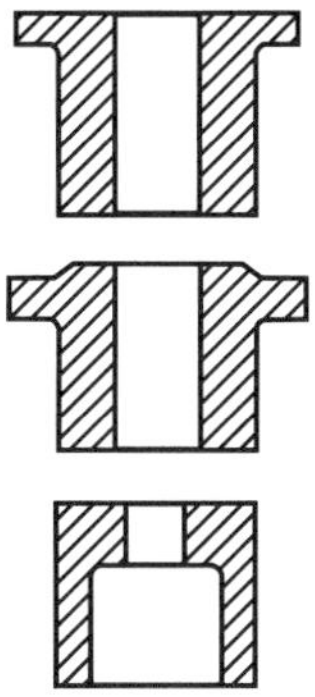

Table 4-8. DFD guidelines and their rationale

Fasteners	Rationale
Minimize the number of fasteners and fastener removal tools.	Most disassembly time involves fastener removal.
Fasteners should be easy to remove.	Saves time in disassembly.
Fastening points should be easy to access.	Awkward movements slow down manual disassembly.
Snap fits should be properly located and able to be dismantled by standard tools.	Special tools may not be available.
Use fasteners of material compatible with the parts connected.	Avoids need for disassembly operations.
Eliminate adhesives unless they are compatible with both parts joined.	Many adhesives cause a contamination of the materials.
Minimize the number and length of interconnecting wires or cables.	Flexible elements are slow to remove; copper contaminates steel, etc.
Product Structure	**Rationale**
Minimize the number of parts.	Reduce disassembly and recyclability costs.
Make designs as modular as possible, with a separation of functions.	Allows options of service, upgrade, or recycling.
Locate non-recyclable parts in one area, which can be quickly discarded.	Speeds disassembly.
Locate parts with the highest value in easily accessible places.	Enables partial disassembly for optimum return.
Design for easy separation, handling, and cleaning.	Disassembly process becomes faster.
Avoid molded-metal inserts or reinforced elements in plastic parts.	Creates the need for shredding and separation.
Access and break points should be made.	A logical structure speeds disassembly and training.

parts being greater than the cost of carrying out the operation. Often, costs can exceed revenue. This is because of the low value that materials have when they are recycled. Material-selection strategy is the first area of the design process that addresses DFD. Broader material options make the high-volume disassembly and recycling of a product more attractive.

In the automobile industry, certain rating criteria are used for recyclability and for the ability to separate the material. The recyclability of a part is considered if the part is remanufacturable (examples include starters and alternators). If the materials in the part are recyclable (examples would be metals), the rating would be two. Materials that can not be recycled would have a recyclability rating of three. If the material can be recycled with additional techniques, the rating would be four, while organic materials (examples would be wooden components) would have a recyclability rating of five.

Similarly, a material separation rating of one would be used if the material can be disassembled easily in a minimum amount of time (for example, a steering column cover). If the material separation is done with a minimum amount of effort in just a few minutes, the rating would be two. If the material is disassembled with more effort and by mechanical means (an example would be an engine), the material separation rating would be three. If the part cannot be disassembled, the rating would be four.

Joining Processes

Joining and fastening are at the heart of DFD. The areas of fastening are well-identified during the process of designing, but an additional step is needed to decide whether disassembly is accomplished through reverse assembly or brute force. In reverse assembly, a fastener, such as a screw, should be unscrewed. If two parts are snap-fit together, they should be snapped apart. Brute force is much less acceptable, but is often the most efficient disassembly method. Parts are pulled or cut apart, depending on the strength of the fastening method. Two questions are noted, regardless of how the joining is done:

1. Which method of disassembly should be used?
2. Given the method of disassembly, are fastening (separation) points accessible?

Threaded fasteners are considered by many people to be inherently contrary to the goals of DFD, one of which is to reduce the amount of parts and fasteners. Screws offer the option of reverse assembly through the utilization of a drill with an attachment to back the screw out. The standardization of screw types and sizes can minimize the number of attachments required to disassemble the product. Brute force disassembly can be achieved by simply pulling screws out of their bosses, but the screw must still be tracked. If it is still attached to one part and not removed, it can cause serious damage to resin reprocessing equipment. Either method, however, requires that the screw be accessible.

Metal inserts, if used in the design, create an additional part for disassembly. As with fasteners, access to the insert is critical. Not only must the fastener be removed from the product, but the metal insert must also be removed. Currently, most inserts are removed with brute force—either punched out with a hammer stroke or cut out (along with some plastic) and thrown away. Ultrasonic inserts may offer some opportunity for reverse assembly, but the cost of fixtures must be offset by high-volume disassembly operations. As with screws, inserts intended for reverse assembly should be standardized to facilitate the disassembly.

Adhesives typically create more problems than they solve. Unless adhesives are water or solvent soluble, brute force is the only efficient method of disassembly.

Thermal methods such as ultrasonic welding, spin welding, vibration welding, hot-plate welding, and hot-gas welding all involve melting thermoplastic to form a strong bond. With current technology, these are not reversible processes because they require brute force for disassembly. However, since thermal methods typically bond similar materials, there may be no reason to separate parts, except for disassembly operations.

Induction welding is one of the quickest thermal assembly processes, even though it leaves metal inside the part. However, it does offer the option of reverse assembly, since the metal is still in the part and can be reheated simply by energizing an electromagnetic field around the metal.

Snap-on fixtures are the ideal fasteners. Removal of a snap-fit latch requires a combination of both reverse assembly and brute-force disassembly. Snap-on connections require no additional parts for separation, contain no additional materials to provide possible contamination, and can be removed quickly and efficiently.

Activity-based-costing Demanufacturing Method

The activity-based-costing demanufacturing method considers three pieces of data that include product, process, and uncertain information. Regarding the product, it considers the abilities to disassemble, reuse, and recycle material. This methodology does not address the issue of potential damage to parts. The program needs to be continuously updated, as recycling information is dependent on the volume of material used and the cost involved (Amezquita, Hammond, Salazar, and Bras 1995).

Life-cycle Assembly, Service, and Recycling

Using the life-cycle assembly, service, and recycling method, a design can be evaluated in terms of its assemblability, serviceability, and recyclability characteristics. Necessary information for the program includes assembly directions, material-type data, price, and cost. Some design constraints on parts may minimize the impact of this method (Ishii, Adler, Barkan, 1988; Ishii, Eubanks, and DiMarco 1994).

Disassembly Analysis

Disassembly analysis is a spreadsheet-based methodology. The design-analysis score is determined by tabulating the difficulty scores for each task. The actual disassembly difficulty is compared to the ideal assembly.

Reverse Fishbone-diagram Method

The assembly fishbone-diagram method supports DFD. It is a way to enhance the product design for ease of assembly by graphically planning the assembly. Drawing the fishbone diagram forces designers to identify the cost of assembly tasks and paths that may lead to defects. This diagram is an essential part of the evaluation of assembly difficulties (Ishii, Lee, and Eubanks 1995).

Figure 4-4 shows a fishbone diagram for automobile design. The central spine in the diagram represents the automobile and each rib represents a specific subsystem or component in the design. Potential problems and flaws are listed along each of the secondary ribs. The question marks in the diagram indicate that more information needs to be added. The reverse-fishbone analysis starts with preliminary analysis of product service information. Areas of the product that deserve critical analysis of disassembly and cost are observed. Prior understanding of the post-disassembly process of component sorting is also necessary. The procedure for constructing the assembly fishbone diagram starts with the main part that other parts are attached to.

Figure 4-5 shows the main idea of a reverse-fishbone diagram using a coffee maker as an example. The diagram schematically describes the disassembly steps for the product and specifies the retirement content for each component. The fishbone diagram promotes DFD by forcing engineers to identify assembly difficulties and then come up with remedies. Each assembly step indicates fixturing needs, reorientation, and insertion directions. The assembly rating is computed based on these values. The diagram can include other symbols indicating time-penalty factors such as the need for inspection and testing.

Knowing the intended fate of each part, the reverse-fishbone diagram can be con-

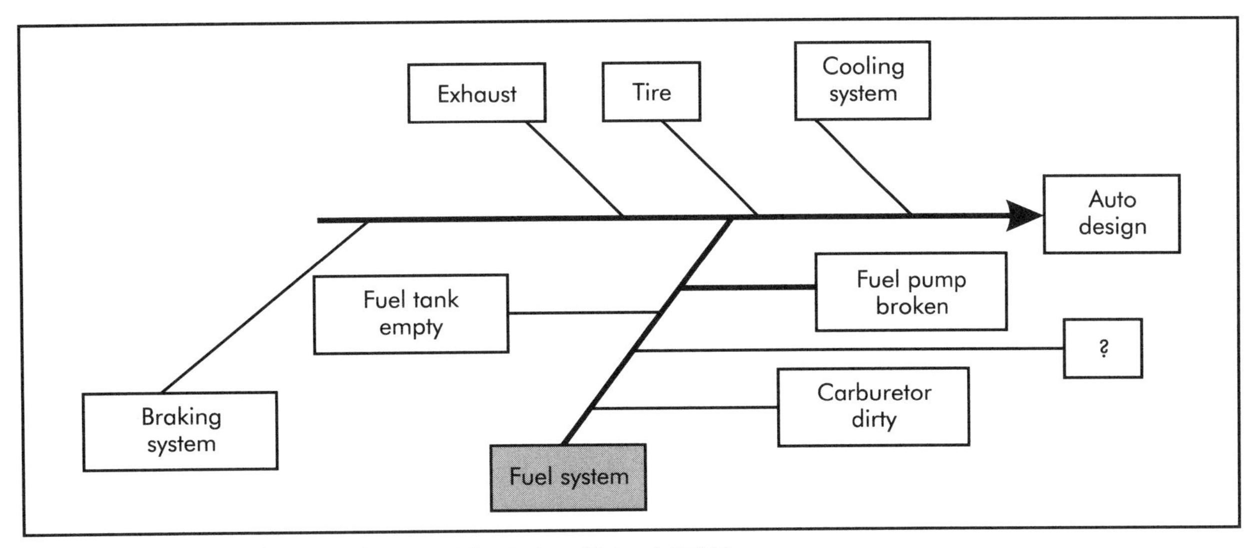

Figure 4-4. Fishbone diagram of automobile design (Voland 1999).

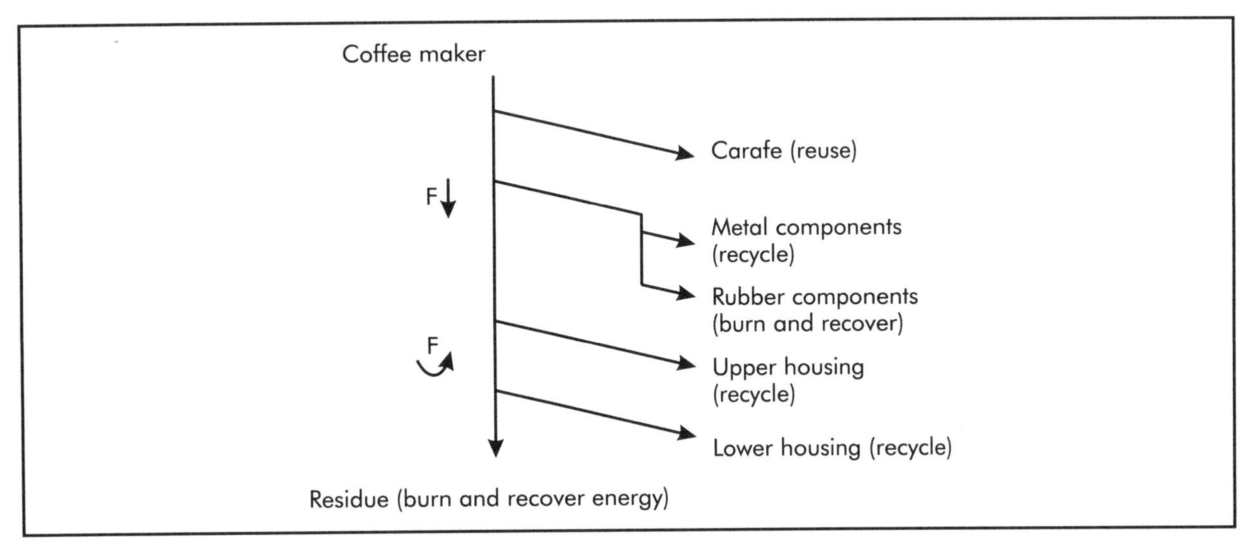

Figure 4-5. Reverse fishbone concept.

structed to determine the sequence dependency of the disassembly process, as well as the parts that need not be disassembled. This procedure allows disassembly to focus only on those items that must be removed to be decomposed into desired parts. A normal approach to constructing the reverse-fishbone diagram is to go from the top to the bottom.

As the designer physically disassembles the product, the reverse-fishbone diagram characterizes each disassembly step by its fixturing needs (the symbol "F" indicates that fixturing is required; reorientation requirements are indicated by a circular arrow; and removal directions are indicated by straight and rotational arrows). The diagram also can include other symbols or markers indicating the component fate category, time-penalty factors, and the connection-separation method (break, pop, and unscrew). The tool requirement and removal difficulty should be included, as these symbols facilitate rapid visual evaluation of the disassembly difficulty. The initial steps of the reverse-fishbone show a short series of disassembly steps as the first set of fasteners and the product skin is removed. Then, the fishbone generates a layer of sequentially independent steps, followed by a series of sequence-dependent steps to the next layer, and so on.

Examination of Reverse-fishbone Results

The designer can generate additional information about the design's performance under different product retirement scenarios on the basis of results. Reverse-fishbone analysis can provide the designer with early guidance in the following areas:

- identification of retirement components;
- matching the retirement component with the market demand for reused components;
- identification of potential improvements in disassembly steps and procedures;
- identification of intercomponent connections that pose disassembly difficulties;
- retirement cost/revenue stream projections; and
- identification of special disassembly tooling and fixturing requirements.

The reverse fishbone graphically characterizes the difference between sequence-dependent and sequence-independent disassembly specifications. Analysis may reveal that additional work is required to make the disassembly process more sequence-independent. It helps the designers to identify the strategic components that are candidates for retirement early in the design process.

The analysis also leads to an estimation of relative volumes of the traffic stream, thus aiding the capacity planning of product retirement facilities. Improvements in the disassembly steps and procedures are another important goal of reverse-fishbone analysis.

Design for Assembly Analysis Method

The design for assembly method described in Chapter 3 also can be applied to DFD. This procedure is based on determining the minimum number of parts required for a given design. The major elements of this process are assembly time and part count. It takes into consideration the minimum number of parts; the part shape, size, and symmetry; handling; orientation; and insertion difficulties. The contribution of this method to identify assembly operations and simplify them through the use of *Z*-axis assembly, self-locating components, and minimum handling also applies to design for disassembly. The methods can be reversed in the disassembly process, leading to similar operational benefits (Boothroyd and Dewhurst 1983).

Part consolidation is a major achievement of design for assembly. The incorporation of multiple parts into a single part reduces the number of materials to be processed in disassembly and recycling. However, a consolidation of parts might lead to such a complex single part that disassembly of associated components would be difficult and time consuming, indicating that part consolidation is counterproductive.

Evaluation-chart Method

Evaluation charts have also been used to evaluate the efficiency of DFD. Ehud Kroll proposed a method based on the degree of difficulty and types of tools used in disassembly. As shown in Table 4-9, the procedure for evaluation is based on a spreadsheet-like chart that uses a database of task difficulty scores (Kroll 1996). The scores are derived from work-measurement analysis of standard disassembly tasks and they provide a

Table 4-9. Evaluation chart (Kroll 1996)

Part Number	Minimum Number of parts	Number of Repetitive Tasks	Tool Type	Task Direction	Tool Required	Tool Accessibility	Position	Force	Time	Special	Subtotal	Total
1	0	4	U	Z	I	1	2	2	1	1	7	28, Screw
2	0	1	F	Z	II	1	1	1	1	2	6	6
3	1	1	P	Z	III	1	1	3	1	1	7	7
4	1	1	C	Z	IV	1	1	3	1	1	7	7
5	1	1	R	Y	III	1	1	1	1	1	5	5

Tasks

P = push/pull
C = cut
R = remove
G = grip
U = unscrew
W = pry out
F = flip

Tools

I = Phillips-head screwdriver
II = large gripper
III = gripper
IV = wire cutter
V = flat-head screwdriver

means of identifying weaknesses in the design and comparing alternatives quantitatively. Time-and-cost estimates provide an idea about the dismantling sequence.

In the evaluation chart, there are columns identifying the task type, the number of repetitions for disassembly to take place, the type of tools needed, and the degree of difficulty to each category, ranging from easy to most difficult. The subtotal adds these values and the total multiplies the subtotal by the number of repetitions that must be carried out for a particular procedure. The overall difficulty of disassembly is identified, telling the designer which area needs to be improved. After changes are incorporated, a second disassembly evaluation chart is prepared.

Design for Maintenance, Disassembly, and Recyclability Method

Another method proposed by Devdas Shetty and Ken Rawolle measures the product for factors such as design for maintenance, disassembly, and recyclability. The method is based on the premise that maintenance, disassembly, and recyclability can easily be integrated. A part that is easier to disassemble will be easier to maintain and recycle. For proper maintenance of a product, users must be able to remove the parts, disassemble them, and recycle unserviceable parts. The method is easy to use and provides consistent results. Analysis can be performed quickly. It also provides information that identifies corrective actions to improve design. This methodology carefully examines the influencing factors for accessibility, recyclability, task, tool, and time (Shetty and Rawolle 2000). Factors include (see Table 4-10):

- Accessibility: Is the location easy to work in with your hands?
- Recyclability: Can the removed part be immediately reused; is reconditioning required or will it go to the landfill?
- Task: Is the task to remove the part difficult?
- Tool: Is tooling required, and if so, how difficult is it to obtain the tooling?
- Time: How long does it take to perform the step?

Once an evaluation has been completed, the method provides an overall design score. The score gives a grade ranging from 0–100%. The perfect score of 100% is very difficult to achieve. There is no right or wrong score; however, a good score should exceed 75%. This can be achieved from a series of ratings that are automatically calculated from the information entered during the design review. There are a total of six criteria that make up the overall design score. They are:

- number of parts,
- time,
- access rating,
- reuse rating,
- damage rating, and
- tooling rating.

It is important to account for potential damage to a part during disassembly. The ability to reuse a part without any rework gives the best disassembly results.

Part score is calculated by dividing the time measured by the total time. If there are 12 steps and a total of 24 parts utilized, the part score is 50%. (This step assumes that, for the best design, only one part is needed for each step.) The weighted average is calculated for each step based on the total number of parts versus the number of parts in that step. The contribution of each step is determined by multiplying the weighted average by the rating and time scores. The rating score is determined by obtaining the sum of the four rating scores divided by 12. (In this case, 12 is considered a perfect score.) The results are multiplied by the parts' weighted average.

Table 4-10. Rating the ease of maintenance, disassembly, and recyclability of a product (Shetty and Rawolle 2000)

A	B (Item)	C Number of Parts	D Damage Rating	E Tooling Rating	F Reuse Rating	G Accessibility Rating	H Rating Score	I Time Measure	J Total Time	K Time Rating	L Estimated Time	Time Score
1	Part 1	1	1	3	3	1	66.7	91	91	1	2	
2	Part 2	2	3	2	3	1	75.0	367	734	2	3	
3	Part 3	2	2	1	1	1	41	349	698	2	1	
4	Part 4	3	0	2	0	0	16	322	966	3	3	
5	Part 5	1	1	2	3	2	66.7	479	479	2	1	
Average		1.8	1.4	2	2	1	52.8	321.6	593	2	2	90%

Damage Rating

3 = Task is easily accomplished with little concern for part damage.
2 = Task is easily accomplished. However, this part is considered fragile and can be easily damaged without proper care.
1 = Typically, the part is damaged during the task and great care is taken during its removal to prevent additional damage.
0 = Destructive assembly is used to remove the part. Part cannot be reused for the same purpose after removal.

Tool Rating

3 = A tool is not required. The task is accomplished by hand.
2 = A common hand tool is required.
1 = Special tooling/equipment are required and there is no delay.
0 = Special tooling is required. There is a delay to acquire the tool.

Reuse Rating

3 = Part can be reused with no conditioning required (part is not damaged).
2 = Part can be reused after conditioning.
1 = Part can be reused after reconditioning and special treatment.
0 = Part cannot be reused because it is too expensive to recondition.

Accessibility Rating

3 = The area is easy to work in. Tools are easily accessible.
2 = There is restricted access and the part can be removed without damage.
1 = There is restricted access and vision, and special care is needed.
0 = It is difficult to access the area and special tooling. Extreme care must be taken to prevent damage.

Time rating is computed by multiplying the time required to perform a step by the required number of parts. For example, one team may be consistently faster or slower than another. This method highlights the parts in a design that consistently take longer to install than others, regardless of who is doing the disassembly. A correction factor is found that accounts for the condition when more than one part is required and where a group of parts is difficult to remove. If the time to perform the step is within 10% of the average, then a value of one is used. If the time to perform the step is between 1.1 and two, then a value of two is used. If the time to perform the step is greater than a factor of two, the ratio of measured time with the average time is used. If it is not possible to measure the time, a value of one, two, or three is chosen, depending on the degree of similarity between this step and other steps.

The information from Table 4-10 is used to calculate the overall design score. It helps to identify whether there are too many parts, or whether damage ratings, tooling, reuse, and access ratings are high. The rating score is the overall average in column H, which is obtained by summing columns D, E, F and G divided by 12. The time score is calculated by dividing the total number of parts by the total time rating. For satisfactory DFD implementation, the rate, time, and part scores should all be greater than 80%.

DESIGN FOR LIFE-CYCLE MANUFACTURE

Major considerations in design for life-cycle manufacture are:

- physical concept of the product,
- part decomposition, and
- total product quality.

Breaking down design for life-cycle manufacture into these components examines the concept of total product quality in terms of quality of concept, quality of design, and quality of ownership (Stoll 1997).

The process can be explained by looking at engineering design. The engineering design process begins with general knowledge of what is required and ends with specific detailed information about how it will be accomplished. Design decisions made during this process determine, in large measure, the product's cost, quality, ease of manufacture, and ease of support in the field. The process begins by conceiving a physical concept as shown in Figure 4-6 for the product based on customer needs and a product specification, and then creating a preliminary layout of the design, which embodies the physical concept. It is preliminary because, at most, only key dimensions and relationships between parts have been specified; actual size, shape, and detail features of parts are, as yet, either undefined or only partially defined.

A preliminary layout is then developed into a completed design. During detail design, the preliminary layout changes and evolves iteratively as questions are answered and uncertainties resolved. The end result is a definitive layout or final design. The layout contains the design information required to fabricate and assemble parts. Combining the concept of design for life-cycle manufacture and the engineering design process yields the following critical observations:

- Physical concept—the physical concept embodies the way in which the product performs or provides its intended function. The key to achieving success in the marketplace often lies in identifying and selecting the right (or best) physical concept for the product. This well-known fact is a primary motivation for creativity and innovation in product design. It is also motivation for many research and development activities conducted by manufacturing firms.

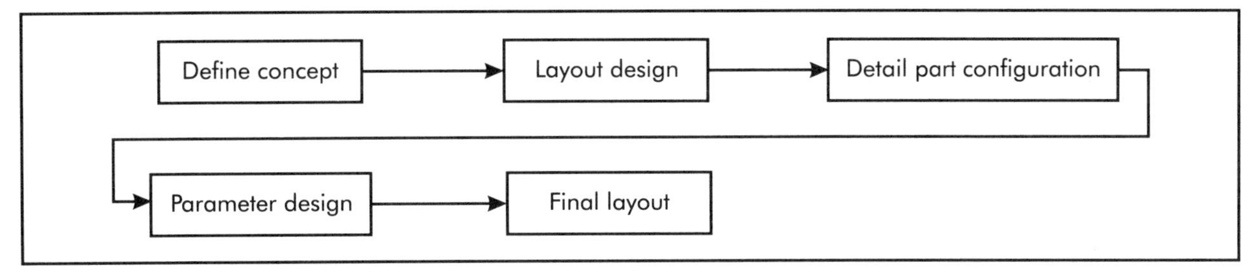

Figure 4-6. The engineering design process.

- Part decomposition—for many products, identifying and selecting the right part decomposition may be just as or even more important than the physical concept itself (see Figure 4-7).

Part decomposition determines the ease of assembly, testability, and serviceability of a product. It also determines the number and complexity of designed parts, which in turn, influence material and manufacturing-process selection, tooling cost, and a myriad of other factors. For many products, therefore, it is decomposition into parts, more than any other factor, which determines profitability. The physical concept and part decomposition together determine product functionality and manufacturability (Ettlie and Stoll 1990).

Step-by-step Approach

The design for life-cycle manufacture methodology is essentially a step-by-step prescription for performing the engineering design process. It focuses on systematically identifying, developing, and evolving the design concept to ensure high total product quality and low total cost and time. The steps are:

1. Develop a physical concept and preliminary layout for the product.
2. Develop and optimize the part decomposition.
3. Develop and optimize the detail design.

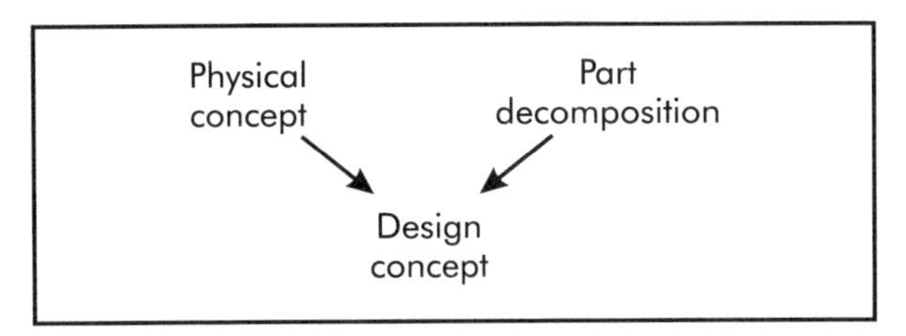

Figure 4-7. Design concept (physical concept + part decomposition).

The general flow of the design-for-manufacture approach is shown in Figure 4-8.

In Step 1, front-end activity is aimed at identifying and selecting the best physical concept for the product. Step 2 focuses on identifying the most appropriate part decomposition from the strategic, assembly, component manufacturing, functional, and quality points of view. Step 3 supplies the detail information required to fully specify a design.

Step 1, which is the most important step, can be further broken down in the general flow of activities as illustrated in Figure 4-9.

Design for Total Product Quality

As discussed in Chapter 2, achieving optimal total product quality requires a conscious and systematic focus on quality as a design objective. To facilitate this, total product quality is resolved into three components. These depend on the product design and can be used to help guide design decisions at each step of the design for life-cycle approach. The components are based on the following key concerns:

- What will cause the customer to select or purchase the product?

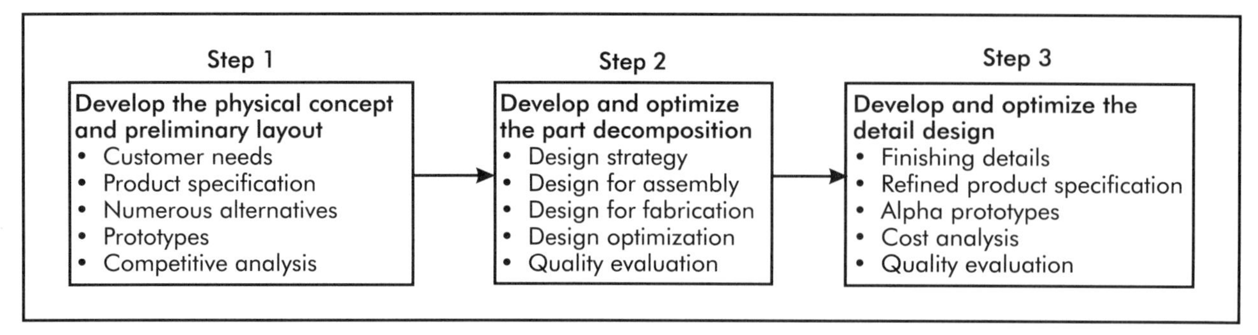

Figure 4-8. Three-step design for manufacture approach.

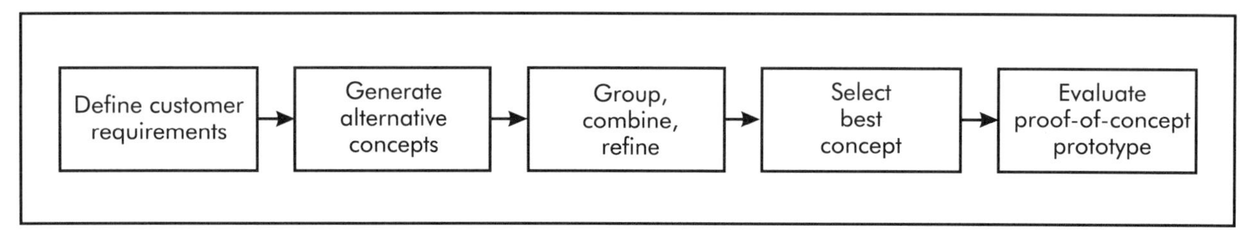

Figure 4-9. Concept development activities.

- What will make it worthwhile for the firm to design and sell the product?
- What will delight and satisfy the customer as an owner and user of the product?

These considerations define the three primary design-related components of total product quality: concept, design, and ownership (see Figure 4-10). Each of these qualities is determined by decisions made during engineering design. By understanding how each quality can be maximized, the design team is able to consciously and systematically design for total product quality. The strategy to define the relationship between engineering design and total product quality consists of:

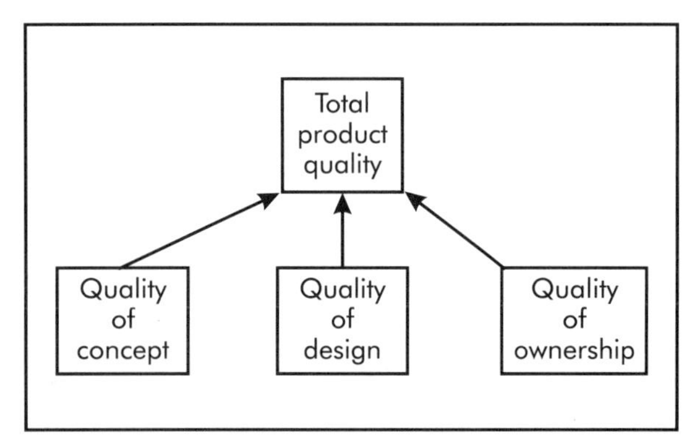

Figure 4-10. Three design-related components of total product quality.

- determining what factors or design considerations affect or contribute to each quality component;
- defining an acceptable quality level for each factor; and
- determining how each factor can be adjusted or controlled by a design to maximize the quality.

Quality of Concept

In essence, quality of concept is a reflection of how well a product satisfies customer requirements. For most products, factors that contribute to the quality of concept include performance, features, aesthetics, ergonomics, and serviceability (see Figure 4-11).

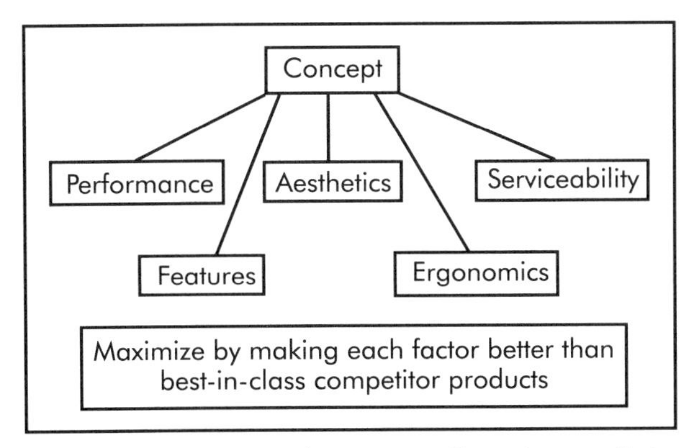

Figure 4-11. Quality of concept reflects how well the product satisfies customer requirements.

Quality of Design

Some characteristics of a successful product are an inherent ease of manufacture and assembly; consistent product characteristics from product to product; insensitivity to hard-to-control disturbances; and minimal scrap rates, rework, and warrantee claims. Such characteristics imply high quality of design (see Figure 4-12).

Good design is when:

1. The part decomposition and associated production planning are appropriate for the quantity required.
2. Trade-offs are possible to achieve the optimum of least possible cost and minimum amounts of time.
3. The quality conformance is acceptable.

Quality of Ownership

Quality of ownership relates to the experience the customer has as a result of owning and operating the product. Considerations such as ease of use, operating cost, reliability, serviceability, maintainability, condition of product when purchased, and customer service all influence quality. High quality of ownership is important because it is what causes a customer to become a repeat buyer and an advocate of the product in the marketplace. Also, quality of ownership contributes to and sustains, over time, the firm's reputation for manufacturing and selling high-quality products.

Quality of ownership depends on how satisfied the customer is with the product, how available the product is over its useful life, and how costly the product is to own and use. These considerations can be grouped to form three primary subcomponents: customer satisfaction, product availability, and operating costs as shown in Figure 4-13.

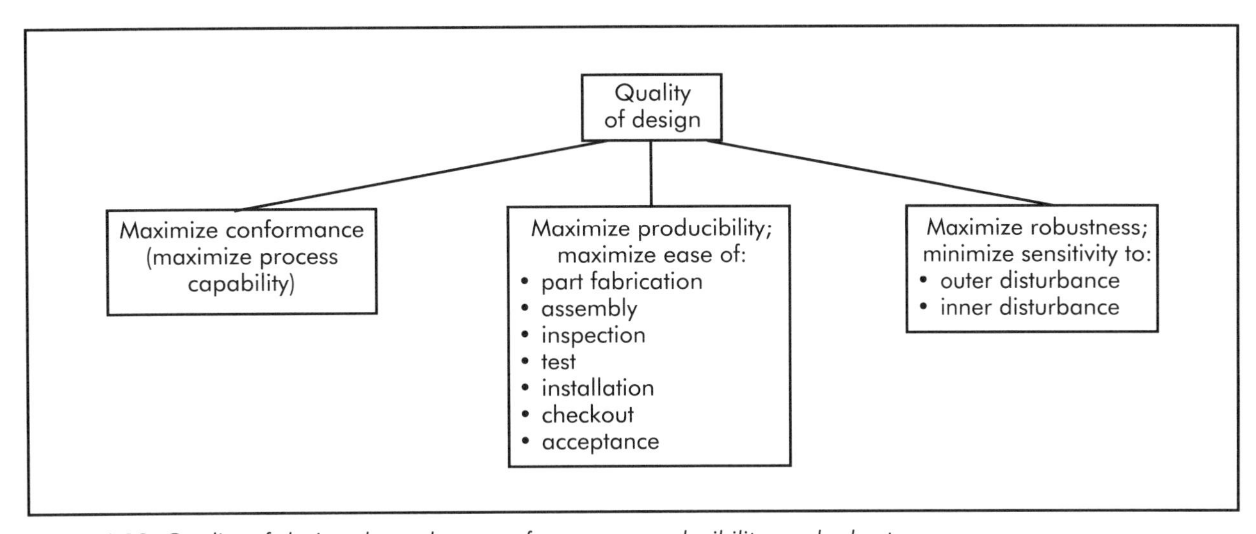

Figure 4-12. Quality of design depends on conformance, producibility, and robustness.

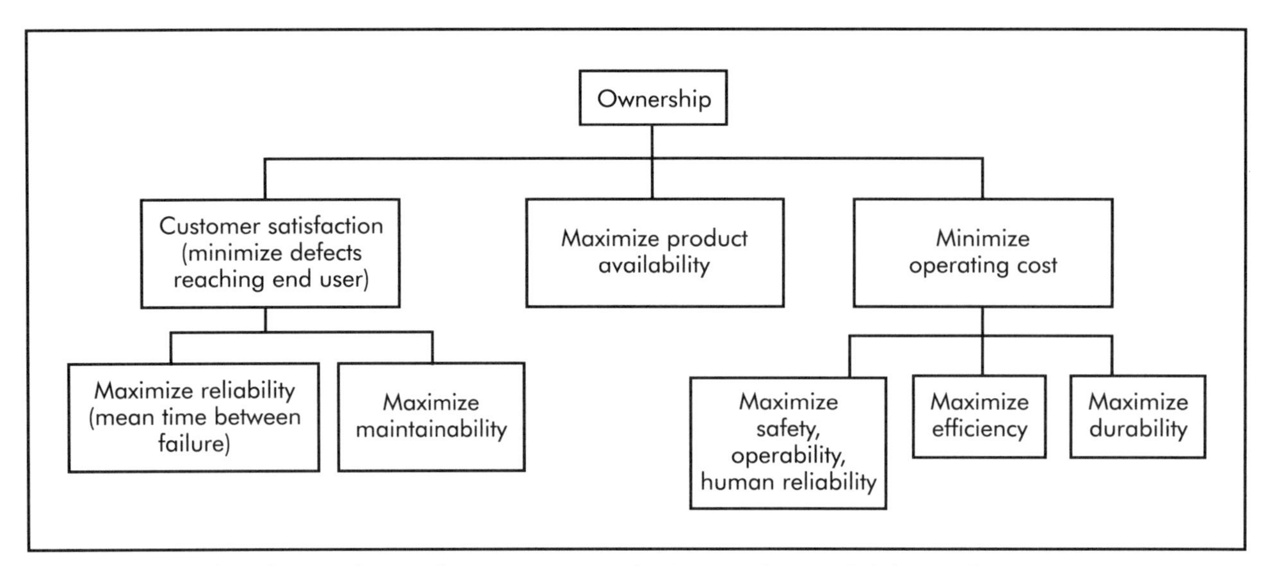

Figure 4-13. Quality of ownership involves customer satisfaction, product availability, and operating costs.

REFERENCES

Amezquita, T., Hammond, R., Salazar, M., and Bras, Bret. 1995. "Characterizing the Remanufacturability of Engineering Systems." ASME Design Technical Conference. Boston, MA: American Society of Mechanical Engineers, September. DE-Vol. 82: 271-278.

Ashby, M.F. 1992. *Material Selection in Mechanical Design*. Oxford, UK: Pergamon Press.

Boothroyd, Geoffrey and Dewhurst, Peter. 1983. *Design for Assembly: A Designer's Handbook*. Amherst, MA: Boothroyd-Dewhurst, Inc.

Bralla, J.G. 1986. *Handbook of Product Design for Manufacturing*. New York: McGraw Hill.

Datsko, Joseph. 1997. *Material Selection for Design and Manufacturing*. New York: Marcel Dekker.

Ettlie, J.E. and Stoll, H. 1990. *Managing the Design Manufacturing Process*. New York: McGraw Hill.

Hundal, M.S. 1997. *Systematic Mechanical Designing: A Cost and Management Perspective*. New York: ASME Press.

Ishii, K., Adler, R., and Barkan, P. 1988. "Application of Design Compatibility Analysis to Simultaneous Engineering." *Artificial Intelligence for Engineering, Analysis, and Manufacturing* (AI EDAM). Vol. 2, No. 1: 53-65.

Ishii, K., Eubanks, C., and Di Marco, P. 1994. "Design for Product Retirement and Material Life Cycle." *Materials and Design*. Vol. 15, No. 4: 225-233.

Ishii, K., Lee, B., and Eubanks, C. 1995. "Design for Product Retirement and Modularity Based on Technology Life Cycle." ASME Winter Annual Meeting Symposium on Life Cycle Engineering. November. New York: American Society of Mechanical Engineers.

Kroll, Ehud. 1996. "Development of Disassembly Evaluation Tool." Irvine, CA: Proceedings of ASME Design Engineering Technical Conference, August. New York: American Society of Mechanical Engineers.

Magrab, Edward. 1997. *Integrated Product and Process Design and Development—The Product Realization Process*. Boca Raton, FL: CRC Press.

Pahl, G. and Beitz, W. 1996. *Engineering Design: A Systematic Approach*. Berlin, Germany: Springer Verlag.

Shetty, Devdas and Rawolle, Ken. 2000. "A New Methodology for Ease of Disassembly in Product Design." *Advances in Design for Assembly*. New York: ASME Press. Also presented at Orlando, FL, ASME 2000 International Mechanical Engineering, Congress, and Exposition, November.

Stoll, Henry. 1997. *Design for Quality and Life-cycle Manufacturing: Concurrent Product Design and Environmentally Conscious Manufacturing*. New York: American Society of Mechanical Engineers. De-Vol. 94 (MED-Vol. 5).

Tanner, John. 1991. *Manufacturing Engineering: An Introduction to the Basic Functions*, Second Edition. New York: Marcel Dekker.

Voland, George. 1999. *Engineering by Design*. Reading, MA: Addison Wesley.

Chapter 5

Tools and Techniques of Product Design

TOOLS OF OPTIMUM DESIGN

Today's productive and competitive atmosphere demands product design optimization after considering all variables that control the design process. The major aim of the designer is to find a design solution that meets the performance requirements of that design, while also satisfying all of the constraints. For a product to be successful, the design needs to be efficient and economical. The process of determining the best solution is known as *optimization*. An optimum design is defined as a design that is feasible and also superior to a number of alternate solutions.

An optimum design can be obtained through an iterative process or by solving an optimization problem. In the iterative process, the design is improved through repeated modification whereby the designer changes the values of the design variables based on lessons learned. Deciding which parameter to change rests with the designer. The second approach is a procedure to find out all design parameters simultaneously to satisfy the constraints and optimize the objectives. Optimum performance can be expressed as objective function or desired criterion. An optimum design is also a measure of product performance and is represented as a performance index or figure of merit.

A common design objective could be cost or weight. An architect designing a building with comfortable surroundings may specify the objectives as minimizing the cost and maximizing the area. A jet engine designer may specify the objective as a high power-to-weight ratio. Typical objective functions considered by product designers are maximizing profit, process yield, and production rate, and minimizing inventory and cost. Design parameters under the control of the designer are called *design variables*. Objectives and constraints of a design have a direct relationship with these variables. The relationship between input variables and the objective function is mathematically represented.

Design Constraints

One step in optimization is the consideration of an objective function with all the design constraints involved. Constraints put a limit on the exploration space of the design by defining the requirements that an acceptable solution should possess. Constraints are considered to be limits on the available design and they define the working region where the objective function is optimized.

Constraints are represented as mathematical inequalities or equalities. They arise from physical laws involved in design and put limitations on individual variables. *Functional constraints*, also called *equality constraints*, specify relationships that must exist between variables. *Regional constraints*, also called *inequality constraints*, specify the uniqueness of a design situation. These functions are mathematical statements of limits between which design parameters must lie. George Dieter and James Siddal have reviewed the development of optimal design methods and classified them into four groups, including optimization by: evolution, intuition, trial and error, and numerical algorithm (Dieter 2000; Siddal 1990). Some of the numerical optimization techniques used are:

- the differentiation method,
- the linear programming technique,
- the gradient search technique (Imai 1986; Groover 1980),
- the LaGrange multiplier method,
- dynamic programming, and
- nonlinear optimization method.

The differentiation method, linear programming, and gradient search techniques are explained in detail in this chapter.

Optimization by Differentiation

While using the optimization by differentiation method, the designer uses differential calculus to determine the optimum. The optimum is determined by solving simultaneous equations found by setting to zero the derivatives of the objective function with respect to each of the parameters.

$$\frac{\partial Z}{\partial x_1} = 0 \qquad (5\text{-}1)$$

$$\frac{\partial Z}{\partial x_2} = 0$$

$$\frac{\partial Z}{\partial x_n} = 0$$

where:

Z = function of variables $x_1, x_2, \ldots x_n$

The optimum point in Equation 5-1 could have either a maximum or minimum value. To check whether it is a minimum or maximum value, the sign of the second derivative of Z with respect to x must be examined. If the curvature is negative, then the stationary point is maximum. The point is minimal if the curvature is positive.

Consider the example of a manufacturing process with two variables, x_1 and x_2. The performance Z of the process is related to the input variables x_1 and x_2 by:

$$Z = 25x_1 - 2x_1^2 + 41x_2 - 5x_2^2 + 4x_1x_2 \qquad (5\text{-}2)$$

What values of x_1 and x_2 maximize the value of Z? The differential calculus approach for two variables involves taking derivatives of the objective function Z, with respect to x_1 and x_2, and setting them so that they are equal to zero. This provides two equations involving two unknowns, which can be solved to find optimum operating conditions.

$$\frac{\partial Z}{\partial x_1} = 25 - 4x_1 + 4x_2 = 0 \qquad (5\text{-}3)$$

$$x_1 = 6.25 + x_2$$

$$\frac{\partial Z}{\partial x_2} = 41 - 10x_2 + 4x_1 = 0$$

$$41 - 10x_2 + 4(6.25 + x_2) = 0$$

$$x_2 = 11, x_1 = 17.25$$

$$Z = 25(17.25) - 2(17.25_1)^2 + 41(11) - 5(11)^2 + 4(17.25)(11)x_2$$

$$Z = 441.125$$

Optimization by Linear Programming

Linear programming is a mathematical method of allocating constrained resources to attain an objective, such as minimizing cost or maximizing profit. This method can be applied in solving problems in which objective function and constraints are linear functions of variables. The objective is to maximize or minimize some linear objective function.

$$Z = a_1x_1 + a_2x_2 + a_3x_3 \ldots a_nx_n \qquad (5\text{-}4)$$

The objective function is constrained by resources. They are shown by constraint equations.

Less-than-or-equal-to constraints

$$b_1x_1 + b_2x_2 + b_3x_3 \ldots b_nx_n \leq b_0 \qquad (5\text{-}5)$$

Greater-than-or-equal-to constraints

$$b_1x_1 + b_2x_2 + b_3x_3 \ldots b_nx_n \geq b_0 \qquad (5\text{-}6)$$

Equal-to constraints

$$b_1x_1 + b_2x_2 + b_3x_3 \ldots b_nx_n = b_0 \qquad (5\text{-}7)$$

A linear programming problem can be solved analytically or graphically. The graphical approach is suitable for two variables, but is difficult to use if there are more than two variables.

Example 1. The linear programming method is applied to a manufacturing engineering situation where the input is raw material and the output consists of two products. The manufacturing process has two stages. Stage 1 is an automated rolling machine into which raw material flows. The output of Stage 1 consists of two base parts for two products (A and B). The two outputs from Stage 1 are fed to the assembly line (Stage 2), where further assembly takes place. The output of Stage 2 results in products A and B, as shown in Figure 5-1.

Profit can be made by either of the products. Unit profit on product A = \$10 and unit profit on product B = \$20.

Total profit can be shown as

$$Z = 10x_1 + 20x_2 \qquad (5\text{-}8)$$

where:

x_1 = number of units of product 1 produced
x_2 = number of units of product 2 produced

In addition, one unit of raw material is needed for one unit of product A and for one unit of product B. The total amount of raw material that can be processed through Stage 1 is equal to 9 units/day, expressed as:

$$1x_1 + 1x_2 < 9 \qquad (5\text{-}9)$$

where:

x_1 = units of 1
x_2 = units of 2

A designer must realize that the maximum profit will not be realized just by maximizing Product B. Rather, more labor resources will be needed for each Product B. One hour of labor is required for each unit of Product A, and three hours of labor is required for each unit of Product B. The total labor hours/day = 15.

$$1x_1 + 3x_2 < 15 \qquad (5\text{-}10)$$

In the graphical method of solving this problem shown in Figure 5-2, the two constraints are plotted as two lines. Since both constraint equations are less than or equal to each other, the useful area is shown to the left region.

$$1x_1 + 1x_2 = 9$$

$$1x_1 + 3x_2 = 15$$

The user's objective is to find the combination of x_1 and x_2 so that the function is maximized.

$$Z = 10x_1 + 20x_2$$

To determine this answer, a series of constant profit lines (shown with dotted lines)

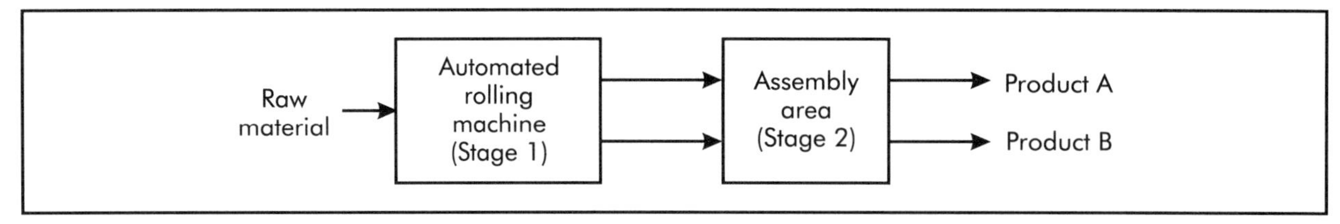

Figure 5-1. Optimization Example 1 (Groover 1980).

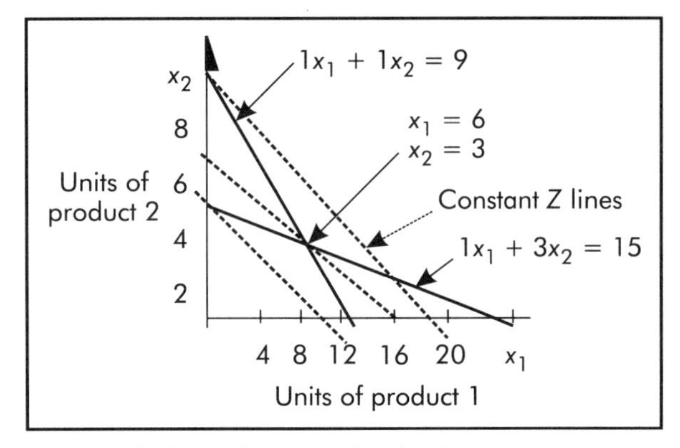

Figure 5-2. Example 1 graphical solution.

are drawn on the same graph as that of the constraints (see Figure 5-2). When the constant Z lines are superimposed on the constraint region, the user sees that the optimum point is at $x_1 = 6$ and $x_2 = 3$, and Z = \$120/day. There is no other combination giving a higher Z value.

Example 2. An assembly department producing printed circuit boards outputs two types of boards: X and Y. Each board requires three operations: component insertion, soldering of components, and inspection of components (see Table 5-1). The time required for each board in each operation and the maximum work hours available per day are plotted in Figure 5-3.

Determine the number of X boards and Y boards that should be produced to maximize output. Assume each X board contributes \$120 profit and each Y board contributes \$60. Determine the number of each of the boards that should be made to maximize the profit.

The constraints can be rewritten as follows:

$$8X + 4Y \leq 80 \tag{5-11}$$

$$3X + 4Y \leq 60$$

$$X + 3Y \leq 24$$

$$X \geq 0;\ Y \geq 0$$

The objective function in this case is maximization of output.

$$Z = X + Y \tag{5-12}$$

Solving these equations shows that the maximum output = 12.8 units. (X = 7.2; Y = 5.6). Since the units in this case are printed

Table 5-1. Example 2

	X Units	Y Units	Man hr/day
Insertion	8	4	80
Soldering	3	4	60
Inspection	1	3	24

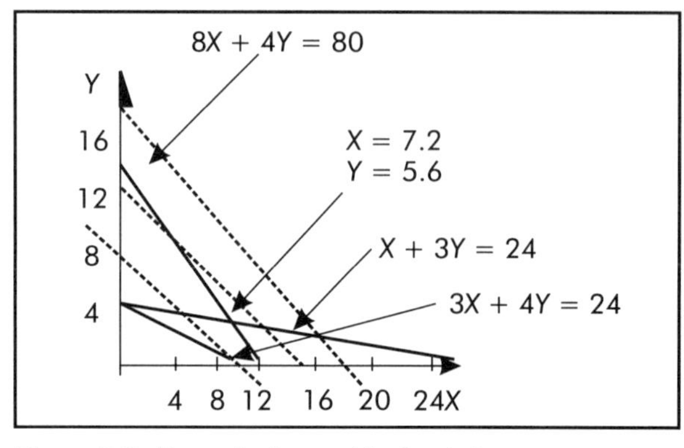

Figure 5-3. Example 2 graphical solution.

circuit boards, they must be taken as whole numbers, $X = 7$; $Y = 5$. The maximum output of 12 units is more realistic.

With the constraints the same as above, assume the objective function to be maximized is the profit. For maximum profit, $Z = 120X + 60Y$. It is obvious the maximum profit will be equal to \$1,200 if $X = 10$ and $Y = 0$. Thus, to maximize the profit, it is best to make board X only.

The *slack-variable method* can be used to maximize the output and the profit. To maximize the output of X and Y boards:

$$\text{Output} = X + Y \qquad (5\text{-}13)$$

To maximize the profit of X and Y boards:

$$\text{Combination profit} = 120X + 60Y \qquad (5\text{-}14)$$

As in Equation 5-11, both objective functions are subjected to constraint functions.

When slack variables S_1, S_2, and S_3 are introduced into the constraint equations, they are modified as,

$$8X + 4Y + S_1 = 80 \qquad (5\text{-}15)$$

$$3X + 4Y + S_2 = 60$$

$$X + 3Y + S_3 = 24$$

In this case, there are $m = 3$ (equations) and $n = 5$ (unknowns). Table 5-2 sets two variables to zero in each instance and shows calculations for the objective functions. There are 10 possible combinations of these where possible combinations are calculated using the formula,

$$\frac{n!}{m!(n-m)!} \qquad (5\text{-}16)$$

The objective functions are optimized if at least $(n - m)$ variables are set to zero.

Item number 9 in Table 5-2 shows the maximum output and profit since none of the variables have a negative number and the output and profit are both positive quantities.

Example 3. Product mix problems occur when several products are produced in the same production plant. For example, a refinery may have two crude oils. Crude A costs \$30 a barrel with 20,000 barrels available. Crude B costs \$36 a barrel with 30,000 barrels available. The company manufactures gasoline and lube oil from crude. The yield and sale price per barrel of the product and market are shown in Table 5-3. How much crude oil should the company use to maximize its revenue?

The objective function is defined as:

$$(0.6X)\$50 + (0.4X)\$120 + (0.8Y)\$50 + (0.2Y)\$120 = Z$$

$$0.6X + 0.8Y \leq 20{,}000 \qquad (5\text{-}17)$$

$$0.4X + 0.2Y \leq 10{,}000$$

$$X \geq 0;\ Y \geq 0$$

$$78X + 64Y = Z$$

where:

X = crude A
Y = crude B
Z = revenue

$X = 20{,}000$
$Y = 10{,}000$

However, this maximizes revenue, not profit. If the goal is to maximize profit, the objective function can be written as,

$$\text{Profit} = 0.6(\$50 - 30)A + 0.4(\$120 - \$30)A + 0.8(\$50 - \$36)B + 0.2(\$120 - \$36)B \qquad (5\text{-}18)$$

$$\text{Profit} = 48A + 28B$$

$$0.6A + 0.8B < 20{,}000$$

$$0.4A + 0.2B < 10{,}000$$

Gradient Search Procedure

The *gradient search procedure* finds the optimum point by using the method of steepest ascent/descent. It is based on the theory that the fastest way of finding an optimum

Table 5-2. Example 2 using the slack variable method

Number	X	Y	S_1	S_2	S_3	Output	Profit ($)	Comment
1.	0	0	80	60	24	0	0	X = 0; Y = 0
2.	0	20	0	–20	–36	20	1,200	S_2, S_3 negative
3.	0	15	20	0	–21	15	900	S_3 negative
4.	0	8	48	28	0	8	480	X =0
5.	10	0	0	30	14	10	1,200	Y = 0
6.	20	0	–80	0	4	20	2,400	S_1 negative
7.	24	0	–112	–12	0	24	2,880	S_1, S_2 negative
8.	4	12	0	0	–16	16	1,200	S_3 negative
9.	7.2	5.6	0	16	0	12.8	1,200	Output and profit are positive
10.	16.8	2.4	–64	0	0	19.2	2,160	S_1 negative

Table 5-3. Example 3

	Yield			
Product	Crude A	Crude B	Sale Price per Barrel	Market
Gasoline	0.6	0.8	$50	20,000
Lube oil	0.4	0.2	$120	10,000

point is to move along the gradient. In other words, the gradient is a vector of directional directive of a given function. The procedure is illustrated in Figure 5-4. The figure shows some of the constant contours of an objective function, $Z = f(x, y)$. The procedure requires the user to choose a starting point. From this starting point, he or she moves in the direction of maximum slope (gradient). After moving a certain distance, the user stops and changes direction to that of maximum slope from that point, and so on, until the optimum is reached.

The basic steps of the gradient search procedure are:

1. Evaluate the objective function at the starting point.
2. Compute the direction of the gradient from the starting point.

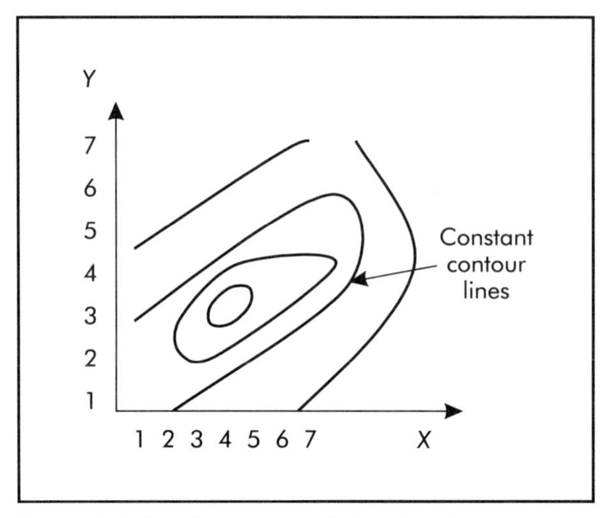

Figure 5-4. Gradient vector of directional directive.

3. Perform a search in the computed gradient direction for determining minimum function along a direction.
4. Evaluate the objective function at a point $(x + 1)$ from the previous step.
5. When the optimum is reached, the search procedure is terminated.

Figure 5-5 shows some of the constant contours of an objective function Z. The map generated by plotting the relationship between the objective function (performance index) and input variables appears like a geographical survey map. The figure shows the plot generated for two variables, X and Y. These constant contour lines are also known as the *response surface*.

The optimum point is the combination of X and Y values at which the objective function is optimized.

Many techniques looking for an optimum are based on gradient techniques. The *gradient* is a vector quantity with components along axes of independent variables X and Y. The magnitude of the component is equal

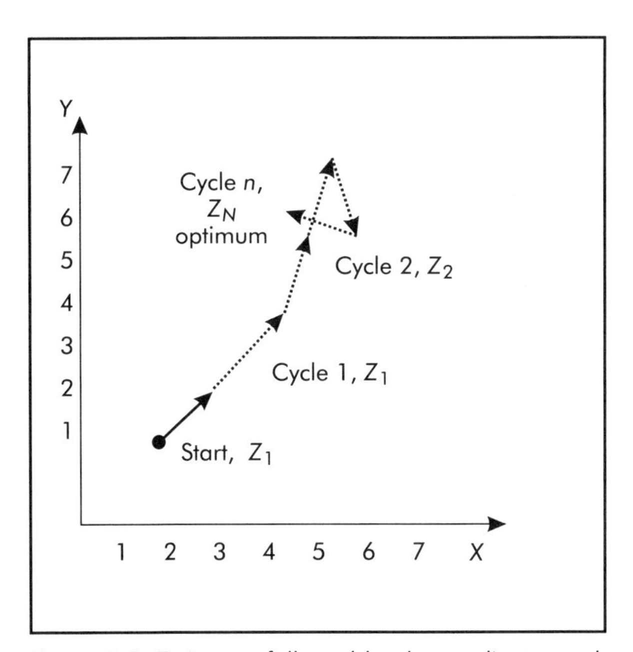

Figure 5-5. Trajectory followed by the gradient search technique.

to the partial derivative of the objective function with respect to the corresponding independent variable. For two inputs, X and Y, components of the gradients are defined as:

$$G_x = \frac{\partial z}{\partial x}$$
$$G_y = \frac{\partial z}{\partial y} \tag{5-19}$$

where:

G_x and G_y = components of the gradients in the X and Y directions on the response surface

$$G = G_x i + G_y j \tag{5-20}$$

where:

i and j = unit vectors parallel to the X and Y axes

The gradient proceeds in the direction of the steepest slope. Moving along the direction of steepest slope is a reasonable strategy to reach the top of the response (objective) surface. The magnitude (M) of the gradient is a scalar quantity given by:

$$M = \left[\left(\frac{\partial z}{\partial x}\right)^2 + \left(\frac{\partial z}{\partial y}\right)^2\right]^{\frac{1}{2}} \tag{5-21}$$

The *magnitude* of the point is defined at a particular point P on the X-Y surface.

The *direction* of the gradient (D) is defined as a unit vector:

$$D = \frac{G}{M} \tag{5-22}$$

As the gradient search proceeds from the starting point toward optimum, the sequence of moves to seek optimum is represented by a trajectory. The definitions of gradient, magnitude, and direction can all be extended to functions with more than two independent variables. As the gradient search proceeds from the starting point toward optimum, the sequence of moves to seek optimum is represented by a trajectory as shown in Figure 5-6.

Method of steepest ascent. Using the method of steepest ascent, the search begins by estimating the gradient at the current operating point. It then moves the operating point to a new position in the direction of the gradient. The gradient is determined at the new position also. The cycle of the gradient determination and step-by-step movement is repeated until the optimum is reached.

In many industrial processes, analytical representation of a response surface is not readily available, which limits the ability to find the gradient at each cycle. Starting from the current operating point, the slope of the response surface is found by making several exploratory moves centered around the current point. Exploratory moves are arranged in the form of a factorial experiment done on neighborhood points. Four experimental points are explored around the current point. At each of those points, the objective function is calculated. Then gradient components are estimated by means of equations:

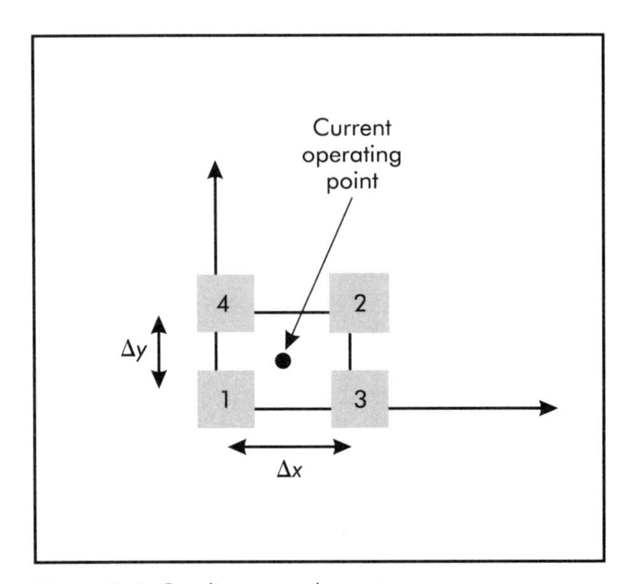

Figure 5-6. Gradient search strategy.

$$G_1 = \frac{(Z_2 + Z_3) - (Z_1 + Z_4)}{2\Delta x} \qquad (5\text{-}23)$$

$$G_2 = \frac{(Z_2 + Z_4) - (Z_1 + Z_3)}{2\Delta y}$$

The gradient is $G = G_1 i + G_2 j$ (Eq. 5-20)

where:

Z_1, Z_2, Z_3, Z_4 = values of the objective function at four experimental points

Δx = difference in independent variable x separating experimental points

Δy = difference in independent variable y separating experimental points

Assuming the current point to be (3,4), Δx and Δy should be considered to be two units each. The coordinates of point 1 will be (2 and 3):

Test point	x	y	z
1	2	3	z_1
2	4	5	z_2
3	4	3	z_3
4	2	5	z_4

Exploratory moves are made for the purpose of determining the gradient. If the gradient is determined, a step move is made to the new operating point. The step move is taken in the direction of the gradient. Input variables x and y are incremented in proportion to components of the direction vector.

$$\text{New } x = \text{old } x + C\frac{G_1}{M} \qquad (5\text{-}24)$$

$$\text{New } y = \text{old } y + C\frac{G_2}{M}$$

where:

C = a scalar quantity that determines the size of the step move

The search continues until optimum is reached. At the optimum value of the objective function, or *performance index*, the gradient has a value of zero. Quite often, the gradient changes direction abruptly indicating values in the opposite direction.

Example. Suppose that the response surface for a certain manufacturing process was defined by:

$$Z = 17x + 27y - x^2 - 0.9y^2 \quad (5\text{-}25)$$

The method of steepest ascent is used to determine the approximate optimum operating point. The starting point of the search should be $x = 2$, $y = 3$ and the step size should be $C = 4$ (Groover 1980).

$$\frac{dz}{dx} = 17 - 2x \qquad \frac{dz}{dy} = 27 - 1.8y \quad (5\text{-}26)$$

$$\frac{dz}{dx} = 17 - 2(2) = 13$$

$$\frac{dz}{dy} = 27 - 1.8(3) = 21.6$$

$$M = \sqrt{13^2 + 21.6^2} = 25.21$$

$$D = 0.51571 + 0.8568j$$

where:

M = magnitude of the gradient
D = direction of the gradient

New $x = 2 + 4(0.5157) = 4.063$

New $y = 3 + 4(0.8568) = 6.427$

For cycle 1, calculate Z_1. At $x = 4.063$ and $y = 6.427$,

$$\frac{dz}{dx} = 17 - 2(4.063) = 8.874$$

$$\frac{dz}{dy} = 27 - 1.8(6.427) = 15.431$$

New $x = 4.063 + 4(0.4985) = 6.057$

$M = 17.8 \qquad \bar{D} = 0.1985i + 0.867j$

New $y = 6.427 + 4(0.8670) = 9.895$

For cycle 2, calculate Z_2. At $x = 6.057$ and $y = 9.895$,

New $x = 6.057 + 4(0.4695) = 7.935$

$M = 10.407 \qquad \bar{D} = 0.4695i + 0.883j$

New $y = 9.895 + 4(0.8830) = 13.427$

For cycle 3, calculate Z_3. At $x = 7.935$ and $y = 13.427$,

$$\frac{dz}{dx} = 17 - 2(7.935) = 1.13$$

$$\frac{dz}{dy} = 27 - 1.8(13.427) = 2.831$$

$M = 10.407 \quad \bar{D} = 0.4695i + 0.883j$

New $x = 7.935 + 4(0.371) = 9.419$

New $y = 13.427 + 4(0.929) = 17.143$

For cycle 4, calculate Z_4. At $x = 9.419$ and $y = 17.143$,

$$\frac{dz}{dx} = 17 - 2(9.419) = -1.838$$

$$\frac{dz}{dy} = 27 - 1.8(17.143) = -3.859$$

There is a change in slope. Reduce the step size to $C = 2$.

New $x = 9.419 + 2(1.43) = 8.56$

New $y = 17.143 + 2(-0.903) = 15.34$

$$z = 17(8.56) + 27(15.34) - (8.56)^2 - 0.9(15.34)^2$$

$$z = 274.64$$

When the vicinity of the optimum is reached, it is better to reduce step size. When the gradient direction changes and the objective function does not change appreciably, it is an indication that the vicinity of the optimum point is located. The objective function tends to fluctuate. The user should repeat the last step until no further improvement of the objective function results.

LaGrange Multiplier Method

The LaGrange multiplier method provides an approach for determining the optimum values in multi-variable problems subject to functional constraints. The method consists of transferring objective and constraint functions into a single constraint-free function called the LaGrange expression. The optimum values of the variable are obtained by solving the equation (Dieter 2000).

Dynamic Programming Method

The dynamic programming method is an approach for optimizing the design systems that are configured in stages. The sequence of design decisions is made in each stage. The word "dynamic" has no relationship to the changes denoted with respect to time. This method is useful in situations involving at least four stages with several design options available at each stage. Dynamic programming converts a large and complicated optimization problem into a number of interconnected minor problems. Each minor problem contains few variables and results in a series of partial optimizations.

Nonlinear Optimization Method

Multivariable optimization of nonlinear problems has been a topic of extensive research and many computer-based methods of accomplishing this process are available. Out of the two types of methods, the direct method relies on the evaluation of the objective function. Indirect methods require information about derivative values to determine the search direction for optimization.

LEARNING CURVE ANALYSIS

When a new model of a product is introduced, learning curve analysis is a useful method to study its economics, especially if the new product has similar work content. As an organization gains experience in manufacturing a product, the resource inputs required per unit of output diminish over the life of the product. Labor times that go into manufacturing the first unit of a new automobile are typically much higher than those needed for unit 100. As the cumulative output of the model grows, labor input continues to decline. It is just like performing a repetitive task with the result that performance keeps improving. The performance time drops off rather dramatically at first, and it continues to fall at some slower rate until performance reaches a constant. This learning pattern applies to groups and individuals as well. If people perform the same operation over and over again, it takes them less and less time to do it.

The general shape of this curve, called the learning curve, is shown in Figure 5-7. This exponential curve becomes a straight line when plotted on logarithmic coordinates as opposed to arithmetic coordinates. In this example, the initial unit requires 60 labor hours to manufacture. As output and experience continue, labor hours diminish to about 23 for unit 20. The general equation for this curve is:

$$Y = Ki^{b} \tag{5-27}$$

where:

Y = labor hours required to produce the ith unit (or the production time per unit after producing a number of units equal to i)
K = labor hours required to produce the first unit
i = ordinal number of unit, that is, 1st, 2nd, 3rd, and so on (number of units produced)
b = index of learning (a constant that depends on the constant percentage reduction)

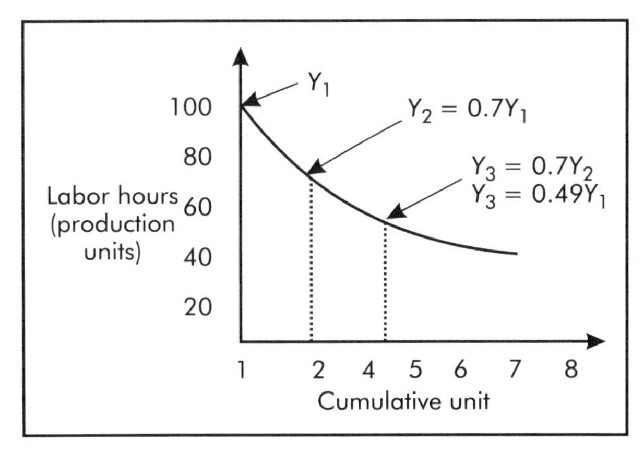

Figure 5-7. Learning curve.

By convention, the learning rate is specified as a percentage. A 90% curve means that each time the cumulative output doubles, the most recent unit of output requires 90% of the labor input of the reference unit. If unit one requires 100 hours, unit 2 requires 90% of 100 labor hours, or 90 hours; unit 4 requires 90% of 90 hours, or 81 hours, etc.

The value of b in Equation 5-27 for a given learning curve (P%) can be found as follows. Each time production quantity doubles, average cost is reduced by a constant percentage, which defines slope. The slope percentage is shown as,

$$P = 2^b x100 \tag{5-28}$$

where:

P = learning curve percentage
b = index of learning

Assume that initially, $Y = Ki^b$ for unit $i = i_a$. For double the unit number, that is, $i = 2i_a$, time is reduced to $Y = PY_a/100$.

Substituting this in Equation 5-27,

$$Y_a = Ki_a{}^b \tag{5-29}$$

$$\frac{PY_a}{100} = K(2i_a)^b \tag{5-30}$$

Dividing Equation 5-30 by Equation 5-29 results in,

$$\frac{P}{100} = 2^b$$

or

$$b = \frac{\text{Log}\left(\frac{P}{100}\right)}{\text{Log}2} \tag{5-31}$$

The exponent, b, is a negative number with an absolute value less than 1, which defines the rate at which the average cost decreases as quantity increases. Exponent values for typical learning curve percentages are given in Table 5-4.

For an 80% learning curve, $y_2 = 0.8y_1$ for $i_2 = 2i_1$.

Example 1

For a product under evaluation, the first unit cost in terms of man-hours is shown as 1,200 hours. In this case, experience has shown that an 88% learning curve can be anticipated. Projected costs for the first 50 units and for the 100 units following the first 50 units are calculated as follows (Tanner 1985).

Using Equation 5-27, $Y = Ki^b$

Taking logarithm on both sides,

Log (Y) = Log(K) + b Log(i).

Users have to determine b, given P = 88%.

Table 5-4. Exponent values for typical learning curve percentages

Learning Curve %, P	Exponent b
65	–0.624
70	–0.515
75	–0.415
80	–0.322
85	–0.234
90	–0.074

b = Log (0.88)/Log2 = –0.1844.

Substituting 1,200 for the first unit K results in,

$Y = 1{,}200(i^{-0.1844})$

Calculate Y at $i = 50$ and $i = 150$.

Cumulative average cost for 50 units $= 1{,}200(50^{-0.1844}) = 583.2$

Cumulative average cost for 150 units $= 1{,}200(150^{-0.1844}) = 496.3$

Total cost for 50 units = 50(583.2) = \$29,160

Total cost for 150 units = 150(496.3) = \$71,445

Cost for 100 units following 50 = \$42,285

Example 2

The first machine of a group of eight machines costs \$100,000 to produce. If a learning rate of 80% is expected, how much time does it take to complete the eighth machine?

Using Equation 5-27, $Y = Ki^b$

For $P = 80\%$, $b = -0.322$ and $K = 100{,}000$

$y = 100{,}000\ (8)^{-0.322}$

$y = \$51{,}200$

Rate of Learning

The rate of learning is not the same in all manufacturing applications. Learning occurs at a higher rate in some applications than others and is reflected by a more rapid descent of the curve. A 90% curve, for example, means that each time cumulative output doubles, the most recent unit of output requires 90% of the labor input for the reference unit. If unit 1 requires 100 labor hours, unit 2 requires 90% of 100, or 90 hours; unit 3 requires 90% of 90 hours or 81 hours, etc.

The use of learning curve concepts as estimating tools involves more than insertion of variables into an equation. The actual behavior of manufacturing cost trends is influenced by a number of key factors. These must be considered in their effect on the actual database and the cost of new or follow-up work.

DESIGN FOR QUALITY AND ROBUST DESIGN

Inspection

Typically, the specifications for a product are intended to guarantee proper assembly of components that are free from manufacturing defects (Schonberger 1982). Maintaining quality assurance is the responsibility of everyone involved in the design and production of a product. There are established methods of sampling a product or process to characterize its correspondence to specifications. *Inspection* is the process of checking conformance of a final product to its specifications. Inspection of variables involves quantitative measurement of characteristics such as dimensions, surface finish, and other physical or mechanical properties. Such measurements are made with instruments that produce a variable result. For highly critical parts, 100% inspection of a process is done with the help of on-line inspection devices.

Another type of inspection is inspection for attributes. The presence or absence of a flaw is an example of *attribute inspection*. Statistical methods extract significant information from large amounts of numerical information. This approach is important in quality control since large quantities of material or product may be involved. Statistical methods are also employed when dealing with variability in data such as in manufacturing processes because no two products are ever manufactured exactly alike (Schonberger 1986).

Statistical Control Methods

There are always variations in dimensions or properties of raw materials, and variation in operation of machines and operator performance. Statistics are important tools for quality assurance because they provide a way of characterizing a production volume by means of a sample. Statistical quality control detects variation in the process. There are two types of variation: natural and assigned. Natural variability in a manufacturing process is inherent, uncontrolled changes occur in the composition of material, performance of the operator, and operation of machines. These variations occur randomly with no particular pattern or trend. In contrast, assignable variability can be traced to a specific, controllable cause. Statistical control methods are intended to distinguish between natural and assignable variability. Ideally, if assignable causes of variability can be identified, the process can be better controlled and defects can be prevented. The systematic method of detecting assignable variability in a process is known as *statistical process control*.

Design of Experiments

Design of experiments is one of the quality techniques used to optimize performance response. It consists of tests on a process where changes are made to input variables or parameters all at the same time so that variables and their interactions that cause output response changes can be observed, identified, and isolated. In traditional experimentation, experimenters change variables one factor at a time (with all other factors constant) to find variables that contribute to response the most.

Robust Design

Robust design is a systematic method for keeping the producer's cost low and delivering a high-quality product. It can greatly improve an organization's ability to meet market windows, keep development and manufacturing costs low, and deliver high-quality products. It is a methodology where quality is brought in concurrently with product design and development. Quality is defined as satisfying customer requirements. The impact of poor product quality is far greater than what appears in the first occurrence. Although it can be expressed in monetary terms, failure to meet customer expectations may also result in direct cost to the customer. The resulting customer dissatisfaction means fewer future sales and a reduction in market share. This, in turn, causes higher marketing and advertisement costs. In addition, direct supplier losses in terms of scrap, rework, inspection, and warranty are passed on to the customer.

Genichi Taguchi, who proposed a unique philosophy for solving quality problems, developed the foundation of robust design (Taguchi 1993). Taguchi's relatively easy-to-understand method of designing a fractional factorial experiment helps to make products/processes more robust and less variable to disturbances. Robust design uses ideas from statistical experiment design and addresses two major concerns:

1. How to economically reduce the variation of a product's function due to the customer's environment.
2. How to ensure that decisions found optimum during laboratory experiments prove to be so in manufacturing and customer environments.

The answers provided by application of robust design to these concerns make it a valuable tool for improving productivity (Chang et al. 1991). Figure 5-8 shows the operational steps in robust design.

To get a full picture of robust design, it is necessary to understand the implications of quality. The ideal quality a customer can expect is that every product delivers target

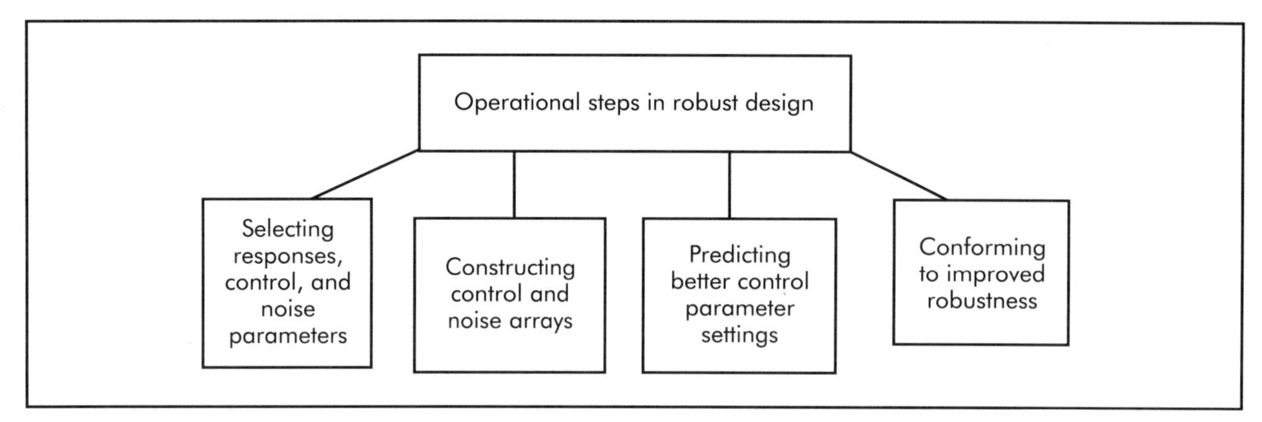

Figure 5-8. Operational steps in robust design.

performance each time it is used under all intended operating conditions, throughout its intended life. The traditional concepts of reliability and dependability are part of the definition of quality. This definition of quality can easily be extended to processes as well as services.

The cost of delivering a product is broken into three areas: operating cost, manufacturing cost, and research and development cost. Operating cost consists of the cost of energy needed to operate the product, environmental control, maintenance, and inventory of spare parts and units. A manufacturer can greatly reduce operating cost by minimizing product sensitivity to environmental and manufacturing variation. Important elements of manufacturing cost are: equipment, machinery, raw materials, labor, scrap, rework, etc. Robust manufacturing can reduce manufacturing cost by reducing process sensitivity to manufacturing disturbances.

Research and development takes time and a substantial amount of resources. Its costs can be kept low by adopting a robust design approach, resulting in more efficient generation of information needed to design products and processes and reduced development time and resources needed for development. Higher quality results in lower operating cost and vice versa.

The following summarizes Taguchi's approach:

- Design quality into the product. Do not use inspection to sort out the poor-quality products.
- Set a target. The cost of quality is the deviation from target values.
- Make the product insensitive to uncontrollable external factors.

These goals are achieved by optimizing product and process designs to make them minimally sensitive to various causes of variation. This is called *parameter design*. However, parameter design alone does not always lead to sufficiently high quality. Further improvement can be obtained by controlling the causes of variation where it is economically justifiable, typically by using more expensive equipment, higher-grade components, better environmental controls—all of which lead to higher product cost, or operating cost, or both. Benefits of improved quality must justify added product cost.

Example

Consider an example where a tolerance specification is given as 0.50 in. ± 0.02 in. In this case, it would be immaterial whether the actual figure is 0.48, 0.50, or 0.52; the specifications would be equally satisfied.

Taguchi defines quality as product uniformity around a target value. Here the target value is 0.50 in., and the actual values achieved closer to the set target value would be better than the values further away from the target.

Conducting a Robust Design Experiment

The first step in conducting a robust design experiment is to identify the factors that influence the process. There are two factors that influence a process: noise and control factors. The number of control factors is determined by the complexity of the process. After both factors have been determined, parameter settings are selected for all factors. Then an orthogonal array (explained in the next section) is constructed using all the factors and levels for each factor that have been predetermined. The experiment is performed according to the parameter settings of each row in the orthogonal array with or without replications.

Replication of the experiment depends on how long the experiment takes, the availability of samples, and financial considerations. After the experiment, analysis by row has to be done first to study the variability and its bias on the data recorded. The detail in the analysis to be studied varies depending on the type of target response. There are three types of target response: larger the better (for example, stiffness), nominal the best (for example, surface roughness), and smaller the better (for example, drag coefficient in cars or planes).

Orthogonal arrays. An orthogonal experiment helps us to understand the simultaneous influence of many factors on product or process quality. Orthogonal arrays need to be constructed or selected for this. They relieve the major burden of designing a fractional factorial experiment and allow the experimenter to evaluate several factors with a minimum number of test runs (see Table 5-5).

The first step in constructing an orthogonal array for a specific case study is to count the total degrees of freedom that determine the minimum number of experiments that must be performed to study all the chosen control factors. For example, a three-level control factor counts two degrees of freedom because, for a three-level factor, the experimenter is interested in two comparisons. In general, the number of degrees of freedom associated with a factor is equal to one less than the number of levels for that factor. Table 5-6 gives examples of commonly used orthogonal arrays.

Table 5-7 shows an example of 4 × 3 orthogonal array for a manufacturing process involving a machining operation. The response to be monitored in this example is surface quality. The control variables are feed rate, spindle speed, and type of tool. Each of the variables has two identified levels that can be tried experimentally. The table shows control factors and the construction of the orthogonal array. Table 5-8 shows the example of a molding machine, where control variables are injection pressure, mold temperature, and set time.

Steps in a robust design experiment. The robust design experiment steps include:

1. Identify the main function, side effects, and failure modes. This step requires engineering knowledge of the product or process and the customer's environment.
2. Identify noise factors and testing conditions for evaluating quality loss. Testing conditions are selected to capture the effect of more important noise factors. It is important that testing conditions permit a consistent estimation of sensitivity to noise factors for any combination of control factor levels.
3. Identify quality characteristics to be observed and the objective function to be optimized.

Table 5-5. Selecting a standard orthogonal array (Padke 1989)

Orthogonal Array	Number of Rows	Maximum Number of Factors	Maximum Number of Columns at These Levels 2	3	4	5
L_4	4	3	3	—	—	—
L_8	8	7	7	—	—	—
L_9	9	4	—	4	—	—
L_{12}	12	11	11	—	—	—
L_{16}	16	15	15	—	—	—
L'_{16}	16	5	—	—	5	—
L_{18}	18	8	1	7	—	—
L_{25}	25	6	—	—	—	6

Table 5-6. Commonly used orthogonal arrays

4 × 3 Orthogonal Array	9 × 4 Orthogonal Array	8 × 7 Orthogonal Array
1 1 1	1 1 1 1	1 1 1 1 1 1 1
1 2 2	1 2 2 2	1 1 1 2 2 2 2
2 1 2	1 3 3 3	1 2 2 1 1 2 2
2 2 1	2 1 2 3	1 2 2 2 2 1 1
	2 2 3 1	2 1 2 1 2 1 2
	2 3 1 2	2 1 2 2 1 2 1
	3 1 3 2	2 2 1 1 2 2 1
	3 2 1 3	2 2 1 2 1 1 2
	3 3 2 1	

4. Identify control factors and their multiple levels. The more complex a product or a process, the more control factors it has, and vice versa. Typically six to eight control factors are chosen at a time for optimization. For each control factor, two or three levels are selected, out of which one level is usually the starting level. The levels chosen should be sufficiently far apart to cover a wide experimental region because sensitivity to noise factors does not usually change with small changes in a control factor setting. Also, by choosing a wide experimental region, the experimenter can identify "good" regions as well as "bad" regions for control factors.
5. Design the matrix experiment and define the data analysis procedure. Using orthogonal arrays is an efficient way to study the effect of several control factors simultaneously. The factor effects thus obtained are valid over the experimental region and the array provides a way to test for inclusion of the factor effects. The experimental effort needed is much smaller when compared to other methods of experimentation like trial

Table 5-7. Control factors of the process and 4 × 3 orthogonal array (three factors each at two levels)

Factors	Level 1	Level 2
A = feed rate	6.6 in./min (168 mm/min)	15 in./min (381 mm/min)
B = spindle speed	85 rpm	290 rpm
C = tool type	High-speed steel	Tungsten coated

Trial Number	A	B	C
1	1	1	1
2	1	2	2
3	2	1	2
4	2	2	1

and error. The choice of using an orthogonal array for a particular project depends on the number of factors and other practical considerations.

6. Conduct the matrix experiment. The levels of several control factors must be changed when going from one experiment to the next in a matrix experiment. Properly setting the levels of various control factors is essential. When a particular factor has to be at level 1, it should not be set to level 2 or 3. Do not worry about small perturbations that are inherent in experimental equipment. Any erroneous or missing experiments should be repeated to complete the matrix.
7. Analyze the data, determine optimum levels for control factors and predict performance under these levels. When a product or a process has multiple characteristics, it may become necessary to make some trade-off. In robust design, however, the primary focus is on maximizing the S/N (signal to noise) ratio. The observed factor effects together with the quality loss function can be used to make a rational trade-off.
8. Conduct the verification experiment and plan future actions. The purpose of this final and crucial step is to verify that optimum conditions suggested by the matrix experiments give the projected improvement. If observed and

Table 5-8. Molding machine example

Variables or Factors	Level 1	Level 2
A. Injection pressure	A1 = 250 psi (1,724 kPa)	A2 = 350 psi (2,413 kPa)
B. Mold temperature	B1 = 150° F (66° C)	B2 = 200° F (93° C)
C. Set time	C1 = 6 seconds	C2 = 9 seconds

	Injection Pressure	Mold Temperature	Set Time	Repetitions		
Exponent Number	A	B	C	1	2	3
1	1	1	1	26		
2	1	2	2	25		
3	2	1	2	34		
4	2	2	1	27		

projected improvements match, the experimenter should adopt the suggested optimum conditions. If not, he or she must conclude that the additive model underlying the matrix experiment has failed, and find ways to correct the problem.

9. The corrective actions include finding better quality characteristics, signal to noise ratios, or different control factors and levels, or studying a few specific interactions among the control factors. Evaluating improvement in quality loss, defining a plan for implementing results, and deciding whether another cycle of experiments is needed are also a part of the final step of robust design. It is, however, common for a product or process to require more than one cycle to achieve the desired quality and cost improvement (Padke 1989).

Case Study: Surface Roughness Analyzer

The product considered for improvement is a new surface roughness analyzer. This instrument works with a new noncontact optical method based on diffraction light principles. When applied to engineering surfaces, it rapidly provides precision surface roughness data on engineering and machined surfaces. The roughness measurement conventionally involves use of a stylus device, which is drawn over the sample to detect and record variations in surface irregularities. Compared to the contact stylus method of inspection, the optical technique is much more preferred.

Roughness measurements are made by utilizing a laser and microcomputer-based vision system to measure intensity of a collimated, monochromatic light source diffracted in the spectral direction and captured by a video system that provides an analog signal to a digitizing system (see Figure 5-9). The

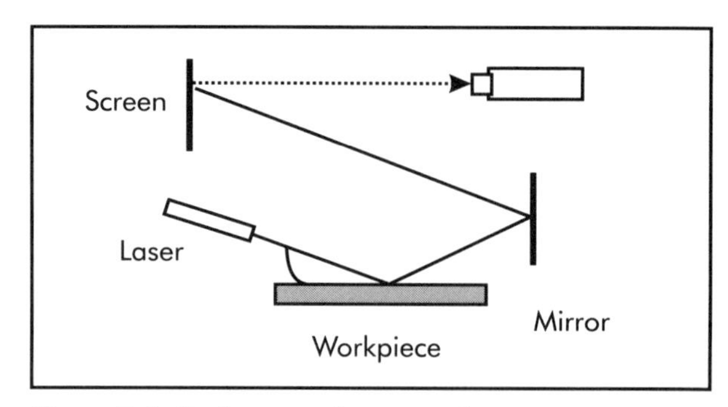

Figure 5-9. Surface roughness analyzer.

information is converted to digital, which is subsequently modified to display the surface roughness value. Intensity is measured as a function of the gray level of the image, processed by the digitizing circuit and compared against a previously defined calibration standard. The method further includes a microcomputer-based procedure, which provides for operator interaction in the form of menu-driven steps that guide the operator through the process, calibration, measurement, and analysis phases (Shetty and Neault 1993).

Improvement using robust design. The objective of a robust design study is to find the optimum recommended factor setting for the surface roughness analyzer to minimize the variability in readings. The instrument relies on the spread of laser light on the workpiece to determine surface roughness. Reliability of the analyzer depends mostly on everything involved with the laser and its path. As an example, four parameter settings, each at two levels, are introduced into the experiment in the form of orthogonal arrays. The parameters are: laser angle, background, distance from laser to workpiece, and background lighting.

Parameter optimization. To optimize the parameters, experiments are done to minimize the effects of the laser angle, distance from the laser to the workpiece, and interactions between these factors. The parameter settings for this experiment are identified in Table 5-9. The optimal settings for the instrument are shown in Table 5-10 (Shetty 1990).

In many industry situations, the robust design method is successfully used to determine the optimum setting of the system. Application of the method in the electronics, automotive products, photography, and many other industries has been an important factor in rapid industrial growth. It proves to be useful and easy to implement in the design process. Furthermore, the method allows the designer to investigate every possible variable and its effects in the application. It also reduces the time required to complete experiments.

FAILURE MODES AND EFFECTS ANALYSIS

Failure modes and effects analysis (FMEA), a methodology used in product development, identifies and prioritizes possible failure modes of a specific system. FMEA is an iterative technique that promotes systematic thinking when a new product or system is developed in terms of what could go wrong with the product, how bad will be the influence of the error, and what needs to be done to prevent failures.

FMEA methodology was first developed by the National Aeronautics and Space Administration (NASA) and later was introduced in industries such as nuclear power and automotive. It has now become a critical step in the design process for many world-class manufacturers and has been incorporated as a design standard. FMEA is used to detail the effects of each individual possible failure mode and evaluate its relative importance. FMEA can be performed on the system, subsystem, assembly, or component levels. The standard FMEA format is used to evaluate single-point failures only. Since a failure in one component may influence the failure of an interconnected component, FMEA analysis can be modified to encompass common cause failures (CCF), in which redundant or interdependent machines are likely to fail within a short period of one another, due to similar reactions to the operating environment. FMEA provides a practical approach well suited for products and systems where there are not many human interface or software-driven concerns.

FMEA should be completed with the input of several disciplines that have knowledge of or experience with the system to be analyzed. Some functional disciplines that would normally be involved are design engineering, quality control, manufacturing, and service. The analysis should be performed so that as many different perspectives are accounted for as is practically possible. The analysis is generally performed because of a desire to improve the reliability of a component or system, to reduce or eliminate failures, to document failure modes for future use, or to use as input for other analyses. Usually, FMEA is performed early in

Table 5-9. Parameter settings for the surface analyzer

Parameter	Level 1	Level 2
Laser angle	20°	30°
Background surface	Glossy	Non-glossy
Distance from laser to workpiece	4.5 in. (114 mm)	5.5 in. (140 mm)
Background lighting	Off	On

Table 5-10. Optimal settings for the surface analyzer

Parameter	Condition
Laser angle	25°
Background surface	Non-glossy
Distance from laser to workpiece	4.5 in. (114 mm)
Background lighting	Off

the design phase of a product, is continued throughout development, and completed prior to final release of engineering production drawings. The FMEA is an active document that parallels the design process and, when used properly, can greatly reduce the possibility of late design changes that can cause significant delays to schedule and have potential to be very costly.

FMEA is commonly run on existing systems to provide a comparison for new development, or to capture possible failure mode causes of systems that have quality problems. One drawback is that FMEA analysis is a lengthy study of the system in question. This means it can take significant levels of effort and time and could be a major concern if schedule is an issue. For these situations, there are other types of analysis that, while not as comprehensive, can be done with relative ease and speed. One such analysis tool is *fault tree analysis* (FTA). FTA is a technique used when the system to be analyzed is complex, with many interconnected functions and human interface situations.

FMEA serves as a summary of the thought process used throughout the design phase. Major benefits of this analysis are that it provides a vehicle for tracking and prioritizing known risks. The analysis does this by taking advantage of the experience and concerns of the people involved (Voland 1999; Womack and Jones 1996).

FMEA Process

The FMEA process is found extensively in quality improvement literature and varies according to its application. One of FMEA's strengths is that it can be adapted for a wide variety of applications. The major steps for completing an FMEA analysis are shown in Table 5-11.

A blank FMEA worksheet is shown in Table 5-12. Organizations can generate their own ranking scales. An example ranking scale is shown in Table 5-13.

FMEA Example: Travel Mug

In this example, a company that produces travel mugs, among other food and beverage containment products, is designing its latest travel coffee mug. It has incorporated FMEA into its design process to help improve quality and reduce rework and scrap, which it has had problems with in its past products. A worksheet such as the one shown in Table 5-12 is completed shortly prior to production startup. The worksheet focuses on the cover for the mug and represents only a portion of the full FMEA, which may also focus specifically on the handle, base, and cup components, either individually or as an assembly.

As the FMEA worksheet is filled out, it soon becomes obvious that this cascading of detail can develop into a somewhat major effort to complete (see Table 5-14). This is one of the reasons that an FMEA is usually a time-consuming task. The advantage, however, is that this format quickly establishes priorities for corrective actions to obtain maximum benefits from early stage corrections and improvements.

Function

For the travel mug cover, after completing the header information and identifying the item(s) to be evaluated, the next step is to list its important functions. This is an important step and it relies on the analysis team's understanding of the product. If a critical function is left out, its failure modes will not be considered, and there will be no conscious effort made to address possible design flaws until after they have been designed in. At that point, the cost to correct the situation will be at least an order of magnitude higher.

Failure Modes

Once functions are detailed, all possible failure modes for each expected function can

Table 5-11. FMEA steps

Step	Task
1	Define the system including subsystems as seen in assembly. Specify the level of the system at which the analysis will take place.
2	Identify all the operational characteristics of the system.
3	Detail the expected functions and outputs of the design. Include expectations and limiting parameters.
4	Determine the environmental profiles. Clarify the type of environment the design will be operating in, the duty cycle, and any external factors that may influence the functionality of the design.
5	Develop a functional block diagram to help clarify the inputs, functions, and outputs of the system.
6	Define the possible failure mode(s) for each hardware item. Determine the effect of each failure mode on the rest of the system.
7	Classify the failure effect's initial severity ranking (SR).
8	Determine the cause(s) of each potential failure mode (use fault tree analysis).
9	Determine the initial occurrence ranking (OR).
10	List the design or test verification tasks that can detect, during development, the failure cause.
11	Determine the failure mode initial detection ranking (DR).
12	Calculate the failure mode/effect initial risk priority number (RPN). $RPN = SR \times OR \times DR$
13	Develop an action plan and responsibilities to reduce failure mode effect severity, failure mode occurrence, and/or change design verification tasks.
14	Implement actions on product.
15	Determine new SR, OR, DR, and RPN after corrective measures have been put in place.
16	Document conclusions and recommendations.

be listed out. Similarly, when failure modes have been defined, the effects of each type of failure are documented. In Table 5-14, notice that a single item can have multiple functions, which can have multiple failure modes, which can have multiple failure effects, etc.

Failure Effects and Severity Ranking

The failure mode of "cover comes off unexpectedly" has two similar, but significantly different effects (see Table 5-14). In the first, "total loss of fluid," the user will be irritated, especially if the failure repeats frequently. Over time, this could result in a loss of customer loyalty to the company, so the severity ranking (SR) for this effect is fairly high; the SR for total fluid loss is an 8. The second effect, "major spill on user of hot fluid," could result in lawsuits against the company and is a safety hazard for the user. The SR assigned is a 10, the highest.

Once the failure effects have all been documented and their severity ratings are assigned, it may be necessary to break away from the FMEA worksheet to determine the causes of the failure modes listed. The information might be gathered from prior experience, test

Table 5-12. FMEA worksheet

Number	Function	Part No.	Failure Mode	Failure Effect	SR	Failure Mode Cost	OR	Verify Design	DR	RPN	Actions	Results			
												SR	OR	DR	RPN

SR = initial severity ranking
OR = initial occurrence ranking
DR = initial detection ranking
RPN = initial risk priority number

Table 5-13. FMEA ranking scales

Severity Ranking (SR)	Severity Effect
1	Minor: no effect on system performance
2, 3	Low: slight effect on system performance
4, 5, 6	Moderate: failure causes some system performance deterioration
7, 8	High: system malfunctions
9, 10	Very high: system safety compromised or noncompliance with codes
Occurrence Ranking (OR)	**Probability of Failure/Event (Quantitative Probability)**
1	Remote: failure unlikely (0.000001)
2, 3, 4	Low: relatively few failures (0.00005, 0.00025, 0.001, respectively)
5, 6	Moderate: occasional failures (0.00025, 0.0125, respectively)
7, 8	High: repeated failures (0.025, 0.05, respectively)
9, 10	Very high: failure is almost certain (0.125, 0.5, respectively)
Detection Ranking (DR)	**Detection Likelihood**
1, 2	Very high: program will almost certainly detect potential failure cause
3, 4	High: program has a good chance of detecting potential failure cause
5, 6	Moderate: program may detect failure cause
7, 8	Low: program is not likely to detect failure cause
9	Very low: program probably will not detect failure cause
10	None: program can not detect failure cause

Table 5-14. FMEA example of a coffee mug (Chicoine 2000)

Item	Function	Failure Mode	Failure Effect	SR	Failure Mode Cause	OR	Design Verification	DR	RPN	Recommended Action	Actions/ Results	SR	OR	DR	RPN
Cover	To restrain fluid from uninten-tional spillage	Cover leaks	Small spills on user	5	Poor fit due to tolerance or design	5	SPC on cup and cover dimensions	3	75	Analyze design and process to improve fit	Temperature control module on molding machine updated for better control	5	3	3	45
					Uneven expansion due to difference in materials	7	Reliability testing of assembly	4	140	Review material selection compatibility	Materials with same thermal expansion coefficient selected		1	4	20
					Wrong material used	5	No incoming material inspection	10	250	Review supplier's QC and alternative suppliers	New supplier certifies material is correct; has SPC in place		1	1	5
		Cover comes off un-expect-edly	Total loss of fluid	8	Poor fit due to tolerances or design	4	SPC on cup and cover dimensions	3	96	Analyze design and process to improve it	Temperature control module on molding machine updated for better control	8	2	3	48
					Uneven expansion due to difference in materials	6	Reliability testing of assembly	4	192	Review material selection compatibility	Materials with same thermal expansion coefficient selected		1	4	32

Table 5-14. (continued)

Item	Function	Failure Mode	Failure Effect	SR	Failure Mode Cause	OR	Design Verification	DR	RPN	Recommended Action	Actions/ Results	SR	OR	DR	RPN
					Wrong material used	5	No Incoming material inspection	10	400	Review supplier's QC and alternative suppliers	New supplier certifies material is correct; has SPC in place		1	1	8
		Major spill on user of hot fluid		10	Poor fit due to tolerances or design	4	SPC on cup and cover dimensions	3	120	Analyze design and process to improve it	Temperature control module on molding machine updated for better control	10	2	3	60
					Uneven expansion due to difference in materials	6	Reliability testing of assembly	4	240	Review material selection compatibility	Materials with same thermal expansion coefficeint selected		1	4	40
					Wrong material used	5	No incoming material inspection	10	500	Review supplier's QC and alternative suppliers	New supplier certifies material is correct; has SPC in place		1	1	10
	To provide smooth, controlled fluid flow for drinking	Air vent or fluid opening blocked or insuf-ficient	Poor fluid flow	4	Mold flashing in hole	6	SPC on cover dimensions	3	72	Analyze molding process and equipment	Molding cavity and process refined with minimal flashing; added mesh-ing removal step as needed	4	2	1	8

Table 5-14. (continued)

Item	Function	Failure Mode	Failure Effect	SR	Failure Mode Cause	OR	Design Verification	DR	RPN	Recommended Action	Actions/ Results	SR	OR	DR	RPN
					Too small by design	1	Calculation and proto-type testing	2	8	No action	No action		1	2	8
	To slow fluid heat loss	Does not retain heat ef-fectively	Fluid cools too quickly	5	Poor material design	3	Reliability testing of assembly	4	60	Review material selection compatibility	Material with acceptable thermal insulation properties	5	2	4	40
					Wrong material used	5	No incoming material inspection	10	260	Review supplier's QC and alternative suppliers	New supplier certifies material is correct; has SPC in place		1	1	5
					Fluid opening and air vent too large	1	Calculation and proto-type testing	2	10	No action	No action		1	2	10

SR = initial severity ranking
OR = initial occurrence ranking
DR = initial detection ranking
RPN = initial risk priority number

data, other types of analyses, such as root cause analysis (RCA) or fault tree analysis (FTA), or from a combination of these and many other sources. Results of this effort are then fed back into the FMEA worksheet in the failure mode cause column. Each of these causes is then rated based on the frequency or probability of occurrence.

In Table 5-14, three failure mode causes are listed twice. Each cause is repeated next to each of the two failure mode effects. This is done for calculation purposes, due to the two effects having different severity ratings. In this particular case, the failure has two possible causes that are design related—one being fits and tolerances—the other material selection. The third cause is process related—the wrong material being used in the cover molding. The occurrence ranking (OR) ratings are based on the knowledge available with respect to frequency of the particular causes.

Design Verification

The ability of the design or process under scrutiny to uncover failure causes prior to the product reaching the end user is addressed under the column "design verification" (see Table 5-14). In this column, the actions, procedures, or checks in place to catch any errors are recorded. These are then ranked according to their ability to successfully detect the flaws in question. At this point, the rankings are multiplied to yield the initial risk priority numbers (RPNs). The failure mode causes with the highest RPNs can then be focused on; improvements in these areas will provide the greatest payback.

The four highest-ranking RPN issues shown in Table 5-14 involve an inappropriate material being used. An action plan was devised to review the supplier's quality and to research other suppliers. For any other issues that were thought to be within control of the design or process, action plans were also developed and addressed in order of the assigned priorities.

Action and Results

The results of the review regarding wrong material were that this was a process problem traced to a combination of a poor supplier and lack of inspection of material upon receiving it. The company changed to another supplier with good quality control methods in place and installed someone who would certify the material on each shipment. With these changes in place, along with additional changes from the other efforts, RPNs were all recalculated. Nothing changed the severity rankings of any of the failure modes, because none of the failure modes were eliminated. However, with the changes described in place, it was determined that the occurrence and detection rankings for some failures were significantly reduced. This resulted in a considerable decrease in the new RPN rankings for wrong material, dropping them to the bottom of the list, essentially implying that the failure mode in question was no longer a concern.

Once the process has been completed for all failure mode causes and each RPN has been recalculated, the process may be repeated until all RPNs are at an acceptable level, or the analysis may be considered complete and documented for future reference. What constitutes an acceptable level is solely the determination of those conducting the analysis.

FMEA can be a highly effective tool for reliability assurance. As seen in this example, it is extremely efficient when applied to analysis of elements that cause a failure of the entire system. Flexibility allowed by the FMEA format makes it especially easy to use in a wide variety of situations. The drawback of this analysis, however, is that it quickly becomes very cumbersome when applied to complex systems with many components or multiple operating modes.

ROOT CAUSE ANALYSIS (RCA)

Root cause analysis (RCA) uses a logical approach to eliminating problems, whether they are defects, process errors, or any other undesired event. RCA aims at finding a solution that prevents the problem from ever occurring again. It requires a persistent, patient, and open-minded search for the initial cause that created the problem, and not acceptance of a perceived cause that is in fact only a symptom, or effect of an earlier cause.

RCA is based on the belief that treating the immediately apparent cause as the only solution to a problem will permit the problem to persist. The driving force is the idea that the cost of solving a problem increases by an order of magnitude for each process it is allowed to proceed through, without correction, beyond where the problem initiated.

RCA is used by a wide spectrum of industries besides manufacturing and engineering. It is used extensively as a risk management tool in the business and medical fields, by the legal profession and insurance companies, and in the environmental and nuclear fields. RCA is deeply ingrained in quality management systems as well. The reality is that RCA can be used in virtually any context where there is a desire to prevent recurrence of a problem. In many industries, human safety and/or financial impact are usually the primary motivators. In the manufacturing industry, major RCA benefits come from reduction of scrap, rework, repairs, and warranty expenses.

Root cause analysis involves the persistent pursuit of the initial controllable event in a chain of events, which leads to a defect or failure and, when corrected, prevents recurrence of the same defect or failure. Key to this definition is the word "controllable." Circumstances are often identified when performing an RCA that are, for one reason or another, beyond a company's ability to control or correct. Acts of nature, pre-existing conditions, or even circumstances that are impractical to consider changing are types of uncontrollable root causes. In the case of an uncontrollable event, there are often additional steps that can be taken to reduce the severity or likelihood of the event causing a failure.

RCA is primarily considered a reactive or after-the-fact tool. In other words, a failure has occurred, and RCA's purpose is to find out why. When used in this manner, RCA often addresses only those failures considered significant, usually instances involving major property or personal damage. There is a problem of missed opportunity with this approach. This is only a narrow perspective of what RCA can do. When coupled with other tools, RCA can improve a company's profitability by identifying the root causes of problems that have commonly been accepted as the cost of doing business.

For instance, if a piece of manufacturing equipment requires resetting or adjustment five or six times a day, and each time it takes 10–20 minutes to perform the operation, it may go unnoticed because each individual occurrence seems like a minor inconvenience. But when the number of occurrences compounds over a year's production, it can add up to a significant cost. Using RCA on these types of events can increase a company's profits considerably by reducing downtime. Another benefit that can be gained from the routine use of RCA is tracking the trends of the analyses.

By recording and trending the types of root causes found, seemingly unrelated failures can point to a common theme. For example, failure in manufacturing a part, filling out required forms, performing a maintenance task, and shipping a product could all lead to the root cause of poor process instructions. This would highlight the need for more focus on creating clear, understandable documentation for these and other tasks.

Regardless of the field or industry, RCA operates on the premise that there are no such things as isolated incidents and that every incident is an indicator of a prior problem. Each incident is viewed as the tip of the iceberg; there are many underlying circumstances yet to be known and understood. Some refer to RCA as making order out of chaos, and assert that everything happens for a reason and usually many errors occur before an undesirable outcome happens. Failures are often regarded not as problems that take away from the bottom line, but as opportunities to improve quality and profitability.

There are numerous ways to undertake an RCA, and there are many tools available to assist in logically proceeding from a failure event to its root cause. Table 5-15 lists some of the more common tools and categorizes them into the basic steps of the analysis. The effectiveness, thoroughness, flexibility, and ease of use of these tools are the main selection criteria for choosing which one to use. Many of these tools come complete with brainstorming exercises, forms, and even final report formats.

RCA Steps

There are six basic steps to performing a successful root cause analysis. The steps are as follows:

1. Identify and define the problem.
2. Preserve and collect data.
3. Analyze the data.
4. Identify and verify the root causes.
5. Communicate findings and recommendations.
6. Implement corrective actions.

Identify and Define the Problem

A clear and complete statement of the precise nature and scope of the failure event defines the focus of the analysis. It is especially important when working in teams that everyone begins with the same understanding of what occurred and where the focus lies.

Preserve and Collect Data

The single most important step of the entire analysis is preserving and collecting data. If a poor job is done, recovery is nearly impossible. By contrast, if a good job of data collection is done, but analysis later proves faulty, a new analysis can be performed on the same collected data. Usually, a failure

Table 5-15. RCA tools and techniques

Generating Ideas
Brainstorming
Force-field analysis
Team forming
Five whys
Prioritizing Data or Action
Histogram
Pareto charts
Solution selection diagram
Nominal group techniques
Action Planning
Storyboarding
Solution selection diagram
Grouping techniques
Nominal group techniques
Finding Patterns and Relationships
Cause-and-effect diagram
Scatter diagram
Failure modes and effects analysis
Event tree analysis
Force-field analysis
Guide data collection
Statistical methods
Storyboarding
Function analysis
Process analysis
Examining Results
Five whys
Root cause test

event will leave parts, data, paperwork, or some other type of physical evidence behind. It is crucial to preserve these items at the onset of an investigation or risk its being lost forever. Interviews with the people involved, review of the procedures used, and reviews of the site of the failure are key elements of good data collection.

Analyze the Data

The type and importance of the failure, as well as the level of effort required for the analysis will determine the specific analysis tools used. If the example considered is that of a hydraulic power transmission system, the cause-and-effect diagram, also known as the fishbone diagram, will be used. What is critical to success of the analysis is to keep digging until the root cause is found. Anything less than the root cause is just a symptom.

Identify and Verify Root Cause

Once the root cause has been identified, it should be tested and verified. When safety and cost allows, this could mean recreating the failure. When safety and/or cost are an issue, modeling, calculation, or other similar means may verify the root cause.

Communicate Findings and Recommendations

The findings and recommendations must be communicated to those in positions of power to facilitate informed decisions. It is important to be clear, concise, and make recommendations that are specific to eliminate the root cause from recurring. General recommendations such as, "operator should use caution when performing this task" should be avoided, as they are ambiguous and impossible to confirm or enforce. A more useful recommendation would be, "operator should confirm that locks are engaged and safety screen is in place before performing this task." Recommendations should relate directly to the root cause, be actionable, verifiable, and should not create unacceptable risks.

Implement Corrective Actions

It may or may not be within the authority of the RCA team to implement corrective actions, which is why it is important to build a strong case for the recommendations made. Occasionally, there will be circumstances where the cost of the recommendation will outweigh the cost of living with the failure event. This becomes a business decision and, hopefully, the cost of prevention will never outweigh the cost of human life or injury.

PRODUCT MODELING USING CAD/CAM

Computer-aided design (CAD) and manufacture is one of the fastest growing areas in engineering. The words computer-aided design and drafting (CADD) are used to emphasize the drafting task. The use of computers to extend the application to manufacturing resulted in computer-aided manufacturing (CAM). The slash between CAD and CAM in CAD/CAM is intended to reinforce the shared functions of design and manufacturing using a common database. Significant technological advances have been occurring in computer-aided design and computer-aided manufacturing, which has resulted in productivity increases. CAD/CAM systems help the designer to design a product by using the speed and efficiency of a computer. The first commercial computer-aided system emerged in the 1960s, when companies like GM, Boeing, and Lockheed developed mainframe computer-based design systems. Until this time, design of a complex part or device was very time consuming. Design iterations and performance testing were impossible to complete on time and under budget (Chryssolouris 1992).

In the automotive and aerospace industries, the design phase represents the most critical aspect of a project. The final product has to be cost effectively built to strict specifications. However, because of the high cost of early CAD systems, only large engineering firms afforded their use until the introduction of mini-computers in the 1970s. In recent years, the introduction of personal computers, workstations (intelligent terminals), and sophisticated software has made CAD available to a broader spectrum of users. Design and engineering personnel are able to use stand-alone systems with a large choice of input/output devices without the expense associated with mainframes. Often the stand-alone systems are interfaced to a central computer, thereby increasing the storage capabilities for CAD software. In addition, such systems improve speed and performance, and result in greater overall productivity. Today, CAD/CAM is used for everything from air-traffic control systems to weapons design and research, and from circuit design to computerized photography. CAD/CAM systems are not only used in electronic design, but in areas such as architecture, civil engineering, and aerospace engineering.

There are different kinds of CAD developers in the marketplace. First, there are subsidiary divisions that sell CAD/CAM technology to different departments or divisions of a larger corporation. The second group is established turnkey CAD vendors. These companies specialize and offer a wide range of CAD systems. Having been involved in CAD/CAM technology for several years, these companies have established reputations in development of new technology. The third are entrepreneurial CAD system developers. These companies tend to be small, young, and very innovative. Although their market share is small, they excel in providing single, high-quality products to a narrow market segment. The entrepreneurial companies often deal with personal computer systems that are useful to customers with specialized design and automation needs. The last group consists of service bureaus. These organizations specialize in performing CAD/CAM work for other companies that have minimal or periodic needs. The number of service bureaus is growing, as they are often the best solution for companies that do not have their own CAD/CAM systems.

CAD Systems

CAD, as used in mechanical design and manufacturing, allows the choice of using two- or three-dimensional (2D or 3D) procedures to create the design. The user inputs 2D drawings of the intended product model, usually needing more than one view to describe the intended 3D product. This means re-entering the same intrinsic data, originally and to accommodate downstream changes. Mimicking the drafting board, designers use simple 2D information on geometry to construct points, lines, circles, arcs, and possibly conic sections together with some type of free-form curves. Two-dimensional systems rarely distinguish a model geometry line from a drafting construction line.

Selecting 3D CAD opens additional possibilities: 3D wire frame, solids modeling, surface modeling, parametric solids modeling, variational modeling, feature-based modeling, or a mix. To create 3D geometry, many systems use 2D views with the third dimension controlled perpendicular into and out of the screen. A given product's lack of any one of the foundation technologies may limit design choices.

3D Wireframe

Building geometry using 3D-wire frame involves representation of the model using lines, arcs, and curves. Although this technology cannot interpret model data for specifications such as area or volume due to difficulties in representation, the user can still benefit from model interaction that closely resembles drafting.

Solids Modeling

Solids modeling is best implemented by defining 3D volumes, both the model's inner material and outer envelope. In solids modeling, the basic techniques are Boolean operations: union (add shape), subtract (remove shape), and intersect (combine shape). Benefits include calculating full mass properties, true cross sections, interference checking between models, realistic graphics, and moving to such applications as finite element analysis, generating numerically controlled machine codes, and stereolithography. New Internet-based applications—including viewing, marking up, and technical documentation—require at least a data-viewing mechanism, which solids modeling provides.

Surface Modeling

Surface modeling deals with defining a product model's outer skin, not its interior. Typically, surface modeling products exist as an adjunct to either wire frame or solids modeling, except for high-end styling products. Moving beyond simplistic ruled surfaces, free-form surface creation allows aesthetic shapes for products such as those used in automotive and aerospace projects.

Parametric Solids Modeling

Parametric solids modeling has its origin in design modeling, which uses the principles of associativity. The process connects an internal programming link from a piece of 3D-model geometry to its 2D drawing representation. With this link, drawings reflect 3D model changes. Actual 3D model changes require explicit model creation and editing techniques, which in turn require the user to directly alter geometric shapes by adding or subtracting other 3D geometry. Parametric systems can represent 3D model dimensions, lengths, or angles as drivers controlling model shape. If you change a driving dimension value, the whole 3D-model shape updates. Non-driving dimensions found only on model drawings merely reflect a resulting shape measurement. Most parametric systems are restricted to solids geometry. Surface models and their mathematical procedures do not offer representations of parametric values, though some products claim limited capabilities. Some parametric modeling products help the user define and remember non-dimensional geometric constraints, such as parallel and perpendicular. The additional features of recognizing geometric constraints add to the overall ease of use while capturing important design considerations. Other hybrid modeling systems allow users to mix explicit and parametric modeling conditions in one part model.

Variational Modeling

Variational modeling covers an important subset of parametric modeling, wherein dimensional and geometric constraints need not follow a sequential definition with one leading to the next. A variational option in parametric modeling lets users add dimensional and geometric constraints randomly.

Feature-based Modeling

Feature-based modeling offers users a more familiar command language interaction using form feature constructs, usually parametrically defined, for such geometric objects as holes, bosses, and rounds. The user interface represents design terminology more closely, helping streamline model creation.

CAD/CAM Systems

Evaluating CAD/CAM software can involve extensive work since inadequacies, inaccuracies, and inconsistencies are very often revealed only after the system has been in use for some time. Therefore, potential users should have clear and reasonable expectations for minimum performance requirements of software function and reliability.

CAD/CAM software can, in general, be divided into three categories: operating systems, graphics software, and applications as shown in Figure 5-10.

The operating system controls the lowest level of system operations. It is divided into three categories:

1. System controls, which include setting job priorities and supervising CPU time;
2. Processing tasks, which include systems utilities such as language translation; and
3. Data management, which controls the organization and access to all on-line data.

The graphics system is that part of the CAD/CAM software with which the user interacts directly. This group of programs creates and manipulates the graphics data on which application of the system depends.

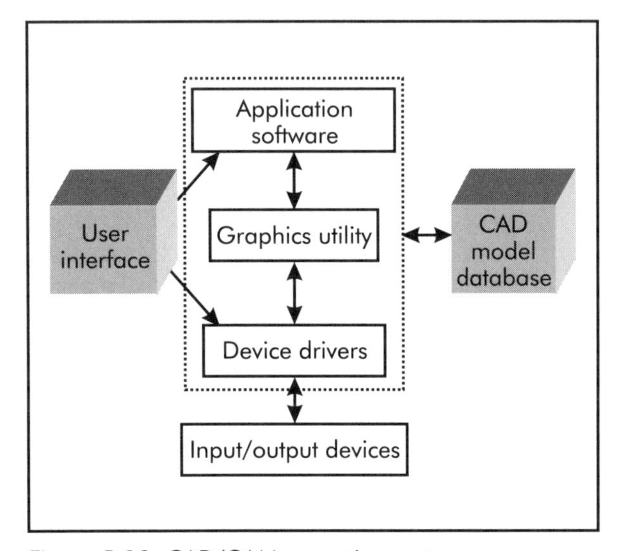

Figure 5-10. CAD/CAM operation system.

Requirements of a good graphics system are to provide a complete set of graphics tools and make those tools easily accessible to the user.

Graphics systems are either menu-driven and/or command-driven. Menu-driven systems are easier for the novice to learn. However, these systems tend to be slower than command-driven systems, especially if each menu is entered through the keyboard rather than through a mouse or light pen. More efficient menu-driven systems are equipped with different levels of menu operation (for example, novice to expert level). Command-driven systems also include a complete set of training aids. Some systems include help and error messages along with complete documentation. Recent systems make extensive use of icons to make on-screen display symbols. Another improvement in the graphics system is the use of solid modeling as opposed to wire-frame modeling.

Wire-frame modeling is completely adequate for two-dimensional representations. For three-dimensional objects modeling, it is rather limited and sometimes produces confusing views, which can be interpreted in many different ways. This problem is overcome if the wire-frame graphics system includes automatic hidden line removal. In solids modeling, the models are displayed as solid objects to the viewer, involving more realism, especially when color is added. Solids modeling requires more computation, which places demands on system memory and speed.

Application software can stand by itself as an independent component or act as an organized set of independent tools, each fulfilling a portion of the overall integrated production plan. Each application is designed to meet the CAD/CAM needs of a defined industry.

In an integrated CAD/CAM system, additional features need to be considered. Design analysis is an important task that can be handled automatically by CAD systems. Modeling is a very powerful tool for conducting product simulation. By use of primitive shapes such as triangles, cones, circles, and rectangles, most models can be constructed and manipulated by the system automatically using features such as shading, scaling, skinning, rotating, cross-section, and merging to emulate virtually any design idea.

Solid modeling results in a complete solid object, consisting of mathematical representations, which can be studied as though it were an actual manufactured object. The designer is able to simulate the product model's physical characteristics and view realistic pictures that can be used in creating exploded views, part details, and manufacturing drawings. Also, solids modeling allows computation of physical factors such as weight, density, center of gravity, volume, and sensitivity to heat and stress, among many other factors. An extremely important engineering objective facilitated by modeling is analysis by the finite-element method. Individual parts are broken up into discrete physical elements, which can be analyzed independently with regard to required stresses and displacements. This process helps to optimize the design and keep part failure to a minimum.

Environment

The engineering design process has been traditionally carried out on drafting tables. The design process consists of arrival at the final shape and size of the desired product, design of its relevant components, preparation of component drawings, and final assembly layout. Similar procedures are followed in structural design, chemical plant design, electrical component design, aerospace and automotive design, etc. As an individual component is designed, its physical dimensions are identified, and its physical properties and characteristics are specified.

In the design process, once the design goal is fixed, a certain know-how is needed to go through the design methodology. The process of designing produces information that can be documented and used in production. Quite often, the environment influences design activity, and the design process is often considered an integral part of a higher-level process. Design is an iterative process that includes steps such as identification of need, problem definition, problem synthesis, analysis and optimization, evaluation, and the presentation of results. It is a series of evolutionary steps along which the designer proceeds from recognition of needs to desires to be fulfilled. Certain product characteristics are determined heuristically at an early stage on the basis of incomplete knowledge about their effect on the design. Synthesis is an attempt to refine the model in such a way that subsequent analysis may produce better results.

In a typical engineering application, a preliminary design is manually synthesized and then subjected to some form of analysis. This analysis procedure involves extensive calculation and human judgment. CAD is a standard tool in many design offices and in operations such as geometrical modeling, engineering analysis, design review, and evaluation. Automated drafting is efficiently performed by the modern computer-aided design-and-drafting system. A graphics-assisted design system, which can present proper tools for image creation and manipulation, can make a designer much more productive. Figure 5-11 shows the progress of the design activity in a computer-aided design environment. A computer-aided design system can be executed through either a large-scale central computer on batch mode or a dedicated microcomputer-based CAD system in interactive mode. The interactive graphics system provides an immediate response to inputs by the user. The system and designer

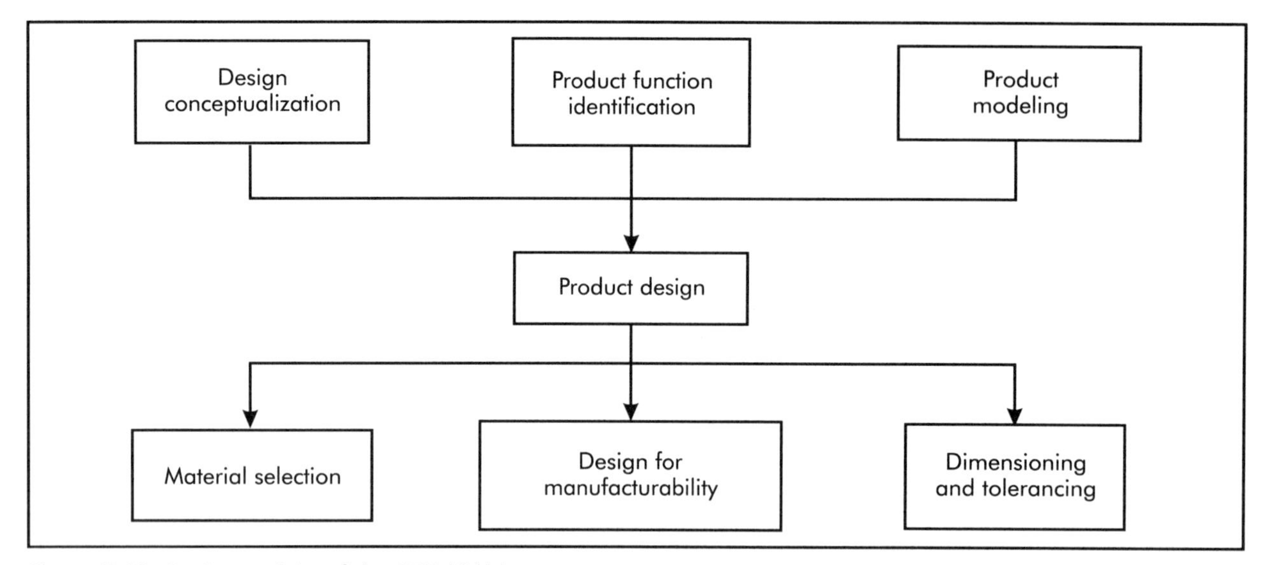

Figure 5-11. Design activity of the CAD/CAM system.

are in direct communication with each other, and the designer uses commands and responds to questions posed by the computer system. CAD systems also handle non-graphic applications such as display of engineering data (Zeid 1991).

Modeling System

One of the issues concerning any application of computer modeling is the representation of geometric data. Most two-dimensional wire-frame models are created from graphical primitives such as points, lines, arcs, and curves used in manual drafting and layout. Two-dimensional wire-frame models, however, have many of the disadvantages of a two-dimensional drawing procedure. At times, they are ambiguous and difficult to interpret and can be inadequate for complicated shapes. The primary limitation is that all lines that define edges (and contoured surfaces) of the model are displayed. If the three-dimensional system has no provision to eliminate hidden lines, the lines that define the edges at the rear of the model show right through the foreground surfaces. It is possible to construct non-imaginable parts using wire-frame models, requiring human intervention to resolve the ambiguity.

A solid model can describe a part completely. The physical attributes of the object consist of solid primitives, such as blocks, cylinders, cones, spheres, and edges. These shapes are considered basic geometric representation. The primitives are combined through the use of Boolean operators of union, difference, and intersection. A tree-like data structure describes how an object is built up from simpler objects. Using these descriptors, the part can be specified as a combination of shapes or features oriented in a specific way.

Boundary representation schemes require the designer to draw the boundary of the object, so a relatively solid image can be developed. Some transformations and editing features are used to modify the model to a desired shape. The boundary representation is especially useful when some unusual shapes are to be represented, which are difficult to create by use of solid modeling techniques. During manufacturing, each part is made from representative part drawings, and assembly drawings are used to create the whole structure. The shape models are classified into surfaces; assembly part shapes define whole shapes. The major requirement of a modeling system for design application is to be able to build and contain unambiguous representations of parts and assemblies. CAD systems currently available are capable of generating three-dimensional models of different parts. The software is available for personal computers and workstations. These programs can generate the pictorial view of the part as an isometric or oblique drawing. They have the additional ability to interact with the manufacturing environment.

Design-related Tasks

As products become complicated and technology advances, the knowledge level expected from the designer increases tremendously. A designer should be able to examine and manipulate the product on a graphical display, interact with the database, and calculate and modify ideas in a short time. This aspect nearly justifies adaptation of CAD systems. CAD systems increase the productivity of the design engineer by helping the designer to visualize the product, component, and subassemblies of the part, thereby reducing the time required for synthesis, analysis, and documentation. CAD systems improve the quality of the design by facilitating explanation of design alternatives and analysis, thereby reducing design errors. CAD systems improve the communication link between the design and designer. They also create a database for manufacturing tasks.

Design-related tasks assigned to CAD systems are:

- geometrical modeling;
- engineering analysis and optimization;
- review of the design and further evaluation;
- synthesis;
- presentation and automated drafting; and
- manufacturing interface.

Handling geometric information about the product is not the sole purpose of CAD systems. Generating the physical form is just one type of information needed in overall CAD integration in engineering. As an example, aerospace designers have to work with the behavior of aircraft wings. The design aspect involves not only geometric data on aircraft wings, but other details such as material properties, material volume, welding behavior of the structure, optimum layout of materials for minimum scrap, etc. Geometric data on the aircraft part can be used in fatigue and crack analysis, aerodynamic computation, etc. Further, the geometric model can be used in an integrated manner with the machining and assembly processes in manufacturing. Successful CAD systems are those that give designers the flexibility to use knowledge anywhere in the design, manufacturing, and planning stages of an operation without loss of information. The current trend is to develop intelligent CAD systems that create computer-based models for conceptual design.

Geometrical modeling is concerned with a computer-compatible mathematical description of objects. Designers convert a graphic image on the screen by three input types. The first set of commands is: points, lines, circles, etc.; the second set: scale, rotation, transformation, etc.; the third set causes various elements to unite into a desired shape. The geometric model involves representation in 2D/3D wire frame, solid model, with hidden line removal, etc.

In any design project, some engineering analysis is required. Analysis may require stress/strain calculations, heat transfer computation, analysis of properties, optimization, etc. Design review and evaluation is facilitated by the ability of the CAD system to perform tasks like automatic dimensioning, tolerancing (designer can zoom into a part), interference, and viewing from various angles. Calculation of materials and volume is also performed in CAD. The automated drafting process includes the ability to print a hardcopy of the drawing from CAD and the creation of a manufacturing database (see Figure 5-12).

CAD systems are used with the intention of:

- increasing the productivity of the designer, which is accomplished by helping the designer to visualize the product, component, and subassemblies of the part, thus reducing the time required for synthesis, analysis, and documentation;
- improving the quality of the design through analysis and a number of design alternatives to reduce design errors;
- improving communication; and
- creating a database for manufacturing (see Figures 5-13 and 5-14).

Feature-based Design and Modeling

Feature-based design and modeling is the process of creating a product model on the basis of features and how they are related. This idea was developed to enhance the productivity of designers who use solid modeling systems. The procedure involves considering the design attributes from the point of view of how the product gets fabricated. Feature-based design and modeling systems have the ability to group entities into form features such as ribs, bosses, flanges, and pockets. This way, the designer is not required to specify the individual primitives needed for a complex product.

The data available in geometric models is of microscopic nature. Lines, curves, and solids represent the general models. Boolean operations are used in constructive solid geometry (CSG) and solid primitive representations. In boundary representation, the models are represented in terms of edges, faces, and vertices. In the conventional geometrical modeling system, there is no macroscopic information or design intent. Parametric modeling systems are a slight improvement. They are dimension-driven models. That is, the geometry is defined using parameters or variables to specify the dimension of an entity. Mathematical relationships between the variables and entities are important. The variable values are solved sequentially or by using simultaneous equations.

A *feature* represents the engineering meaning or significance of the geometry of

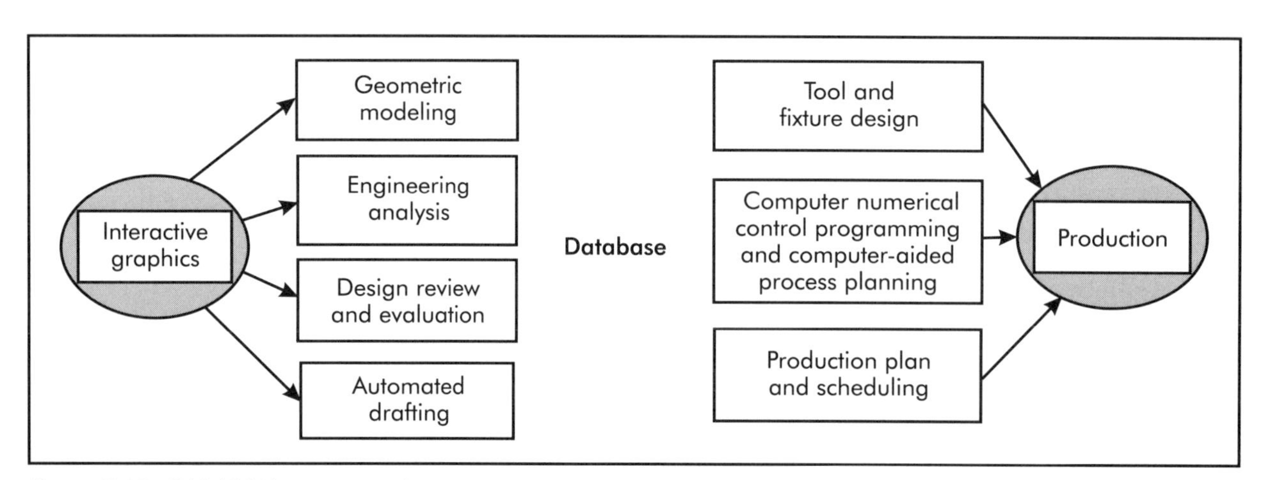

Figure 5-12. CAD/CAM representation.

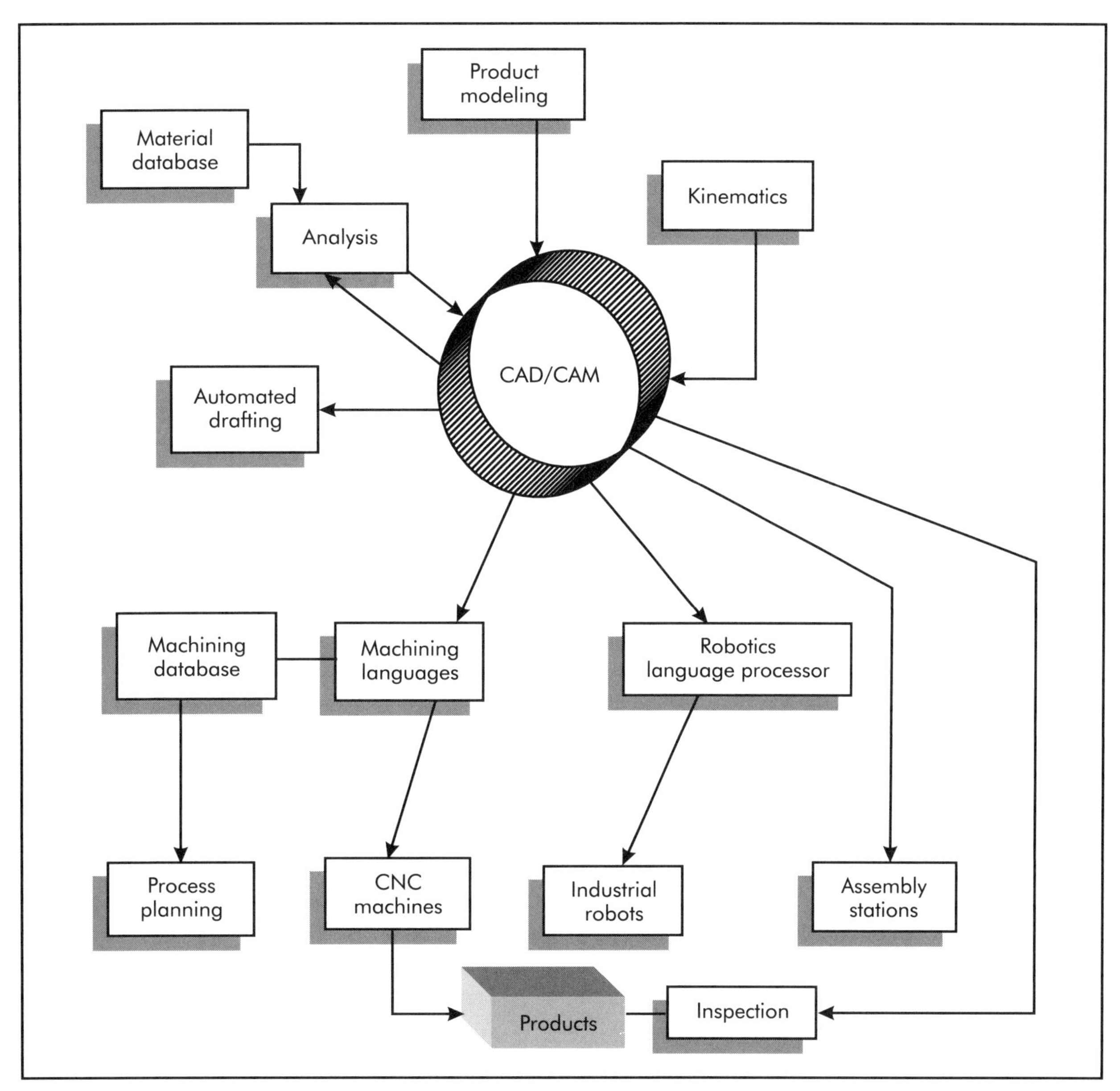

Figure 5-13. The CAD/CAM production system.

a part or assembly. A feature is any perceived geometric or functional element, or property of an object useful to understand its function, behavior, or performance. A feature can be mapped to a generic shape and has engineering significance. It has predictable properties. Feature-based design and modeling systems allow users to define their own sets of form features. A feature model is a data structure that represents a part or an assembly in terms of its component features. Features can be broken down into form features, tolerance features, assembly features, functional features, and material features.

Form features are portions of a part's geometry that keep happening again and

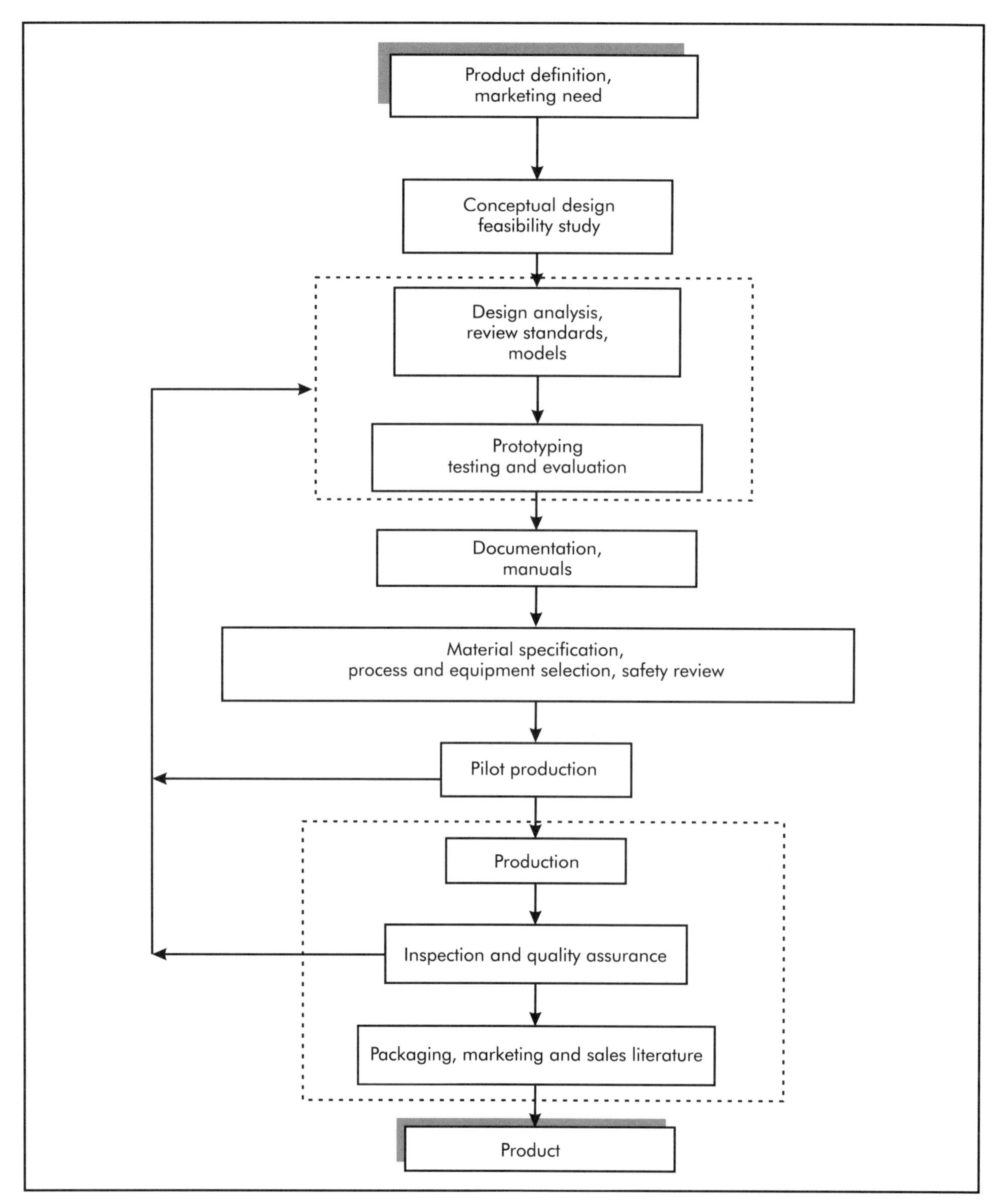

Figure 5-14. Product concept to realization.

again. Tolerance features are deviations from the norm; assembly features represent a grouping of some feature types to define assembly relations; functional features are related to specific function or performance; and material features are related to the composition and condition of materials.

REFERENCES

Chang, T., Wysk, R., and Wang, H. 1991. *Computer-aided Manufacturing*. Englewood Cliffs, NJ: Prentice Hall.

Chicoine, Troy, 2000. "Failure Mode and Effects Analysis of Hydraulic Elevator Power Transmission Systems." Hartford, CT: University of Hartford. Masters Thesis, May.

Chryssolouris, George. 1992. *Manufacturing Systems—Theory and Practice*. New York: Springer-Verlag.

Dieter, George. 2000. *Engineering Design: A Materials and Processing Approach*, 3rd Ed. New York: McGraw-Hill, Inc.

Groover, Mikell P. 1980. *Automation, Production Systems, and Computer-aided Manufacturing*. Englewood Cliffs, NJ: Prentice-Hall, Inc.

Imai, Masaaki. 1986. *Kaizen: The Key to Japan's Competitive Success*. New York: McGraw-Hill, Inc.

Padke, Madhav. 1989. *Quality Engineering: Using Robust Design*, AT&T Bell Laboratories. Englewood Cliffs, NJ: Prentice-Hall, Inc.

Schonberger, Richard. 1986. *World Class Manufacturing: The Lessons of Simplicity Applied*. New York: The Free Press.

——. 1982. *Japanese Manufacturing Techniques: Nine Hidden Lessons in Simplicity*. New York: The Free Press.

Shetty, Devdas and Neault, Henry. 1993. United States Patent. "Method and Apparatus for Surface Roughness Measurement Using Laser Diffraction Pattern." Patent Number: 5,189,490; Feb. 23, 1993.

Shetty, Devdas, et al. 1990. *New Experiments on Non-contact Inspection of Ground Turbine Blades*. New York: ASME.

Siddal, James N. 1990. *Expert Systems for Engineers*. New York: Marcel Dekker.

Suzaki, Kiyoshi. 1987. *The New Manufacturing Challenge: Techniques for Continuous Improvement*. New York: The Free Press.

Taguchi, G. 1993. *Taguchi on Robust Technology Development*. New York: ASME Press.

Tanner, J.P. 1985. *Manufacturing Engineering: An Introduction to the Basic Functions*. New York: Marcel Dekker.

Voland, George. 1999. *Engineering by Design*. Reading, MA: Addison Wesley.

Womack, James P. and Jones, Daniel T. 1996. *Lean Thinking: Banish Waste and Create Wealth in Your Corporation*. New York: Simon & Schuster.

Womack, James, Jones, Daniel T., and Roos, Daniel. 1990. *The Machine that Changed the World: The Story of Lean Production*. New York: Harper Perennial.

Zeid, Ibrahim. 1991. *CAD/CAM Theory and Practice*. New York: McGraw-Hill, Inc.

Chapter 6

Streamlining Product Creation

WORKPLACE DESIGN

The choice of a workplace structure depends on the design of parts to be manufactured, the lot sizes of parts, and market factors such as required responsiveness to market changes. In industrial practice, there are five general approaches to structuring the process area: the job shop, the project shop, the cellular layout, the flow line, and the continuous system approach.

Job-shop Layout

In a job-shop layout, as shown in Figure 6-1, machines with the same or similar material processing capabilities are grouped together. They are usually general-purpose machines, which can accommodate a large variety of part types. In this structure, the component moves through the system by visiting different work centers, according to the part's process plan. Material handling must be very flexible to accommodate many different part types, which is why it is usually done with manually controlled implements such as forklifts and handcarts. This is advantageous for a number of reasons (Chryssolouris 1992), including:

- Each operation can be assigned to a machine that yields the best quality or best production rate.
- Machines can be evenly loaded.
- The problem of machine breakdown can be addressed easily.

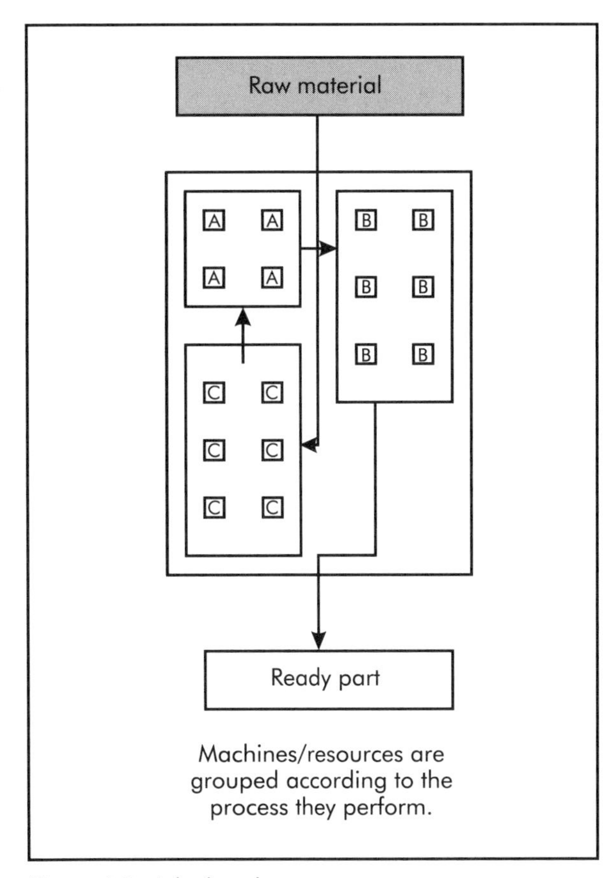

Figure 6-1. Job-shop layout.

Project-shop Layout

In a project shop (see Figure 6-2), a product's position remains fixed during manufacturing because of its size and/or weight. Materials, people, and machines are brought to the product as needed. Facilities organized as product shops can be found in the aircraft and shipbuilding industries and in bridge and building construction.

Cellular Layout

In manufacturing systems organized according to a cellular plan (see Figure 6-3), equipment or machinery is grouped according to the process combinations that occur in the family of parts. Each cell contains machines that can produce a certain family. The material flow within the cell may differ for different parts of a part family. Intracellular material flow takes place either automatically or manually.

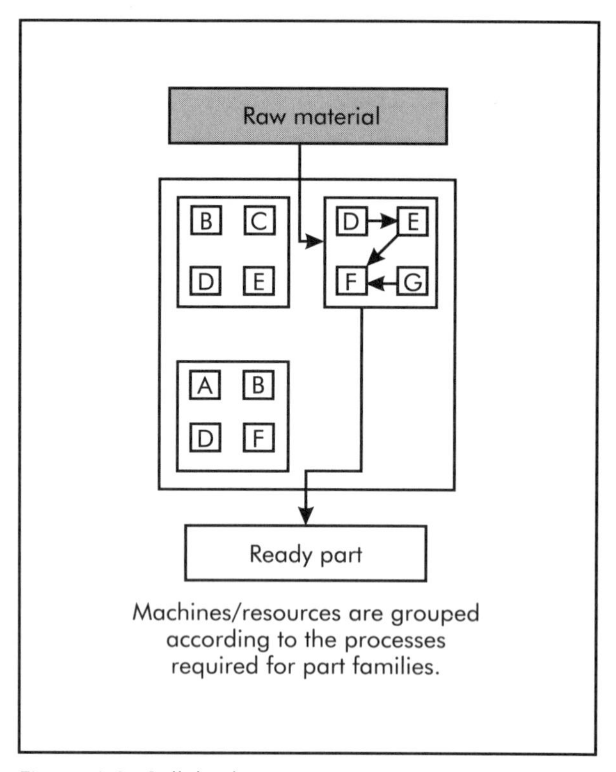

Figure 6-3. Cellular layout.

Flow Line

Another way of structuring a manufacturing system is to create a flow line in which machines and other equipment are ordered according to the process sequences of parts to be manufactured (see Figure 6-4). Typical examples of the flow-line manufacturing system are: a transfer line used in automobile assembly, a car wash, and a television set assembly line. A transfer line consists of a sequence of machines typically dedicated to one particular part, or at the most, a few very similar parts. Only one part type is produced at a time.

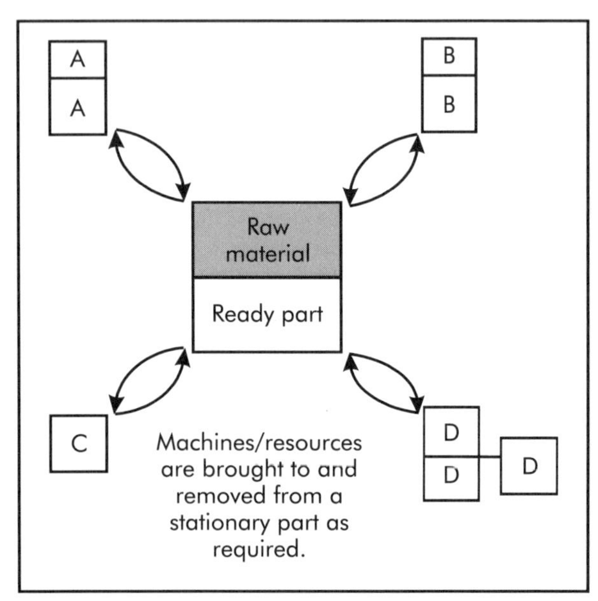

Figure 6-2. Project-shop layout.

Continuous System

In contrast to other types of systems that manufacture discrete parts, continuous systems (see Figure 6-5) produce liquids, gases, or powders. As in a flow line, processes are arranged according to the processing sequence of products. The continuous system is the least flexible type of manufacturing system.

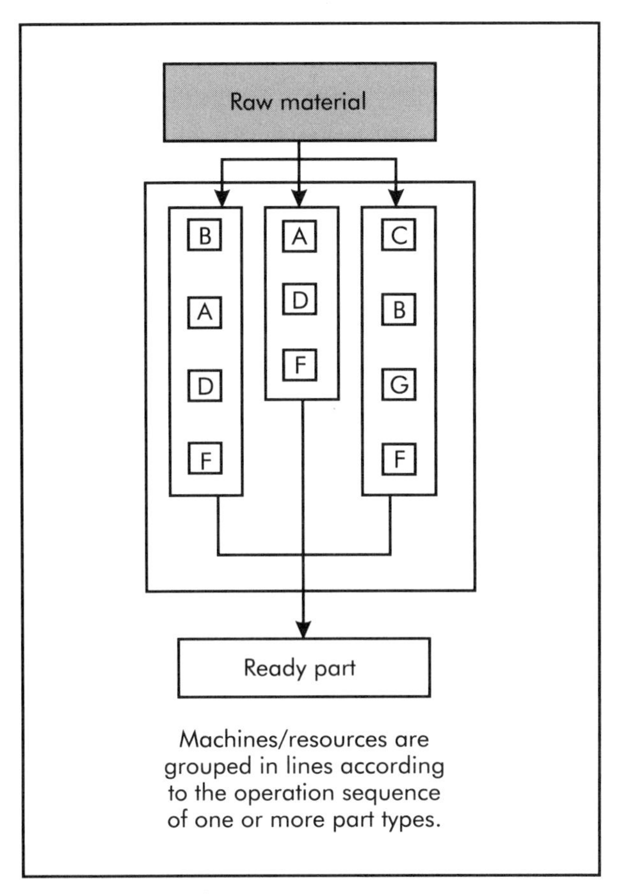

Figure 6-4. Flow line.

PRODUCTION CAPACITY MODEL

Equations can be developed to determine the number of workstations required on a production line. When designing a single model line to satisfy annual demand for a product, the required hourly rate of the line will be

$$R_P = \frac{D_a}{50SH} \qquad (6\text{-}1)$$

where:

R_P = actual average production rate (units/hr), assuming 50 work weeks/year
D_a = demand rate per week
S = number of shifts/week
H = number of hours/shift

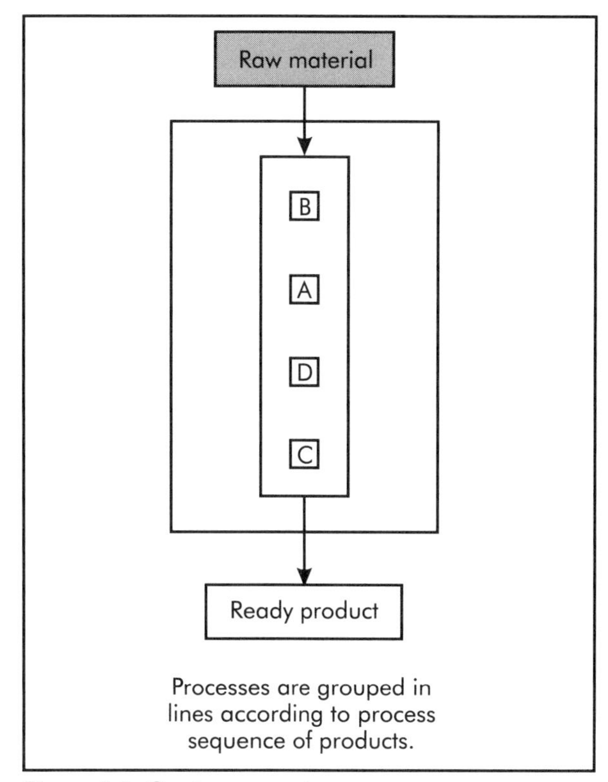

Figure 6-5. Continuous system.

The average production time, per minute (min/unit), T_P, is shown as,

$$T_P = \frac{60}{R_P} \qquad (6\text{-}2)$$

Production capacity is generally known as the maximum rate of output that a manufacturing facility is able to generate under normal operating conditions.

Production capacity, P_C, is expressed as,

$$P_C = WSHR_p \qquad (6\text{-}3)$$

where:

W = number of work centers

CHOOSING THE PRODUCTION METHOD

A production line consists of a series of workstations arranged so that the product moves from one station to the next. At each location, a portion of the total work is performed. The slowest station determines the line production rate. This bottleneck station limits workstations with a faster pace. Transfer of product along the line is usually accomplished by a mechanical transfer device or conveyor system, although some manual lines simply pass the product by hand between stations. Production lines are associated with mass production. If product quantities are very high and work can be divided into separate tasks assigned to individual workstations, then a production line is the most appropriate manufacturing system.

There can be many combinations of design, materials, and manufacturing, even without one best route from the design stage to inspection and sales. If one or even two of these three elements are fixed (for example, a fixed design with fixed materials), there may still be a number of acceptable alternative manufacturing routes. This may depend on the number of parts to be made, the availability and quality of raw materials, the number and capabilities of existing machines in the factory and, of course, the resulting cost per part.

The design of a product can be influenced by the process of manufacturing and by the production route it takes in a manufacturing facility. In some cases, the desired properties of a product can be attained through the use of a specific process. For example, the forging process is one of the most important methods of manufacturing items for high-performance uses. It changes the shape of a piece of material by exerting force on it. One of the characteristics associated with cold forging is that it increases the yield strength of the component.

Sometimes it may be nearly impossible to manufacture an article using particular materials due to production difficulties. Certain materials can cause severe problems if they are extremely difficult to machine or weld. Clearly, there must be some interplay between what the designer specifies and what certain processes can actually achieve economically. Specifications of tolerances and surface finishes required must be considered when different manufacturing methods are compared as alternative production methods. Linked to such considerations is the question of overall economics. Savings arising from one or two operations early in a production sequence may be lost altogether if expensive finishing operations are required later. The number of parts produced, design complexity, capital and labor costs, and parts per assembly play important roles when deciding production methods and assembly modes.

Job shops are more suitable for low-volume production of multiple, but dissimilar, part types. They possess general-purpose machines and a flexible, manual material handling system, which is ideal for this situation. Cellular systems are most suitable for manufacturing in low-to-medium production volumes and lot sizes of part types with enough similarity to be clustered into part families. Flow lines are best suited to the high-volume, high-lot-size production of a single-part type or few very similar part types. This is a consequence of having dedicated machines and material handling equipment.

FLEXIBILITY IN MANUFACTURING

Flexibility is a cornerstone and key concept used in the design of modern automation. It can be defined as a collection of properties of a manufacturing system that support changes in production activities or capabilities. The changes are due to both internal and external factors. Internal changes could be due to equipment breakdowns, software failures, worker absenteeism, and variability in processing times, etc. To absorb uncertainties due to product-design changes, the manufacturing system must be able to produce a variety of part types with minimal cost and lead times.

The types of production systems are:

1. high-volume, low-variety (H-L) production systems (transfer line);
2. stand-alone computer numerical control (CNC) machines; and
3. mid-volume, mid-variety production systems:
   a. manufacturing cell;
   b. special manufacturing system; and
   c. flexible manufacturing system.

Two extreme production situations are H-L and low-volume, high-variety (L-H). Between these two extremes, there is an important mid-volume, mid-variety (M-M) production situation. Figure 6-6 shows volume-variety relationships used to categorize production systems.

H-L Production Systems

An example of a H-L manufacturing system is a transfer line. It can also be referred to as a fixed automation manufacturing system, where dedicated processing and material-handling equipment are used. The parts produced are limited to one or two varieties. Some of the features are:

- Machines are dedicated to the manufacture of one or two product types; this system permits no flexibility at all.
- There is maximum utilization and a very high production volume.
- Direct labor involvement is minimal.
- There is a low unit cost for production.

Stand-alone CNC Systems

Stand-alone CNC machines can produce a variety of parts. Processing requirements of parts should be within a machine's capability. An L-H production system normally consists of a CNC machine augmented by a part buffer, a tool changer, a pallet changer, etc. Such an augmented system is also known as a *flexible manufacturing module*. Some of the features are:

- It has the highest level of flexibility; any job can be processed, provided it is within the process capabilities of the CNC machine.
- There is low utilization and production volume.
- The unit cost of production is much higher than for similar products manufactured on a transfer line.

Mid-volume, Mid-variety Production Systems

Between the extremes of one or two part types produced on a transfer line and a large variety of parts produced on a stand-alone machine, there is an important category of mid-volume, mid-variety parts, which constitute approximately 75% of discrete-parts manufacturing. The simultaneous requirements of flexibility and production volume place more emphasis on system integration and automation. Mid-volume, mid-variety parts systems are classified into the following types:

- manufacturing cells,
- special manufacturing systems, and
- flexible manufacturing systems (FMS).

Manufacturing Cell

The design of manufacturing cells is based on the concepts of group technology. The objective is to process some families of parts on a group of CNC machines within a cell so that intercellular material-handling effort is minimized. Selection of parts for processing on machines may be both sequential and random. In a typical cell, the CNC machines are often linked together by a direct numerical control (DNC) system. Some of the features are:

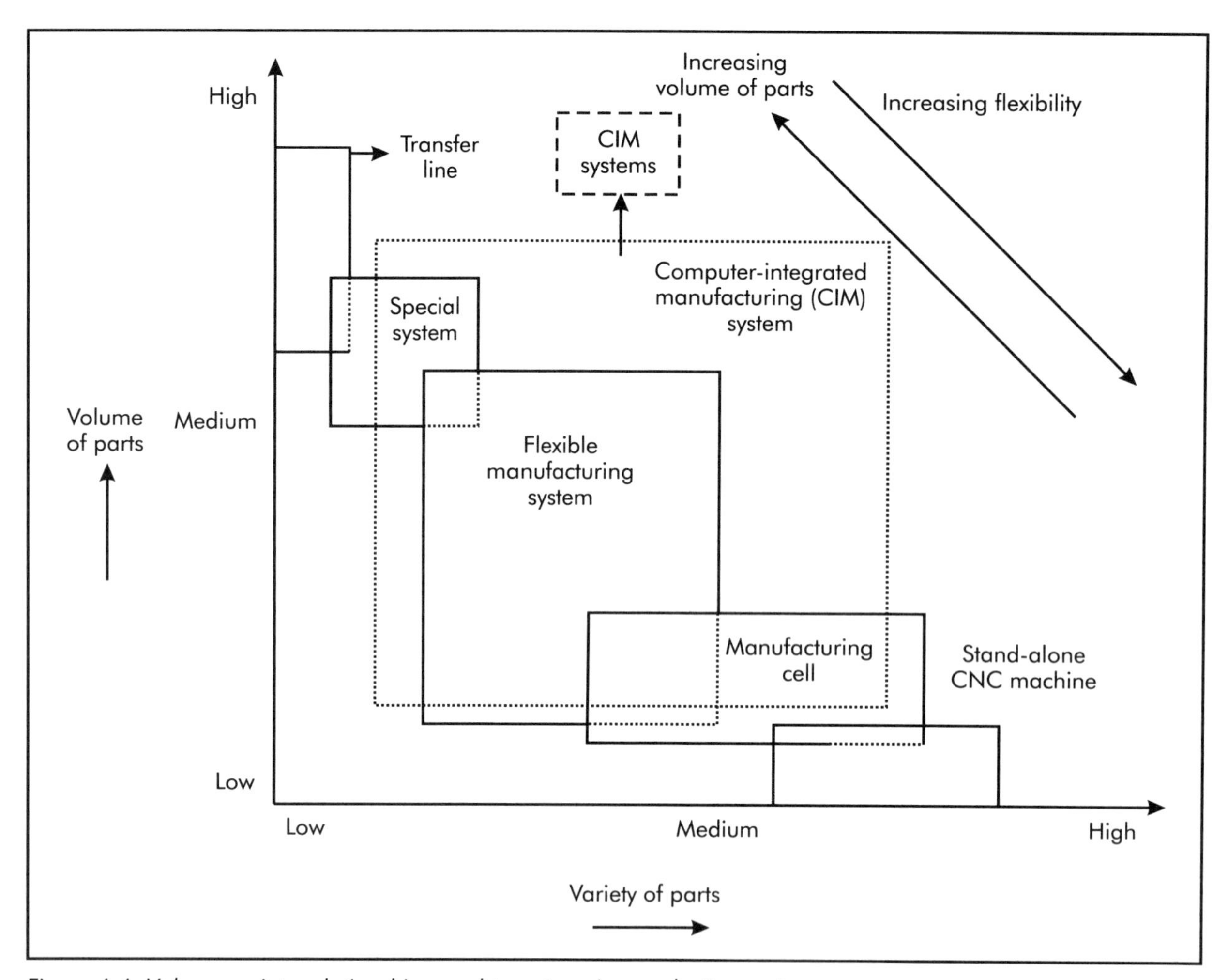

Figure 6-6. Volume-variety relationships used to categorize production systems.

- The system handles a low-to-mid volume of parts.
- A variety of parts is manufactured in batch mode.
- A manufacturing cell is a FMS without a central control.
- A cell is more flexible than a FMS, but has a lower production rate.

Special Manufacturing System

In a special manufacturing system, machines are laid out to manufacture a family of parts based on a sequence of operations. Parts move on the material-handling system in sequence from machine to machine. This system also has a high production rate. Some features are:

- A fixed path material-handling system links machines together.
- This is the least flexible category of computer-integrated manufacturing (CIM) system.
- This system uses multi-spindle heads and a low-level controller.
- This system has a high production rate and a low unit production cost.

Flexible Manufacturing System (FMS)

A flexible manufacturing system lies between the two extremes of a manufacturing cell and a special manufacturing system. It is a true mid-volume, mid-variety manufacturing system, having a higher production rate than a manufacturing cell and much more flexibility than a special manufacturing system. Some of the features are:

- A FMS is an automated, mid-volume, mid-variety, central computer-controlled manufacturing system. It covers a wide spectrum of manufacturing activities such as machining, sheet metalworking, welding, fabricating, and assembly. Families of parts with similar characteristics are processed in a FMS.
- Group technology (GT), and consequently cellular manufacturing, are significant parts of the system.
- A FMS consists of a series of flexible machines, an automated material-handling system, an automated tool changer, and other equipment such as coordinate measuring machines, part washers, etc., all under a high-level centralized computer control.
- The system permits both the sequential and random routing of a wide variety of parts.
- A FMS has a higher production rate than a manufacturing cell and much higher flexibility than a special manufacturing system.

Physical components of a FMS are:

- potentially independent NC machine tools capable of performing multiple functions and having automated tool-interchange capabilities;
- an automated material-handling system to move parts between machine tools and fixturing stations;
- components (machine tools, material-handling equipment, and tool changers) that are hierarchically computer controlled; and
- equipment such as coordinate measuring machines and part-washing devices.

A FMS consists of two subsystems:

1. a physical subsystem, and
2. a control subsystem.

The physical subsystem includes the following:

- workstations consisting of a NC machine tool, inspection equipment, part-washing devices, a loading and unloading area, and a working area;
- a storage retrieval system consisting of pallet stands at each workstation and other devices, such as carousels, used to store parts temporarily between the workstations or operations; and
- material-handling systems consisting of powered vehicles, towline carts, conveyors, automated guided vehicles (AGVs), and other systems to carry parts between the workstations.

The control subsystem required for optimum performance of a FMS includes:

- control hardware, including computers, programmable logic controllers, communication networks, sensors, switching devices, and many other peripheral devices such as printers and mass-storage memory equipment; and
- control software, consisting of a set of files and programs used to control physical subsystems.

The types of flexibility in a FMS include:

- *Machine flexibility* refers to the capability of a machine to perform a variety of operations on a variety of part types and sizes. It also represents the ease of parts changeover on a machine. The changeover time, which includes setup, tool changing, part-program transfer, and part-transfer times, is an important measure of machine flexibility. Computer numerical control (CNC) machining centers are normally equipped with an automatic tool changer, part-buffer storage, part programs, and fixtured parts on pallets. Machine flexibility also encompasses routing and mix flexibility.

- *Routing flexibility* means that a part or parts can be manufactured or assembled along alternative routes. Alternative routes are possible if manufacturing or assembly operations can be performed on alternative machines, in alternative sequences, or with alternative resources. They are used primarily to manage internal changes resulting from equipment breakdowns, tool breakage, controller failures, etc. Routing flexibility can help increase throughput in the presence of external changes such as product mix, engineering changes, or new product introductions. These changes could alter machine workloads and cause bottlenecks. Routing flexibility is one way to achieve mix flexibility.
- *Process flexibility*, also known as mix flexibility, refers to the ability to absorb changes in product mix by performing similar operations or producing similar products or parts on multipurpose, adaptable, CNC-machining centers.
- *Product flexibility* refers to the ability to change over to a new set of products economically and quickly in response to market or engineering changes, or even to operate on a make-to-order basis.
- *Production flexibility* refers to the ability to produce a range of products without adding major capital equipment, even though new tooling or other resources may be required. The product envelope is the range of products that can be produced by a manufacturing system at a moderate cost and time, which is determined by the process envelope.
- *Expansion flexibility* refers to the ability to change a manufacturing system with a view to accommodate a changed product envelope. In the case of production flexibility, there is no change in major capital investment. In the case of expansion flexibility, there are additions as well as replacements of equipment. These changes are easy to make because such provisions are made in the original manufacturing system design.

VALUE STREAM MAPPING℠

Value Stream Mapping℠ provides a clear view of the procedures involved in the manufacture of a product. It deals with all the value-added and non-value-added activities involved in production. The essential process flow includes movement of raw materials through delivery to the customer. Value Stream Mapping, used as a tool, provides several benefits to a process. The first main benefit is enabling people to see the processes as a whole, rather than just as a set of individualized processes. This also makes evident the sources of waste in the value stream. With a clear process flow and identified sources of waste, decisions needed to improve the flow are apparent. Another feature of Value Stream Mapping is it utilizes a format that provides a common language for the manufacturing process, which ties together lean concepts and techniques. Value Stream Mapping is also considered a tool that provides a link between information flow and material flow.

Comparison to Other Methodologies

There are a few techniques that are comparable to Value Stream Mapping (Cudney and Shetty 2000):

- value analysis,
- group technology,

- theory of constraints, and
- line balancing.

Value Analysis

Value analysis is a method to identify and remove unnecessary costs without compromising quality or reliability of design. It seeks to improve the relationship between the function of a component or product and its associated cost. The product or component is studied to determine a better design, material, or manufacturing method. This technique can be applied to a spectrum of problems, but is generally associated with material selection.

Value analysis asks several questions:

- How can a given function of a design system be performed at a minimum cost?
- What is the value of the contribution that each feature of the design makes to the specific function that the part must fulfill?

To be successful, a team of engineers and managers from different backgrounds should carry out value engineering with support from top management.

The value analysis job plan illustrates the tasks and functions necessary to perform the study (see Figure 6-7). This ensures that all important aspects are considered and itemized, and it provides a written record of progress. The first step is to gather basic information. The second is to speculate on alternative means for accomplishing basic functions. Step three involves selecting promising alternatives for further analysis and definition. A complete plan for implementation is developed in step four. The best alternative is then selected for implementation and several alternatives are selected as backups. The final step is to present the best alternative for approval.

Evaluation of design function is a key step in value analysis. Answers are provided to questions regarding what the design does and its implications. The function is then divided into the basic and secondary functions. The basic function defines a performance feature and receives all the value and attention. After establishing the functions, dollar values are assigned to each of them. The worth of a basic function is determined by comparing the present design with other methods of attaining the same function.

The next step is to determine the cost of the method to carry out the function. It is important to identify and focus on high-cost elements of the design. Pareto's Law states that about 80% of the total effect of any group will come from only 20% of the components of that group (see Figure 6-8). The total unit cost is broken down into material, labor, and overhead. One method for calculating cost is to develop the cost of each element in the design for each step in the manufacturing process, from raw to finished materials. The final step is to determine the value of a design or system. Value can be expressed as a ratio of the cost to the worth. This is referred to as the *value index*. Large value indices give a signal that the part under consideration is a target for cost reduction.

Group Technology

Group technology is the grouping of parts into families and then making design and manufacturing decisions based on these family characteristics. Parts are typically grouped together based on shapes, sizes, material types, and processing requirements. Group technology is an aggregation process that achieves a standardized part number and the standard specifications of purchased parts. Arranging production equipment into cells to facilitate work flow creates manufacturing efficiencies. Group technology also provides advantages to product design by

Step 1 Information				
What is it? What does it do? What does it cost? What is it worth?	**Step 2 Speculation**			
Get all the facts.	*What else might do the job?*	**Step 3 Analysis**		
Get information from best source.	Seek new information.	*What does that cost?*	**Step 4 Development**	
Get all available costs.	Eliminate the function.	Put money on each main idea.	*What will satisfy the users' needs? What is needed to implement it?*	**Step 5 Presentation and follow-up**
Work on specifics not generalities.	Simplify.	Evaluate by comparison.	Use specialty vendors and processes.	*What is recommended? Who has to approve it? What was done? How much did it save?*
Define the functions.	Blast and refine.	Evaluate by function.	Use specialty products.	Use good human relations.
	Use creative techniques.	Use experts.	Use standards.	Spend the organization's money as you would your own.
			Use your own judgement.	Monitor progress of review and implementation.
			Substantiate conclusions.	
			Prepare implementation plan.	

Figure 6-7. Value analysis job plan.

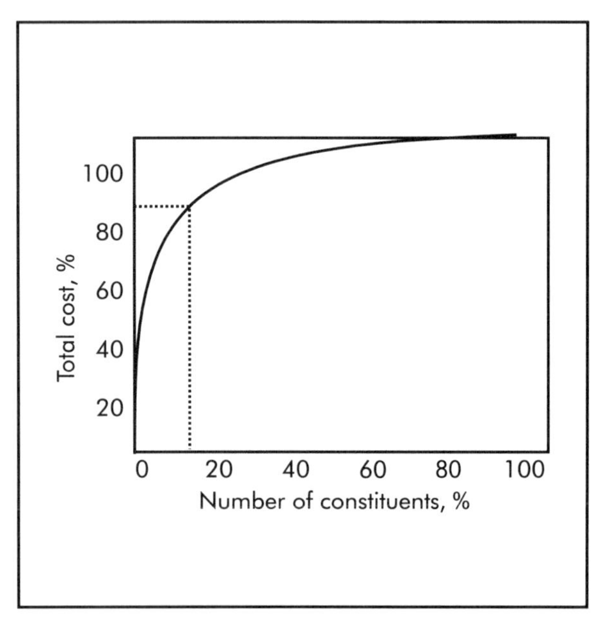

Figure 6-8. Pareto's Law of distribution of costs.

simplifying the classification and coding of parts.

Parts are organized by design attributes and manufacturing attributes. In Figure 6-9a, a process-type layout for batch production is shown. Machines are grouped by function (turning, milling, drilling, heat treatment, and assembly). As shown by the product movement, there is a significant amount of material handling, a large in-process inventory, a higher number of setups, longer lead times, and higher costs. Figure 6-9b shows an example of a group technology layout where machines are arranged into cells. Each cell is arranged to manufacture a specific part family. This reduces material handling, setup times, work-in-process, and lead times. Group technology promotes standardized part numbers, specifications of

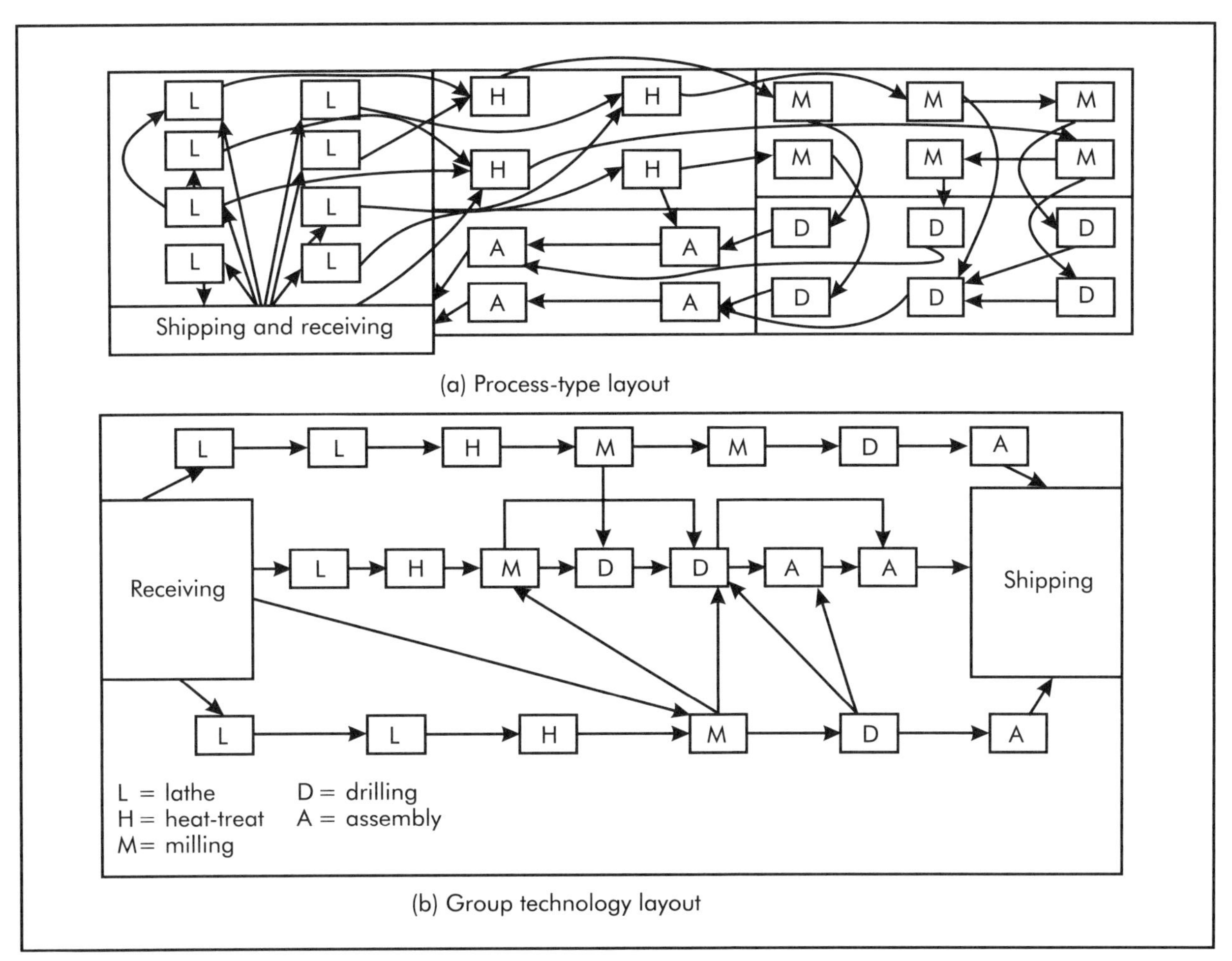

Figure 6-9. Layouts.

purchased parts, and process selections. Benefits are also typically found in design, tooling and setups, materials handling, production and inventory control, process planning, and employee satisfaction.

For design purposes, existing parts with matching codes can easily be retrieved and modified, if needed, rather than redrawing a part. In group-technology environments, tooling is designed for group fixtures and jigs to accommodate different members of a parts family. Since there is a similarity in the parts on fixtures, there is typically little to any setup required for changeover. Material handling is reduced in group technology environments, because layouts provide for efficient flow of material through the facility. Production scheduling is also reduced due to cells, since there are less production centers to schedule. Manufacturing lead times and work-in-process are reduced because of fewer setups and efficient material handling procedures. By setting up a solid part classification and coding system, an automated process-planning system can be put into place. Another key aspect to group technology is employee satisfaction. Operators can visualize the contributions of their cell, leading to improved attitudes and higher levels of job satisfaction.

Theory of Constraints

The *theory of constraints* (TOC) is a methodology that focuses on profit. The basis of the TOC is that every organization has at least one constraint that limits it from getting more of whatever is the goal, typically profit. The TOC defines a set of tools that can be used to manage constraints. Most organizations can be defined as a linked set of processes that take inputs and transform them into saleable outputs. The TOC models this chain of linked processes. This system is based on the theory that a chain is only as strong as its weakest link. Eliyahu Goldratt and Jeff Fox defined a five-step process where a change agent can be used to strengthen the weakest link (Goldratt and Fox 1992).

Step 1 in the TOC is to identify the system constraint. A *constraint* is anything limiting a system from achieving a higher performance level. This link can be either a physical or a policy constraint. Step 2 is determining how to exploit this constraint. The change agent should obtain as many capabilities as possible from the constraining link. As with any other type of continuous improvement, these changes should be inexpensive measures. Step 3 is to subordinate non-constraint components. This allows a constraint to operate at a maximum level of effectiveness. The overall system should then be reviewed to determine if the constraint has moved to another component. If a constraint is eliminated, the change agent will skip Step 4 and continue on to Step 5. Step 4 is to elevate and eliminate the constraint. This action may include major changes to the existing system. Step 5 is a return to Step 1. The TOC is a continuous improvement process.

The TOC also defines three essential measurements to drive changes. *Throughput* is defined as the rate money is generated through the sale of a product or service. This represents all of the money coming into an organization. *Inventory* is money invested in a product or service that an organization intends to sell. Inventory, according to Goldratt and Fox, includes facilities, equipment, obsolete items, raw material, work-in-process, and finished goods. *Operating expense* is defined as money used to turn inventory into throughput. An operating expense can include items such as direct labor, utilities, consumable supplies, and the depreciation of assets. An improvement effort should be prioritized by how it affects the three measures. The formula for implementation maximizes throughput while minimizing inventory and operating expenses.

Line Balancing

In production flow, there are typically several different processing and assembly operations that invariably restrict the order that operations must be sequenced. These restrictions are called *precedence constraints*. There is usually a production rate set to meet demand. The performance of a system is dependent on production scheduling, reliability, and line balancing.

The objective of line balancing is to assign equal amounts of work to each station. Ideally, the goal is to balance the workload so that all of the station times are equal. If these times are unequal, the slowest station determines the production rate of the line. Line balancing promotes the efficient use of labor and equipment. When a line is unbalanced, operators at slower stations may experience a loss of morale because they must work continuously to keep up with the flow.

Step 1 of line balancing is to create a table of work elements, as shown in Table 6-1. Work must be divided into component tasks to distribute the job among its stations. The minimum rational work elements (T_e) are the smallest practical tasks that a job can be divided into. This is considered to be a constant. The total work content (T) is the

Table 6-1. Table of work elements

Number	Element Description	T_e (sec)	Must be Preceded by
A	Rough drill—machine 1	10	—
B	Rough drill—machine 2	15	—
C	Finish drill	15	A, B
D	Tap	20	C
E	Tap	15	C
F	Counterbore	10	D
G	Counterbore	10	E
H	Assemble	5	F, G

sum of all of the work elements to be completed. Workstation process time is the sum of the times of work elements performed at a station. A workstation consists of one or more individual work elements. Cycle time (T_c) is the time interval between parts coming off of the line.

A precedence diagram is a graphical representation of a work element sequence, as shown in Figure 6-10. The sequence is defined by the precedence constraints. If the work elements can be grouped so that all of the station times are exactly equal, there will be a perfect balance on the line and smooth production flow can be expected. The goal in the line balancing procedure is to distribute the total work load on the assembly line as evenly as possible among workers. Three methods used to solve line balancing problems are:

1. largest candidate rule,
2. Kilbridge and Wester method, and
3. ranked positional weights method.

These methods are heuristic and mostly based on common sense and experimentation rather than on mathematical optimization.

History

Value Stream Mapping is based on a Toyota Production System (TPS) method called "material and information flow mapping." The TPS uses this method more as a means of communication between individuals who are learning through hands-on experience. It is used to illustrate current and

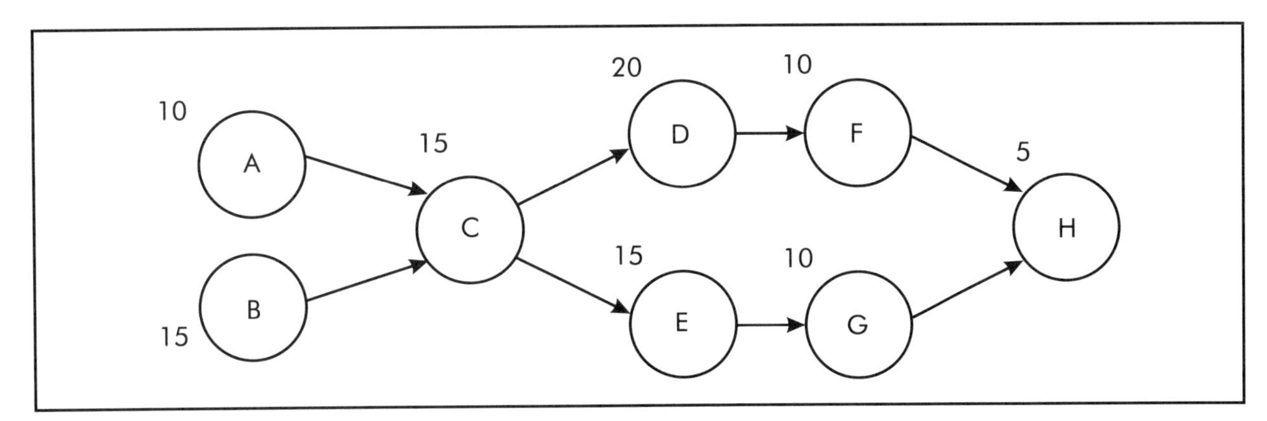

Figure 6-10. Precedence diagram.

future states of processes needed to implement lean systems. The focus at Toyota is to establish a flow, eliminate waste, and add value. Toyota teaches that there are three types of flow in manufacturing. These types include:

1. material flow,
2. information flow, and
3. people/process flow.

Value Stream Mapping is based on the first two.

Implementing Value Stream Analysis

To implement value stream analysis, the following steps are taken:

1. select a product family,
2. draw a current state map,
3. plan,
4. identify standard work,
5. draw the future map, and
6. standardize.

Step 1—Selecting a Product

In step 1, a product that needs to be streamlined and process mapped is selected. The selection process can consider a diverse range of products with a range of benefits. For instance, results from previous automotive industry case studies have outlined well-established material and information flow lines resulting from the implementation of value stream mapping.

Step 2—Current State Map

Step 2 of a value stream analysis is to draw a current state map. In evaluating the current state of the process, several improvements will be made, including implementation of: cellular manufacturing, one-piece flow, automation, and waste minimization. The processes used to attain the improvements are then mapped to help determine the future state of the process.

The purpose of a current state map is to make a clear representation of production by drawing a map of material and information flows. The current state map depicts the process initially present when the study began. The map begins with a shipment from the supplier. Once material is received, it flows through possible routes for the first few stations.

Step 3—Plan

The goal of lean manufacturing is to plan ahead. It is important to plan so that one process makes only what the next process needs, and only when it needs it. The customer defines the value here because the product is delivered at the right time and price.

The principles of concurrent engineering are applied systematically for each process to plan a lean value stream. Areas to be addressed include:

- standard work,
- takt time,
- pull system (see Figure 6-11),
- Kaizen philosophy,
- one-piece flow,
- cellular manufacturing, and
- cycle-time reduction.

Step 4—Standard Work

Identifying standard work is the process of determining how to achieve the maximum performance with the minimum amount of waste. Standard work eliminates variability from a process by establishing a routine for work. By implementing standard work, problems are exposed and waste is identified.

Takt time. *Takt time* can be defined as how frequently products must be finished to meet customer requirements. It sets the rhythm for standard work. Takt time is calculated by dividing the available time by the customer demand.

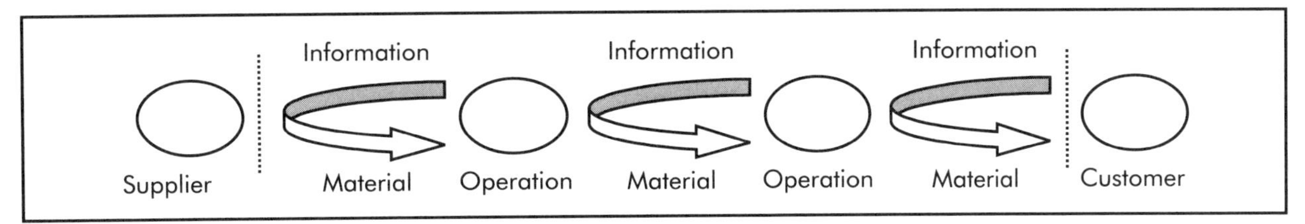

Figure 6-11. Pull system.

Pull system. In a pull system, instructions for the production and delivery cascade from downstream to upstream activities. The product is not sent upstream until a downstream activity signals the need. The purpose of a pull system is to provide accurate production requirements for the upstream process. It is also a method to control the production between the flows of two processes. This eliminates a need to predict demand or to schedule the upstream process. The pull system is based on a concrete order of customer requirements. Production control is visible and disciplined, typically with the use of a communication device called a *kanban*. Material is controlled with kanbans to replenish the system and determine the proper inventory levels. With systems that control inventory, the operator's indirect work becomes manageable. Leadtime is based on the customer, rather than the process.

Kaizen. In lean thinking, perfection is the complete elimination of an activity that uses resources, but does not add value (also called *Muda*). *Kaizen* is a continuous incremental improvement that drives the reduction of waste and adds value. *Kai* means change; *zen* means for the good. There are seven forms of waste:

1. overproduction ahead of demand;
2. inventories above the absolute minimum;
3. unnecessary transportation of material;
4. overprocessing of parts due to poor tool and product design;
5. unnecessary movement by employees during the course of their work;
6. waiting for the next processing step; and
7. production of defective parts.

Overproduction occurs when a product is produced and pushed forward, regardless of the needs of the downstream process. Since this material is not yet needed, it must be handled, counted, and stored. This causes waste in the process. With the production of inventory between processes, defects can not be seen until a product is actually used in a process downstream. This makes it more difficult to determine the root cause of a defect. Overproduction also increases the total time for a product to go door-to-door, even though the value-added time to produce it may be small. Excessive inventory creates a need for more manpower, equipment, and floor space to transport and stock it.

One-piece flow. Flow is the progressive movement of a product in a value stream without stoppages, scrap, or back flow. This may be a product taken from design to launch, from ordering to delivery of the product, or from raw materials to the customer. Continuous flow in lean thinking means producing one piece at a time. The product proceeds immediately from one process task to the next without stopping. Operators are multi-skilled workers who run several machines or perform several steps. Lean production may incorporate cellular design. However, flow production is based on the customer order. Another characteristic of one-piece flow production is flexible setups. One-piece flow in a cellular layout significantly reduces the amount of transportation needed. It also reduces the inventory and waiting time. At the same time, one-piece flow significantly improves quality, delivery, and cost. Other benefits of one-piece flow include:

- shorter lead times,
- better product distribution,
- lower scrap and rework,
- easier scheduling,
- better utilization of floor space,
- reduced material handling,
- better labor utilization and productivity, and
- exposure of problems.

Cellular manufacturing. *Cellular manufacturing* involves a group of machines or processes connected by a process sequence in a pattern that supports efficient production flow. The process determines the layout. Quality is designed into each production step, rather than at the final inspection. A cell is flexible; only one operator runs the entire cell. Layout is a U-shaped design that flows counter-clockwise. Cellular manufacturing provides many benefits to production—including less material handling, occupying less space, and better communication. It also increases the flexibility of the line, allows for the production of smaller lots, and increases throughput and quality. On the other hand, there may be building constraints that do not allow cellular manufacturing. Machine size and the size of the product may also work against U-shaped cells. Investment costs may be too high to allow a change in layout or to justify buying duplicate machines to eliminate backtracking. The use of cellular manufacturing is dependent upon the volume and variety of products, as shown in Figure 6-12. Fixed position layout is ideal for products such as a space shuttle or a ship. These products are special orders; therefore, variety and volume are one. Product layout

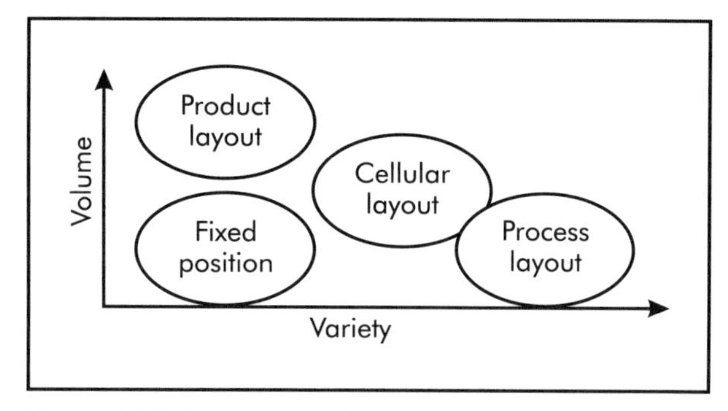

Figure 6-12. Layout as a function of volume and variety.

is used for items with high volume, with little or no variety. Process layout is used for products with low volumes, but with a large variety between products.

Cycle-time reduction. Cycle-time reduction has become a key competitive parameter in the product development process. Any attempt to improve cycle time will involve investigation and improvement of the development process. A company that is good at developing new products can use this advantage to gain market share. The reduction in cycle time for a product's development can open up new market opportunities and improve the company's responsiveness to the customer. This also lessens the market risk by reducing the time between product specification and delivery. Many companies have experienced improvement in quality and profits.

Step 5—Draw the Future Map

The purpose of the value stream plan is to break the future-state concept into reasonable steps. This is done by dividing the future state map into segments or loops. Determining the material and information flows between pulls can identify the segments. The value stream plan then has objectives, as well as the planned steps to achieve those objectives.

Several guidelines can be followed to achieve a goal of producing only what the next process needs, and only when it needs it. The first guideline is to know the takt time. It sets the rhythm for production based on sales.

Continuous flow should be developed wherever possible. This means producing one piece at a time and passing it immediately from one process step to the next, without stagnation. This is the most efficient way to produce. However, continuous flow sometimes should be limited, because combining flows into a continuous flow can merge all lead times and downtimes. First In, First Out (FIFO) inventory methodology also can be used to maintain the flow between two processes.

The next guideline is to send the customer schedule to only one production process: the pacemaker process. This process point then sets the pace for all upstream processes. Material transfers from the pacemaker process to the finished goods downstream. This needs to occur as a continuous flow. Continuous flow may be difficult upstream from the pacemaker process, because other processes may be far away, and may operate at faster cycle times. Also, there may be a changeover for multiple value streams, or they may have long lead times. A pull system should be installed where continuous flow is interrupted, so the upstream process can still operate in a batch mode. Another solution is to level production mix evenly over time at the pacemaker process. By leveling the mix, the upstream storage can be much smaller. This will reduce total lead time.

Step 6—Standardize

The first pass at implementing a future-state value stream should not consider the inherent waste from product design, the current processing machinery, or the location of some activities, since these changes may require a great deal of work and will not change immediately. These conditions should be addressed in later iterations.

Waste is identified and eliminated in the future-state value stream. Several questions are asked to improve the process flow. These questions address takt time, continuous-flow processing, the production process that sets the pace, pull systems, production mix at the pacemaker process, increments of work at the pacemaker process, and process improvements.

Sample questions and answers include:

Q. What is the takt time based on the available working time of the downstream processes closest to the customer?

A. Takt time is calculated by subtracting non-working time from available working time, and dividing this value by customer demand.

Q. How can a company use continuous flow processing?

A. Continuous flow can be achieved by rearranging equipment and/or responsibilities so that the cycle time for each operator's total activities is under the takt time ceiling.

Q. What production process needs to be scheduled?

A. The pacemaker process is the first process downstream to be scheduled. All subsequent process steps are downstream of the pacemaker process and must occur in a flow.

Q. What process improvements will be necessary for the value stream to flow as the future-state design specifies?

A. The purpose of this final question is to start on process improvements with a clear understanding of why they are needed.

The basic steps to be investigated to create a future state map are shown in Figure 6-13.

The value stream plan shows what a user plans to do and when; it also has measurable goals, checkpoints with deadlines, and a reviewer.

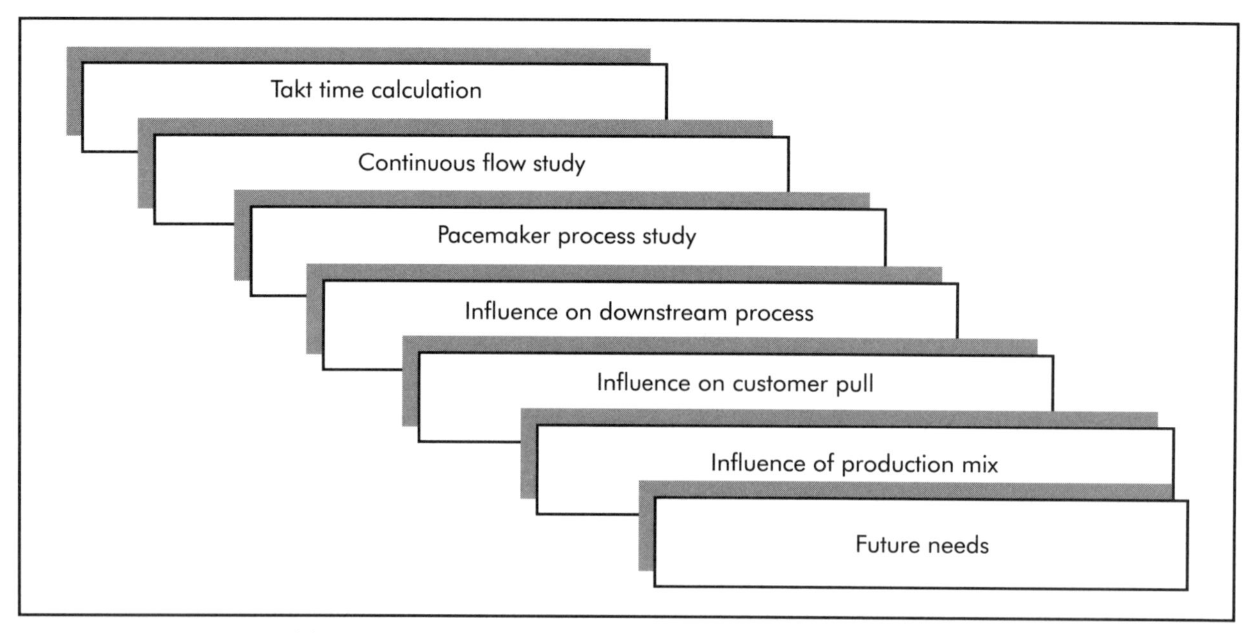

Figure 6-13. Basic steps of future state mapping.

As stated earlier, a value stream plan helps to break the implementation of a future-state concept into reasonable steps (see Figure 6-14). The future state value stream map can be divided into segments or loops. It lists objectives and measurable goals for each objective. A typical pattern for implementing improvements is to start with developing a continuous flow based on the takt time. This tends to be the simplest place to start and typically provides the biggest bang for the buck by eliminating Muda and shortening lead times. The next step is to implement a pull system. The pull system will provide production instructions for continuous flow. Once the order for implementation has been decided, the elements need to be written down for the yearly value stream plan. The value stream chart looks like a variation of a Gantt chart (see Figure 6-15). The key to successfully implementing the value stream is incorporating it as a part of normal business practice.

Macro Value Stream Mapping

Macro value stream mapping extends beyond plant-level maps. Macro mapping can be done after creating current and future state maps for the facility. These maps are created for several reasons. First, a large portion of costs can be attributed to purchased materials. Downstream inconsistencies can threaten leanness inside the facility. Added costs downstream can also negate internal cost savings. This can affect whether or not sales grow. The whole picture allows a user to identify major asset reconfigurations

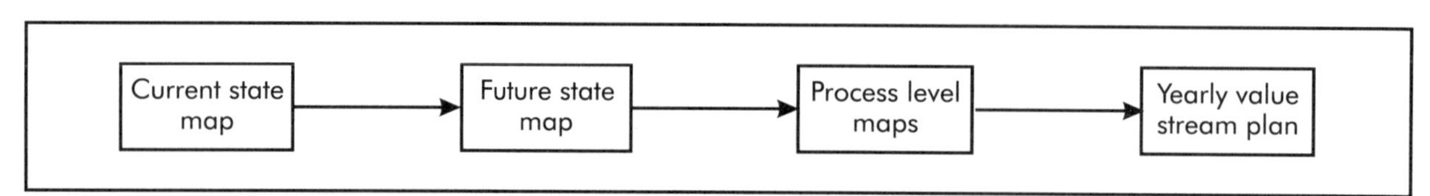

Figure 6-14. Value stream plan.

Value Stream Manager			Monthly Schedule														
Plant-level Objective	Value Stream Objective	Goal Measurable	Jan	Feb	Mar	Apr	May	Jun	Jul	Aug	Sep	Oct	Nov	Dec	Responsibility	Team	Reviewer (Date)

Figure 6-15. Value stream chart.

by showing who does what, where, and with what tools.

Facilities closest to the customer should be mapped first. Information such as frequency, distance, cost, processing time, lead time, inventories, cost per unit, daily volume, shift data, variation, frequency variation, and demand variation, should be collected. In the ideal state of macro Value Stream Mapping, all activities should be located in the exact process sequence.

CASE STUDIES

Automotive Industry (Cudney 2000)

One automotive component supplier plant located in Michigan had not made a profit in 30 months. Due to consistent financial losses, customer demands for a better product, and an internal need for space and growth within the plant, the company decided to implement Value Stream Mapping. Several improvements were made, including:

- implementing the kanban system,
- reducing setup times,
- implementing takt time,
- implementing U-shaped cells,
- implementing standard work, and
- creating a central market.

After implementing Value Stream Mapping:

- The defect rate decreased from 27,000 parts per million (PPM) to 178 PPM.
- The paint yield increased from 60–65%, to 80–85%.
- The lead time decreased from 29.4 days to 8.5 days.
- The raw inventory turns increased from 31 to 38.
- The work-in-process (WIP) turns increased from 88 to 400.
- The finished goods inventory turns increased from 90 to 120.

Aerospace Industry

An aerospace component manufacturer located in New York decided to implement Value Stream Mapping because of increased space requirements, and issues with profitability and survivability. Improvements to the current state included rearranging the factory layout to make it functional to the process, collocating support functions, and locating all assembly processes in one building. After implementing these improvements:

- The number of employees decreased from 153 to 88.
- The support factor decreased from 69% to 47%.
- The unit cost decreased from baseline to 69%.
- Quality improved from baseline to 37%.
- Takt time decreased from 9 days to 8 days.
- Assembly lead time decreased from 64 days to 55 days.

SUMMARY

Value engineering addresses how a given function can be performed at the minimum cost. It also addresses the contribution each feature makes to the function that a part must fulfill.

In group technology, parts are grouped together into families, based on part shapes, part sizes, material types, and processing requirements. Design and manufacturing decisions are then based on family characteristics.

The theory of constraints assumes every process has at least one constraint limiting it from reaching its goal. It encompasses a five-step process for a change agent to strengthen the weakest link.

The objective of line balancing is to assign equal amounts of work to each station; otherwise, the slowest station will determine the production line rate.

A value stream includes all value and non-value-added activities required to manufacture a product. The purpose of Value Stream Mapping is to identify waste and eliminate it. By creating the current state map, future state map, and the value stream plan, material and information flow can be established throughout the process. Value Stream Mapping is a complete view of the process; therefore, the entire flow is apparent, as is waste in the process. It is a valuable tool to identify waste and areas for improvement. As with other lean methodologies, Value Stream Mapping should be an ongoing effort. However, without a clear understanding of its techniques and philosophies—such as the Toyota Production System, pull systems, kanbans, just-in-time, standard work, one-piece flow, etc.—Value Stream Mapping will not provide benefits.

REFERENCES

Chryssolouris, George. 1992. *Manufacturing Systems—Theory and Practice*. New York: Springer-Verlag.

Cudney, Elizabeth. 2000. "Investigation into Value Stream Mapping and its Application." Masters Thesis. Hartford, CT: University of Hartford.

Cudney, Elizabeth, and Shetty, Devdas. 2000. *Value Stream Mapping*. Proceedings of the 16th International Conference on CAD/CAM, Robotics and Factories of the Future. June. Port of Spain, Trinidad: Institution of Electrical Engineers.

Goldratt, Eluyahu and Fox, Jeff. 1992. *The Goal: A Process of Ongoing Improvement,* Second Edition. New York: North River Press.

Groover, Mikell P. 1980. *Automation, Production Systems, and Computer-aided Manufacturing*. Englewood Cliffs, NJ: Prentice-Hall, Inc.

Rother, Mike and Shook, John. 1998. "*Learning to See: Value Stream Mapping to Add Value and Eliminate Muda.*" Version 1.1. Brookline, MA: Lean Enterprise Institute.

Chapter 7

Product Creation: Aligning for Design and Business

WORLD-CLASS PRODUCT DEVELOPMENT

Views presented in this section provide an initial step toward the development of a world-class product methodology. The effectiveness of these elements depends on both how and when they are used in the development cycle. Some of the lessons learned from the Japanese include integrating engineering with manufacturing and building long-term partnerships with subsidiaries and vendors. The Japanese have simplified manufacturing processes enabling them to understand, control, and manage processes more effectively. Japanese companies emphasize employee involvement by broadening factory worker responsibilities and dramatically reducing the need for outside professionals.

Studies of successful product development cases reveal common themes that contribute to superior attributes of a product. These essential elements have been extracted from several product development examples that involve complex, innovative products. Significant technology development was required in many of the cases before the products could be designed and manufactured (Hall 1987; Lee 1999).

PRODUCT DEVELOPMENT PROCESS

The development process originates with product ideas and ends with product manufacturing. Product and process technologies are developed concurrently, as are product and process designs. At this very general level of consideration, there are some basic essential elements that greatly affect the outcome of the development process. Some of them apply to the entire development process, while others apply to particular phases (Ulrich and Eppinger 1991).

Three of the essential product and process technology elements are so fundamental that they affect the entire product development process. These elements integrate many diverse product development activities into a coherent, focused process. The elements are:

- the single-team approach;
- user-oriented product development; and
- the convergence of information at the product's definition.

Single-team Approach

The element with the most far-reaching effects is the selection of a single development and manufacturing team to control the project. Companies like United Technologies and IBM commission product development teams, called independent business units, to develop new and innovative products with complete autonomy.

Hewlett-Packard used a team approach to develop a color printer. Many major companies competing in the international market have found that organizational structure, with its accompanying lack of responsibility for the product, is not a means of being competitive when pitted against global competition. In some cases, the product delivery systems can be very inflexible and create bottlenecks in product delivery. A single development team is an essential element for world-class product development.

Since the single team is responsible for a product throughout its development, it must be comprised of members with the proper skill mix and experience to complete the job effectively. Various team members must have design, manufacturing, marketing, testing, and other skills needed to develop a product successfully. An effective team controls all aspects of a project, including definition and specifications, from technology selection through the first six months of manufacturing.

User-oriented Product Development

Another essential element affecting the entire product development process is the proper determination of customer needs, or user-oriented product development. In a development context, proposed products must provide a competitive solution to customer needs at a future time period, starting with the product introduction and continuing at least until the product development costs are recovered. Customer-needs projections developed from this analysis are used during the product-definition stage to establish product specifications. A customer-needs analysis must project future needs, along with providing current needs statements from the marketing department (Womack 1996).

Development teams should participate in the analysis of customer needs. Encouraging development teams to participate in analysis can enhance their creative contributions by enabling opportunities that might not be seen by merely reading a report from a distant market analysis group. For example, one data-storage firm established a customer advisory board to provide direct customer input into the design process. The board was composed of technically astute customers, such as data-center directors and systems engineers. The firm provided advance notice of new products and solicited suggestions for product changes and improvements from the advisory board. Design engineers attended meetings to answer questions and receive direct feedback from customers (Wilson and Kennedy 1989).

Convergence of Information at Product Definition

Convergence of marketing, engineering, and manufacturing information and goals is essential in the process of creating an adequate product definition (see Figure 7-1). This information convergence assures that marketing, engineering, and manufacturing issues are considered simultaneously as a product is defined. Simultaneous consideration of these issues enables project leaders to agree on a common set of product goals and action plans. This, in turn, enables parallel product and process development to occur with a minimum of conflict. If common goals and plans are not developed, simultaneous engineering of the product and process is likely to cause divergence, resulting in major product and process reworking late in the product's development.

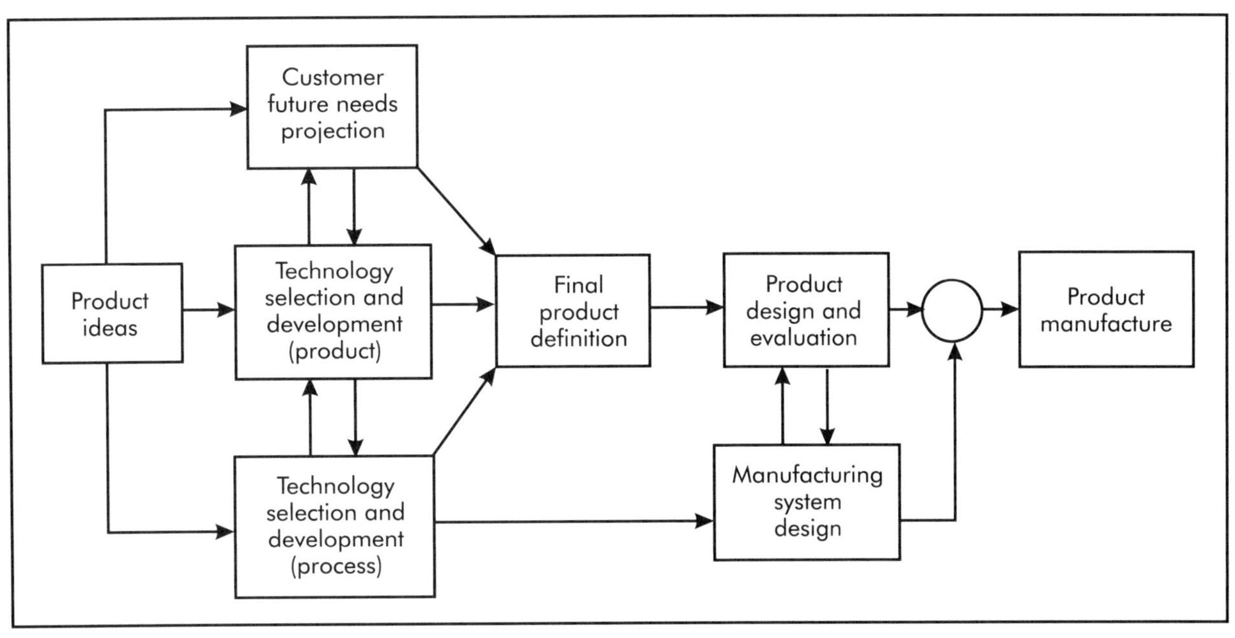

Figure 7-1. World-class development process for an innovative product.

TECHNOLOGY SELECTION

Many essential elements apply specifically to particular phases of the product development process. These elements emphasize the use of the most appropriate engineering methods at the most appropriate times, and they aid in discovering problems at the least damaging stage of development. One essential element of the technology selection and development process is the evaluation of the ability of a selected technology to accomplish its intended purpose. The failure to assure the feasibility of a technology can lead to a poor technology choice, which, in turn, will ultimately result in product failure (Singh 1996; Usher, Roy, and Parsaei 1998).

Seven Rules of New Product Introduction

1. Define the customer requirements, or opportunities. This phase is normally led by the marketing representative and involves an intensive understanding of the present market and ways to explore sales opportunities. The purpose of this phase is to create an objective assessment of what the company can and cannot do. This often leads to a matrix sheet detailing an entire range of capabilities.
2. The product concept, definition, and specification phase determines how requirements can be met. Design engineering and marketing personnel lead the team. The group uses brainstorming, storyboarding, and quality-function deployment to match market opportunities with company capabilities. The group determines how product specification requirements can be met. If a product concept looks feasible, the group will need to scope out a product specification. The marketing representative ensures that a product meets the perceived needs of the market. Creating a product specification also forces a definition of what the company will offer to the market. This minimizes the chance

of making an offer that is difficult to fulfill. It also tends to preclude any misunderstandings between customers and vendors about what constitutes the successful delivery of a product.

3. Produce engineering drawings and specifications (design specifications). After the abstract phase is complete, the team needs to create a first iteration of a design that is relatively producible. Simulation techniques may avoid some trial and error, but real manufacturing with real production equipment is still needed to test the process.
4. Define a method for manufacturing the product (build a prototype and test it). Here, the team will be heavily influenced by the capabilities of the company. For instance, what type of equipment does it possess and have strength in using? Companies strong in metalworking would naturally favor metal, rather than plastic, for the base material. A group will develop a bill of materials and routing to determine the total cost. This is the phase where economic viability for full production is determined. A prototype is structured for manufacturing and necessary vendor supply-chain relationships are established.
5. Evaluate the capacity to make this product, along with all other products requiring the same capabilities (manufacturing capacity development). This step, often overlooked, is led by manufacturing engineering staff. The team has a design, a process, plans for prototypes, and an idea of sales potential. Many companies, liking what they see, will plunge here into selling a product and disregard all previous commitments that require factory resources. If the capacity is insufficient and customers are time-sensitive, another grievous error can occur when the company needs to commit capital resources to obtain capacity. The group should compare its capacity with the needs of the customer. This is usually done concurrently with the internal evaluation of capacity. New products need to be introduced in a timely manner to take advantage of market opportunities.
6. Determine product cost. Profits have to be realized for a company to remain in existence. The team, led by marketing and finance representatives, takes all data, calculates potential profit, and verifies that a new product will match the company's strategic plan. It takes corrective action to achieve a required margin and still meet customer requirements. If unsatisfactory results are indicated by profit calculations, the team must reconsider all of its options to see if the project can be saved.
7. Launch the product. The product is released for manufacturing, after all of the internal and external requirements are determined to be achievable. Senior management may decide to stop, hold, refine some more, or commence a launch. The team monitors the production and sales progress, and it takes corrective action as required. Each member needs to monitor the early progress of commercialization to ensure that everything is running smoothly.

MANAGEMENT PHASES OF PRODUCT DEVELOPMENT

Successful product development requires effective project management that can result in high quality, low-cost products by making the best use of time, money, and other resources.

The integrated product development (IPD) phases are as follows:

1. product feasibility,
2. planning and specification,
3. development,
4. preparation for product release and qualification,
5. pilot production and feedback, and
6. full release.

The initial step is to form the IPD by selecting the dedicated core team whose members represent key functional areas necessary for the product. A typical IPD team consists of representatives from engineering, marketing, supply management, manufacturing, field, finance, code, and intellectual property departments. The best practice is to hold a team-building and kick-off meeting early in the project to facilitate relationship building within the team.

Product-feasibility Phase

The product should be evaluated for market opportunities, strategic fit, financial viability, etc., using a small team and the best-known assumptions. The feasibility-study phase considers the technical aspects of conceptual alternatives and provides a firm basis to decide whether to undertake the project. The purposes of the feasibility phase are to:

- plan the project development and implementation activities;
- estimate the probable elapsed time, staffing, and equipment requirements; and
- identify the probable costs and consequences of investing in the new project.

The end result of the product-feasibility phase is a comprehensive business-case report of 5–15 pages.

Planning and Specification Phase

For the planning and specification phase of the project, the formal IPD team is selected to verify earlier assumptions. The team represents the key functions necessary for the project. It typically includes engineering, marketing, supply management, manufacturing, and finance. These members must clearly define the product and its functionality. The manufacturability of the proposed product also must be determined, as well as a strategy/plan for the supply chain and inventory.

The key technical and project reviews for the phase should be scheduled and planned at the beginning of the phase to ensure availability of key team members, reviewers, and appropriate facilities. Having firm review dates also puts emphasis on maintaining the project schedule. The core team should meet on a regular basis to maintain open communication, update one another on the progress, and key in on accomplishments, issues, and risks (Rivera 2001). The necessary information required includes:

- a statement of work,
- project specifications,
- project milestones, and
- a work breakdown schedule.

If the project is expected to have many changes, a project-control team is established that includes a project manager and members from key functional areas. The project-control team looks into the aspects of prioritizing project changes and identifies resource needs for completing all tasks. The team refines the draft schedule and finalizes planned deliverables and milestones. The information will help the team avoid overlap and inconsistencies.

Key deliverables at the end of this phase are:

- a validated business-case study;
- clearly defined product and requirements;
- a program and phase plan; and
- a product-cost model.

Development Phase

A detailed design can be formulated by using the technical product concept, the requirements, and the component characteristics. Models are used to confirm that allocated performance requirements are achievable. Development begins with supporting processes such as manufacturing, service, and the supply chain. The component requirements to be implemented can be decided based on various engineering disciplines. It is important to verify that all interfaces, timing, and boundary conditions for the separate assemblies have been met. The component model should be updated to track changes. In addition, the performance requirements for various assemblies must be assessed.

The component factory lead times, manufacturing planning, and process development status must be documented. Using the product issues and test reports, the design can be modified to meet the component requirements.

The manufacturing process for each component must be defined and manufacturing standard times, along with equipment and tooling for each component, must be established. A manufacturing layout plan needs to be developed next, along with the space and manpower requirements. Based on the manufacturing standard time estimates, the manpower requirements should be finalized.

Key deliverables at the end of the development phase are:

- a working prototype;
- the release drawings;
- a program and phase plan;
- a release plan; and
- an updated product-cost model.

Preparation for Product Release and Qualification Phase

The preparation for product release and qualification phase includes the process of conducting an initial component and system quality audit. Product testing must be completed before preparing for volume production. Other steps include:

- finalizing the launch plan;
- getting supporting processes approved and qualified;
- identifying parts and assemblies requiring certification by external organizations;
- preparing for necessary data collection; and
- submission for the timely notification of certification results.

Key deliverables at the end of this phase are:

- a program and phase plan;
- an updated product cost model;
- completed product testing and verification;
- completed launch and rollout plans; and
- approved and qualified supporting processes.

Pilot Production and Feedback Phase

A pilot product can be evaluated and improved based on the feedback from testing results. This phase includes the processes of updating the product cost model, preparing for sales deployment, and planning for operational personnel and maintenance support. In addition, the requirements of spare and repair parts and related inventories must be calculated. Maintenance of test and support equipment must be planned for as well. The deliverables at the end of this phase are:

- the program and phase plan;
- an updated product-cost model;
- successful supply chain deployment; and
- the continuation of an engineering plan.

Full Release Phase

A project's status can be reviewed in terms of the progress achieved. The cost, model, and product design should be analyzed for areas of improvement and further modification. The important activities in this phase are:

- the project retrospective,
- the completion of project records,
- documentation of the results, and
- field deployment.

The key deliverable at the end of this phase is the final documentation.

WEB-BASED PRODUCT DEVELOPMENT

With the Internet available in many countries, web-based technology provides many new product development opportunities. For market research on new products, the resources available from the web are unlimited. This technology can also help a product get to the market quickly.

In the past, the computer technique and simulation tools used for product design were largely static in nature. Advances in software and hardware, however, have made it possible to animate prototypes and create design alternatives that add a whole new dimension of utility to the practice of virtual prototyping.

VIRTUAL PROTOTYPING

Virtual prototyping is a novel approach in product development, especially in the early stages of the development process. It is a software-based procedure that entails the modeling of a mechanical system, simulating and visualizing 3D-motion behavior under real-world conditions, and optimizing a design through iterative design studies. Virtual prototyping simulates product features with a degree of functional realism comparable to a physical prototype. This concept is important during the critical step in new-product development of selecting from multiple product concepts one that will carry the firm forward into the market place (see Figure 7-2).

Depending on the cost of developing each concept into a customer-ready prototype, it is optimal to carry multiple-product concepts into the prototyping and testing phases to select the best design. The ultimate goal is to enable a fully digital, front-end process for product development. This increases overall innovation in terms of product design because it shortens the product development cycles; reduces development costs; and improves the accuracy and quality of development to meet the needs of the customer and market. Virtual-prototype analysis can lead to earlier detection of flaws that would otherwise be expensive to correct at later stages of development.

INTERACTIVE PRODUCT SIMULATION

Product designers and manufacturers can increase efficiency with the use of interactive product simulation. This is a procedure that assists the company's development efforts by providing earlier access to prototypes; faster updates than with physical models; distribution of information in an easy-to-understand format; support for existing processes; and long-term value that extends beyond the finalization of a product design. These gains are based on the involvement of a greater number of people who are typically involved in the life cycle of a product (Ghee 1987).

While CAD software is typically designed for non-real-time modeling, the interactive product-simulation software (IPS) is a real-time visualization and interactive system. CAD/CAM geometry is exported to IPS software that comprises two-core components: large-scale assembly visualization and navi-

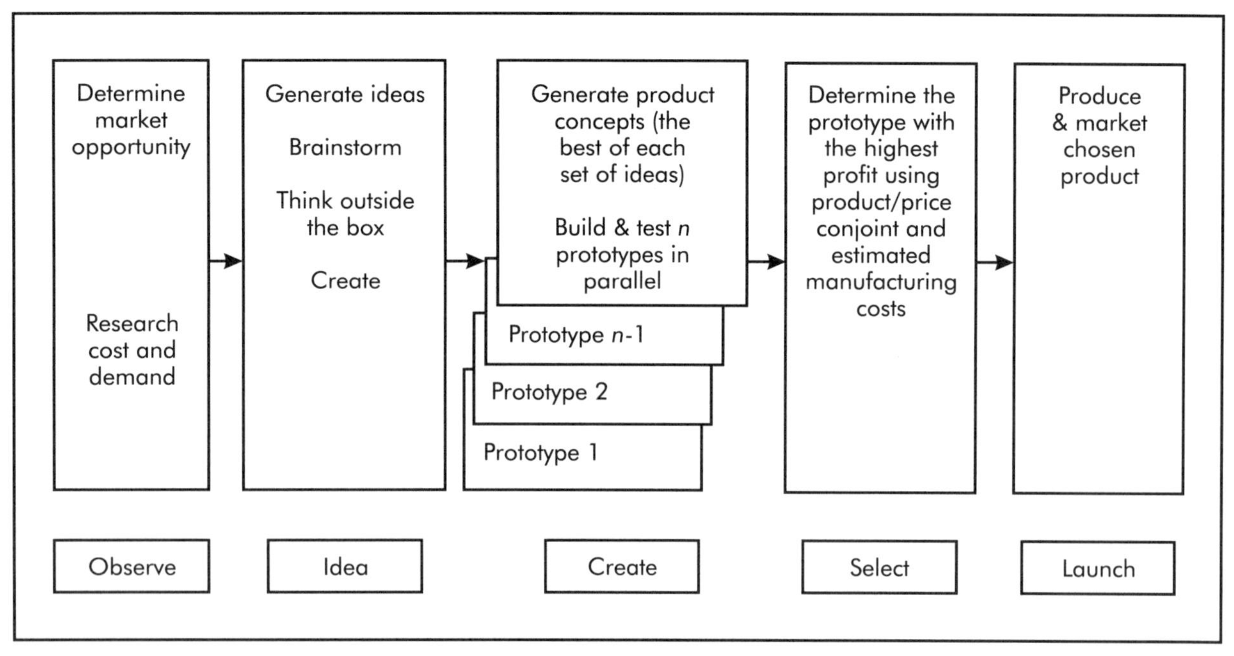

Figure 7-2. New product development process (Dahan and Srinivasan 2000).

gation, and the ability to simulate a product's functionality or behavior. These two components provide a communication medium that can be shared across networks, manipulated, and analyzed. The technology provides significant returns during the entire product life cycle, enabling designers, engineers, customers, and others to not only visualize and navigate design geometry, but to also interact with functional characteristics of a prospective product. Increasingly, this type of software has been integrated into system infrastructure to create visual databases of CAD/CAM geometry. A visual database enhances product understanding at all stages: concept, detailed design, analysis, manufacturing, maintenance, supply chain, project and enterprise management, and personnel training. By providing an intuitive interface to select parts and retrieve part information with the help of a product data-management system (PDM), IPS has become an indispensable and inexpensive communication medium.

Concept Phase

IPS has many applications in the concept phase, where one of the most critical challenges for manufacturers is the economic evaluation and frequent review of multiple high-level concepts and configurations. There are typically two objectives in this phase: the assessment of multiple configurations to make trade-off decisions and the clear communication of a design and its functionality. Trade-off considerations usually include comparisons of cost, weight, aesthetics, ergonomics, manufacturability, serviceability, and more; while the discussion of functionality is intended to emphasize a product's utility for executives, customers, and other non-engineers to generate more effective feedback.

Detailed Design Phase

In the detailed design phase of product development, when a primary need is to conduct frequent design reviews, IPS tools integrate work from many different people and processes for group review and feedback. Users can add real-time animations to the virtual product and edit them to indicate part paths and sequences, as well as check for collision and clearances. Cross-disciplinary collaborative design reviews can quickly evaluate individual ideas in the context of overall product assembly, maintenance, and usability.

The use of IPS models in early focus-group sessions can yield more effective feedback from potential customers, leading to higher market acceptance for the ultimate product. To accomplish this, designer, engineers, and marketers collaborate in the development of a virtual product to be evaluated by potential customers during focus sessions.

A concern of product manufacturers today is the coordination of disparate design teams. Traditional communication and collaboration methods often result in significant expenses for travel, personnel relocations, and telecommunications. IPS addresses most of these needs by enabling real-time linking of multiple sites in one virtual product. The result is a far more efficient use of time than traditional meetings.

The Internet provides an extremely inexpensive and widely distributed format for product feedback. Distributing an IPS model via local networks, wide-area networks, or the Internet is very easy. Web access is simple and intuitive with the help of the standard browsers available today. IPS software uses standard editors (for example, Hypertext Markup Language [HTML]) to add text and graphics, and it can convert animations and camera views with simple procedures. Examples of new technologies available on the web include systems that allow visual, auditory, and tactile information to be distributed and retrieved. The low cost of distributing product designs via the Internet makes assembly and disassembly sequencing, training, and feedback on testing an attractive alternative to traditional methods for product designers.

Examples of web-based product development tools include iMAN™, ProductVision™, oneSpace™, and FirstSpace™. Some of the advantages and disadvantages of a web-based system are listed below.

Advantages of web-based product development tools include:

- shortens the product-definition phase involving the customer, marketing, and design engineering;
- provides instantaneous access to knowledge and experience from different locations and disciplines during the design and development stages to achieve high-quality results quickly;
- on 3D models, allows instant online model changes, design optimization, and lower manufacturing costs through immediate understanding of impact and implications of design alternatives;
- explores and builds on ideas in the product implementation stage involving trade-off decisions between manufacturing, tooling, and design engineering; and
- improves design coordination with remote and non-remote partners.

Disadvantages of web-based product development tools include:

- Internet security,
- bandwidth, and
- exposed proprietary information.

The emerging trend in web-based product development has created more focus, not only on functional and behavioral simulation, but also on Internet security. New software upgrades constantly grow in size and provide more functions. Simulation of product functionality and behavior is becoming an increasingly important feature of software. Future products will have multi-modal user interfaces to use the complex functionality provided by such software. This means that the design and testing of user interfaces will require powerful, flexible tools for simulating and testing multiple design alternatives.

One of the main advantages of virtual prototyping is that it can produce a large number of consecutive prototype versions very rapidly, unlike conventional prototyping based on physical prototypes. This means that the simulation of virtual prototypes of complex products consisting of multiple components have to be executed in a distributed manner. At the same time, valuable simulation models should be downloadable over a network, without compromising proprietary information.

REFERENCES

Dahan, Ely and Srinivasan, V. 2000. "The Predicting Power of Internet-based Product Concept Testing Using Visual Depiction and Animation." *Journal of Product Innovation Management*, 17:99-109.

Ghee, Steve. 1987. "The Virtues of Virtual Products." *Mechanical Engineering*, June, pp. 60-63.

Hall, Robert. 1987. *Attaining Manufacturing Excellence: Just-in-Time—Total Quality, Total People Involvement*. Homewood, IL: Dow Jones-Irwin.

Lee, Kunwoo. 1999. *Principles of CAD/CAM/CAE Systems*. Boston, MA: Addison Wesley.

Rivera, Jim. 2001. Discussion on PDP Guidelines. Otis Elevator Co. United Technologies Corp., May.

Singh, Nanua. 1996. *Systems Approach to Computer-integrated Design and Manufacturing*. New York: John Wiley and Sons, Inc.

Ulrich, Karl T. and Eppinger, Steven D. 1991. *Product Design and Development*. New York: McGraw-Hill.

Usher, John, Utpal, Roy, and Hamid, Parsaei, ed. 1998. *Integrated Product and Process Development: Methods, Tools, and Technologies*. New York: John Wiley and Sons, Inc.

Wilson, Clement and Kennedy, Michael. 1989. "Some Essential Elements of Superior Product Development." Paper 89-WA/DE-7. San Francisco, CA: American Society of Mechanical Engineers (ASME), Winter Annual Meeting.

Womack, James P. and Jones, Daniel T. 1996. *Lean Thinking: Banish Waste and Create Wealth in Your Corporation*. New York: Simon & Schuster.

Chapter 8

Building Successful Product Work Groups

The group approach is a successful tool of concurrent engineering for product design purposes. This chapter examines the importance of groups for product development and it identifies group characteristics. It also reviews the stages a team must go through as it matures. There have been several studies on the effectiveness of the group approach. This chapter identifies the methodologies and techniques that have contributed to its success in design and development.

WHAT IS A PRODUCT DESIGN GROUP?

A *product design group* is defined as a team of people working together to achieve a common goal by using their combined skills, talent, and knowledge. A group may consist of members of the same organization, or it may represent members with different backgrounds from other units of the organization. A cross-functional group is created when goals cover more than one department. For example, if a company is making preparations for the acquisition of a major computer-aided design/computer-aided manufacturing (CAD/CAM) system, the situation might require the creation of a company-wide team from different divisions. Members of this group would be selected from departments that are primary users of the CAD system to ensure that these departments' requirements are addressed.

Product design groups can be classified as either core groups or work groups. A *core group* is the lead group assigned to a new design project. The function of the core group is to direct project planning and implementation. For larger projects, *work groups* are formed for follow-up. Generally, the core group is responsible for the direction and implementation of the overall project, and the work group is responsible for specific portions of it, taking guidance from the core group. Group members, of both the core and work groups, serve dual roles. They not only serve as members performing the tasks of their respective groups, but they also operate as employees of the functional department from which they were selected. These dual roles, as well as an association with the parent department, represent one of the challenges of successful group performance. Very large projects may require auxiliary groups to direct program portions.

Group versus Individual Decisions

Groups tend to make better decisions than individuals, especially in situations that are relatively broad-based for the company—such as strategic planning or issues that apply to everyone. Diversity in background, experience, and skill levels of group members allows a group to consider a wider range of alternatives to solving problems and to be more creative. In group meetings, members work to achieve consensus with other group members. Unfortunately, reaching a consensus as a group takes longer than individual decision-making. This is not necessarily bad, however, because issues assigned to a group generally require more careful and detailed analysis. When dealing with sensitive issues, a group has more potential power than any one of its constituents.

Within a group, members are usually more comfortable expressing issues and concerns openly. Management will view the issue in question as a group opinion, rather than associating it with an individual.

A lot of financial investment and risk-taking are involved when a company goes ahead with new product development. Groups tend to make more systematic decisions than individuals, who tend to be more concerned with the consequences of decisions. Groups are more comfortable with the decision-making process since they have a broader knowledge base. Individuals that comprise a group are less fearful because they have support from fellow team members. Thus, the quality of the group's decision is better. A higher, more sustained energy level results from the support that group members give one another, despite that the work of product groups is more exhaustive. Overall, management support plays a crucial role in decision-making.

Building Effective Groups

Various studies have identified several points as crucial to the success of product development groups. The characteristics of an effective group are:

- There must be strong leadership within the group.
- There must be clearly defined goals.
- It should have budget and staffing priority.
- It should begin with a small, manageable problem.
- There must be a balanced representation of appropriate disciplines from within the company.
- Members should have good communication skills and professional respect for all of the group members.
- There must be an atmosphere that is conducive to free thinking.
- Management should select a group leader and set objectives.
- The leader should participate in the selection of the group members.

Group composition is important and should include all those involved in the decision process. Typical product design groups include motivated individuals from various company areas who have never worked together, but are responsible enough to carry out an important project. Group members with challenging and opposing views can create an atmosphere of true innovation; however, if too much time is spent arguing, less work will be achieved. Once basic objectives are known, the group defines design criteria and sets goals.

Group Selection

Group success depends to a large extent on the selection of players who will serve on the team. Groups with the right combination of individual skills, attitudes, and leadership have a positive impact on the outcome of a project. It is also important to have a good understanding of how groups can work together to meet goals.

Group Membership

Selection of the right group leader is critical to a project's success. A group leader has a major influence on the performance of the entire group. The leader needs a combination of knowledge and qualities that bring out the best in the group.

When selecting group members, it is necessary to consider the knowledge and skills that are important to the process from the functional areas of design, production engineering, purchasing, sales and marketing, field service, fabrication, assembly, quality, suppliers, cost accounting, and other specialties. The main question is, "What knowledge and skills are needed for the group?" The answer will determine the functional departments that should be represented. Members should be chosen based on the size of project and the skills needed during product creation. Group members should be chosen in a fashion similar to the selection of their leader. The selected leader should play a major part in interviewing and choosing other members.

Characteristics of good group leaders and members are:

- good people skills;
- strong written and oral communication abilities;
- assertiveness;
- good listening skills and an open-minded attitude;
- excellent analytical abilities and a good understanding of product design and costing functions;
- determination and persistence;
- an understanding of manufacturing processes and costing;
- an understanding of the organization and the roles of various functional groups; and
- knowledge of the specific product design procedure.

Once members have been recruited, various group training activities should be considered. Depending on the size and resources of an organization, the group leader could perform selected training in team dynamics. Alternately, external consultants, an internal education department, or other qualified personnel could do the job.

Allocation of Activities between Group Members and Others

Group work comes from the combined contributions of the group, and contributions from other organization members. The roles of group members are discussed in the following sections.

Group leader. The role of a group leader includes the following:

- helps a group focus on continuous improvement;
- reminds the group of the long-term outlook;
- stays uninvolved in the daily work of the team; and
- uses periodic reviews with the team to help it focus on measures.

Facilitator. The role of the facilitator includes the following:

- could be a member of the group;
- participates in all meetings;
- does work that is part of what the team does;
- is a person trained in group dynamics who assists the group with problem-solving tools; and
- communicates, coordinates, and facilitates various activities within the team and between the team and others.

Business leader. The role of a business leader includes the following:

- leads development of 1–5-year plans for company growth, machinery investment, and people development;
- develops and enforces policies to guarantee fair treatment to all;
- presents overall company measures in a timely fashion to the leadership team; and
- sets up annual goals.

Product coordinator. The role of a product coordinator includes the following:

- produces results for the company's product line;
- measures monthly results;
- presents results achieved, problems, opportunities and action items to the leadership team;
- works with teams to develop action items to improve results for product lines;
- leads implementation of continuous improvement for product lines;
- focuses yearly goal setting on the product line;
- assures that orders are ready for production; and
- reports results monthly to the project coordinators.

Support team members. The roles of support team members include the following:

- participates on preparation review and allocates tasks to teams as needed;
- works continuously to improve processes;
- reports results of work through the facilitator;
- reviews preparations for production; and
- reports activities and results monthly to the product coordinators.

Group Interaction

Working in a group is generally more fun than working individually. Groups accomplish important tasks by:

- working together;
- knowing what needs to be done;
- splitting the work among group members;
- doing what is essential;
- measuring progress;
- monitoring measures on a regular basis; and
- acting together to correct performance problems.

A support group can provide a manufacturing group, which is primarily responsible for making parts, with information on what needs to be produced. The manufacturing process is then identified. The group makes parts using materials provided to them by another group. Support groups work with these groups on reports and disseminating information throughout the organization. The support group prepares information on the next set of needs and the group evaluates its own work. Other support groups become involved when the main group needs help.

Groups are responsible for helping to define and improve how they work. This step-by-step work method is a process. It usually involves looking at success measures over a longer period of time, identifying areas of opportunity for improvement, and working together to redesign a process to achieve better results. Good ideas are easily lost when not noted and remembered by the team. Follow-through is essential to promote accountability.

Short-group Meeting or Huddle

A "huddle" is a very short meeting that has the same agenda each time. The purpose of a huddle is to review goals and measures, identify needed actions, and agree to an action plan. A team leader or designated member leads the huddle. For a production team, this involves working with the planning/scheduling person. The team reviews the progress and follow-up agenda. Where results fall short of expectations, root-cause problem solving is used to address why. Action items are agreed upon for the team and others, such as those in engineering or maintenance. Action items from previous huddles are reviewed. The group agrees how it will meet demands.

Any team can use huddles as part of its communications plan. It is important to agree on frequency, length, timing, and a normal agenda for the huddle.

Guidelines for successful group meetings are as follows:

- A meeting facilitator should be appointed.
- An agenda should be followed.
- Everyone should be encouraged to participate; structured brainstorming should be used.
- Ground rules should be set and leaders empowered to use them.
- The leader should set the process and follow-up mechanism.
- Decisions should be recognized.
- Detailed notes should be kept.
- Progress from last meeting should be reviewed.
- Who will do what by when should be decided.
- The meeting should end by agreeing to the next steps.

Monitoring the Group

Working in groups may be a new concept for many. As a result, there will be some degree of uncertainty along the way. Here, cooperation goes a long way. Team members should focus on treating each other with respect. Group work also needs to be monitored. The following questions should be asked: "When will this team meet next?" "Who will lead this meeting?" and "How should the team prepare for the meeting?" The team should document its decisions as it goes. It should measure, track, and present its results on a regular basis.

Daily communication on customer needs and product flow is useful. As a team is responsible for action, measurement, and monitoring (control) of what it does, someone needs to be responsible for continuous improvement. The continuous improvement team should be a diverse group, comprised of people in marketing, sales, manufacturing, and other departments.

Steps for Improvement of Group Work

A team should:

- Prepare a list of what is being done well now.
- Ask what is working; this can build on basic ideas.
- Be responsible for action, measurement, and monitoring what is done.
- Step back from daily actions and evaluate the team's work.
- Look at its success over a long period of time.
- Identify areas that need to be improved.
- Work together to redesign the process and get better results.
- Seek clarity and then follow through.
- Record good ideas and use them as steps for improvement.

CHARACTERISTICS OF SELF-DIRECTED WORK GROUPS (SDWG)

Among the most successful product groups are those that are self-motivated. An important aspect of the motivated group is its ability to understand and focus on major goals, including improving quality, reducing cost, and delivery. Each group monitors what it does best on a daily basis to affect quality, cost, and delivery. Some companies form SDWGs and emphasize training and education for these groups. The success of SDWGs results largely from continuous changes taking place in the company.

Work-group-oriented companies recognize manufacturing tasks have evolved at a rate requiring high skill levels for entry-level employees and skill upgrades as experience levels increase. In many industries, SDWGs have evolved from employee-involvement programs. For example, the techniques of lean manufacturing, concurrent engineering, just-in-time production and maintenance, robust design, and computer-aided design/computer-aided manufacturing (CAD/CAM) are some key enablers of productivity. SDWGs have evolved out of ongoing training programs and other programs that encourage employee participation.

The backgrounds and capabilities of each team member and their abilities to achieve overall team goals must be studied. An array of matrices categorize skills as operational, administrative, and technical-functional. This allows the team to monitor its inventory of specific skill levels and then prioritize future training programs, which results in almost total self-direction. Each functional skill level is classified into progressive levels of competency for use as a tool to measure improvement toward team goals. Support teams work with SDWGs to develop systems that help increase autonomy for them and provide information needed to make sound decisions.

Group members must understand the importance of sharing the leadership position. Understanding the role of encouragement in the success of overall work efforts is equally important. Since humans naturally seek approval during goal-accomplishing periods, members must recognize each other publicly for a job well done. Compliments must become commonplace and negative comments should never occur. Management must provide necessary resources and structures for employee success. A focused performance by all results in group success. Senior management must set examples and follow the same formative structures they expect their employees to follow. To create a successful team, management and employees must consider certain elements: a common purpose, shared measures, and shared rewards. Without these elements, success becomes difficult for any team to accomplish.

START-UP PROBLEMS

With most companies adopting a philosophy of lean thinking, there has been emphasis on concurrent engineering. However, start-up problems are usually anticipated regardless of whether concurrent engineering is used to develop a new product or process. If people in various functional departments are comfortable with traditional design approaches, the switch to a group system may initially be difficult. In the traditional approach, each department has well-defined goals, and product progresses from one department to another, adding value every step of the way. Under the group approach, using the concurrent engineering philosophy, communication at various levels is important. Lack of communication among group members and other parts of the organization can hamper progress. Converting from a non-integrated approach to a more open and interactive group environment requires considerable adjustment.

Breakdown of Departmental Walls

In many institutions, departmental walls severely hamper cross-communication. Only rarely have people been required to interact and share ideas with each other. In large product-oriented companies, people from different departments may be complete strangers to each other. Individuals often mirror the attitudes of their respective divisions. People create their own walls, particularly if they previously have learned to act defensively. Leaders of interdisciplinary groups are faced with breaking the ice and optimizing group performance. Adequate managerial preparation is important in terms of convincing each department about teamwork advantages and the potential for successful product creation.

Poor communication is a major contributor to ineffective product development. In such cases, individuals do not receive information they need to do their jobs well. Product success depends to a large extent on management's ability to communicate strategic plans and objectives, so that everyone focuses in the same direction. Employees should keep management informed about their progress, achievements, problems, and concerns. Effective communication channels must exist between and among team members, teams, and functional organizations. Barriers to good communication can exist horizontally and vertically within a company. Communication between peers and peer organizations (horizontal barriers) can be as bad or worse than communication between management layers and between management and the work force (vertical barriers).

Companies with a strong functional structure sometimes build up significant interdepartmental problems. Meetings and joint activities can become adversarial proceedings. In such cases, departments may communicate with each other only when they have no alternative. Introduction of cross-functional teams can be effective in removing these barriers. It is important to note, however, that team members will experience some new challenges as well. They may find themselves torn between allegiances to their functional managers and other team members. An individual used to working alone or in relative isolation can feel uncomfortable working closely and interacting on business and personal levels with team members. Thus, the transition may be difficult for some.

Several common types of barriers to good communication are listed in Table 8-1.

Table 8-1. Communication barriers

Horizontal Barriers
• Departmental walls
• Poor communication
• Physical separation of employees
Vertical Barriers
• Poor downward communication
• Fear of reprisals

Horizontal Barriers

There are numerous barriers to good horizontal communication. These barriers restrict the interchange between peers, and result in decisions without sufficient information. Intelligent decision-making and plan execution depends on the complete knowledge of related facts and issues. Peer communication provides, in particular, information too detailed and technical to be communicated normally through vertical lines and over departmental boundaries.

Departmental Walls

Concurrent engineering philosophy for new product design encourages breaking down departmental walls. As with virtually all other communication barriers, top management must take the lead. Without its assistance and strong encouragement, such cultural changes cannot be successfully implemented. For many reasons, individual departments tend to become isolated and remote from each other. They erect communication walls between them as depicted in Figure 8-1. Information exchange tends to move up the departmental chain, until it arrives at the department head, over the departmental barrier, to the head of the other department, and then back down the chain until it reaches the intended recipient. As a result of this circuitous route, information gets delayed, distorted, and embellished along the way.

Managers may focus heavily on their department's goals, even when doing so is

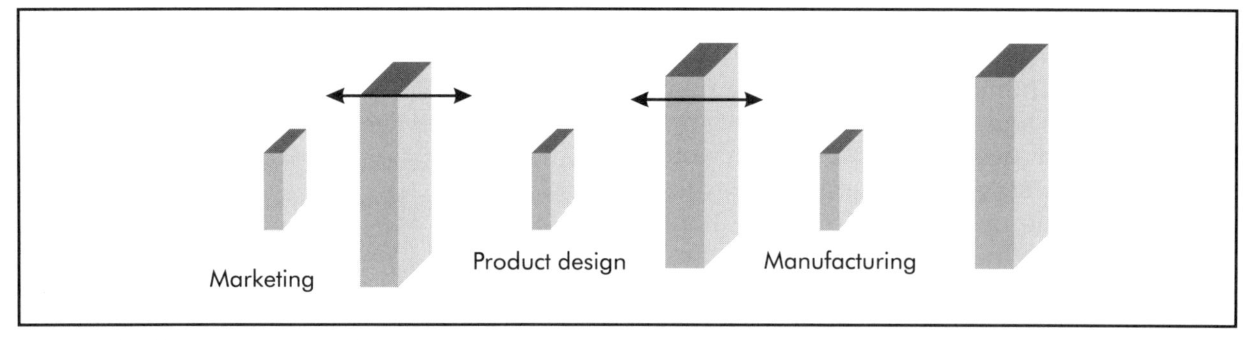

Figure 8-1. Walls between functional departments.

clearly to the detriment of the goals of the corporation or other departments. They may jealously guard their internal resources and assets. Frequently, they keep their problems to themselves. By hiding problems as long as possible from peer organizations, they increase the impact these problems may have on the company.

The following are some causes of departmental walls:

- fear of embarrassment;
- lack of interdepartmental trust;
- organization emphasis on the individual, rather than group or team goals;
- compensation system based on individual performance, rather than group accomplishments;
- interpersonal conflict between peers;
- fear of management reprisals for unexpected problems; and
- belief that problems will be solved before the rest of company needs to be informed.

The Lone Wolf

Some individuals prefer to work alone, rather than as part of a group. Such workers can be poor communicators, resisting requests for information or task sharing. Throwing a "lone wolf" into a team structure or culture without proper preparation and training can be disastrous to the individual and team. Lone wolves must be made to appreciate the benefits of team dynamics and activities. Their innate fear or resistance must be overcome. If these individuals cannot be convinced of the merits of teamwork and cooperation, it may be better to place them in positions where individual contributions are desired. Not everyone can be made to function successfully in a team environment.

Physical Separation of Employees

The effectiveness of communication between employees is related to the relative distance between their offices. The frequency of communication, especially face-to-face, decreases rapidly as distance increases. Dependency on e-mail, telephone, and written communications subsequently increases. It is important to have frequent face-to-face contact between team members. When team members work closely together, they develop social as well as business relationships. It is much easier to work and cooperate with someone who is known personally. Physical closeness fosters such relationships.

Vertical Barriers

Good vertical communication lines are important to the success of any company. They provide vehicles for management to inform the work force of company objectives, goals, policies, programs, status, etc. In addition, managers use vertical channels to give work direction, provide performance feedback, and improve departmental communication. Similarly, employees need good vertical communication lines to management. These channels are used to give feedback on employee concerns such as morale, reaction to policy changes, suggestions for job and work improvement, and ideas for new products and services. Employees should regularly communicate to management on job assignments, problems encountered, and concerns.

Poor Downward Communication

Inadequate communication from top management is a frequent cause of missed plans and objectives. Management must make its desires and intentions clear so that energies are properly tailored to meeting the right goals and objectives. If the company has a well-defined roadmap, it should be properly disseminated within the organization. For example, how can management expect the core team to make the right decisions in selecting products for development if management has not clearly stated its financial and marketing objectives?

Planned periodic management communication meetings can help create a company culture characterized by free and open communication. Management communication meetings should extend throughout the entire work force. Two-way communication during these meetings should be strongly encouraged. The objective is to create an environment in which employees view open bilateral communication as routine and expected.

Breaking Down Barriers

There is no magic formula to creating an environment that fosters good communication. The first step should be to determine the barriers that exist in a company. Managers and employees alike should be surveyed to obtain their personal perceptions. Internal (human resources personnel) or external (outside consultants) resources can conduct these surveys. It is important that the people used are unbiased, and that they feel free to report facts as they find them. Issues concerning fear as a barrier to open communication are particularly pertinent. For this reason, many companies elect to use outside consultants for fact-finding and reporting.

Having determined the issues and concerns, a program must be developed to address them. A program must address all key issues. For example, the practices of top-level management may be among the major deterrents to good communication. Addressing changes here will require the support and commitment of top management people. Changes in management style and culture cannot be effected overnight. They require continuous and dedicated effort. Periodic progress checks should be made and programs developed to address these issues and provide for perpetual modification. Management commitment should be clear and well communicated. Staff and individual contributors must understand and accept the objective of free and open communication.

A team focus on major goals produces substantial gains in product yields and reductions in cycle times. Teams result in greater customer focus, more work force cooperation, greater participation and involvement, and accelerated improvement. Employees on teams must receive proper training in group dynamics (social training) and problem-solving techniques (technical training). This training is essential to develop a high-performance product team.

In addition to traditional approaches to training, industry also uses another successful technique called experiential learning. *Experiential learning* focuses on learning from experience. This training technique begins with the formation of a group of employees, usually an established team or natural work group. A training facilitator assigns the group a problem-solving task, also called an initiative, with various resource constraints and a timeline for initiative completion. An initiative begins following an instructional period when the facilitator informs the group members of constraints for that initiative. Learning begins during the debriefing periods via reflection on what has transpired. This reflection on experience allows for self-discovery to occur. The overall training method has been effective in developing desired behaviors in employees. Group members must subsequently apply skills they learn and then develop them continuously to create a successful work environment.

CONCLUSION

Product design groups can be beneficial to problem-solving. A group consists of people using their combined skills to work toward a common goal. Diversity in the composition of group members makes for a wider range of alternatives and the solutions are naturally more creative. Groups make better decisions than individuals, who are usually more concerned with decision consequences.

APPENDIX: BELIEFS OF PROACTIVE PRODUCT GROUP MEMBERS

- Members of a group believe in respectful treatment of all people.
- Members of a group believe communication is critical to success, even when it is clearly a challenge.
- Group members approach change as a necessity. They recognize that the risk involved in change can create fear.
- The people in a group believe that those who do the work should be involved and empowered to improve processes.
- Members of a group know that a sustainable and competitive advantage can only come through the utilization of the unique skills, the work habits, and the inventions of its people.
- Group members encourage everyone in the organization to do their daily tasks and contribute to the company's ongoing and continuous improvement.
- They recognize that time pressures create tension.
- Members of a group recognize that although there is sometimes a need for dramatic action such as a redesign, steady improvement is a better, less risky approach.
- Group members believe that most people want to do a good job.
- Those in a group believe in positive goals, clear measures, and performing honest work.
- Members of a group believe that people should measure and report on their own results.

REFERENCES

Dieter, George. 2000. *Engineering Design—A Materials and Processing Approach*, 3rd edition. New York: McGraw Hill.

Clemson University. 1997. "High Performance Work Teams—Advanced Team Strategies." Conference Proceedings. Orlando, FL: Clemson University, Aug.

Caltech Industrial Relations Center. 1999. "Creating Breakthrough Products." Conference Proceedings. Pasadena, CA: Caltech Industrial Relations Center, June.

Chapter 9

Case Studies

PRODUCT AND PROCESS IN AEROSPACE (CASE STUDY 1)

In the last 10 years, thousands of firms have adopted modern techniques of product development. The aerospace industry has been in the forefront in applying modern techniques in product development and concurrent engineering to improve the processes used to create marketable products. This case study presents an ideal benchmark solution that may serve as a model for future work. It shows how a product progresses from design, manufacturing, and testing to customer delivery. Close observations of the development of a new aircraft shows how business-process reengineering has influenced the way other products are now designed, managed, manufactured, and marketed.

Product Development Techniques

The large airframe and jet engine industry is arguably among the toughest fields in which to compete. Due to the high development costs involved with designing an airplane or a jet engine, aerospace companies have always been keen to win government contracts. Their products take billions of dollars to develop; therefore, the processes used to develop and manufacture these products have to be efficient. In recent years, companies like Boeing, Pratt & Whitney, and General Electric have redesigned their processes (Norris and Wagner 1996, 1997; Sabbagh 1996).

Back in the 1950s, when commercial jets were on the verge of taking over piston-powered planes, product development was significantly different than today. It was rare to have design teams that worked side by side with manufacturing teams to develop a new airplane. Once designers completed a project, the design was passed on to manufacturing and tooling engineers who established fabrication and assembly sequences, job instructions, and processing and inspection requirements.

A modern technique Pratt & Whitney uses is to manufacture products with Kaizen continuous improvement philosophy. The aim is to reduce cost and the time from product concept to market, and to prevent problems with quality and reliability. The success of a product is not only based on the quality of the product, but also on the development costs and timeliness of the product launch.

Methods used to design and develop products have evolved over time. In a product's development there are typically three stages: design, process planning, and manufacturing. In the initial design stage, a small fraction of the cost to produce a product is committed. The material, capital, and labor to manufacture the product consume about 90% of the cost. The design process consists of the inclusion of many design decisions that go into the final design, affecting a major part of the manufactured cost of the product. The decisions made beyond the design phase can influence only about 25% of the final cost of the product. If there is any problem with the design just before it is launched into the market, it will cost a great deal of money to correct it.

Concurrent engineering is a concept that enables those impacted by design to have early design access and the ability to influence the final design and identify and prevent future problems. With concurrent engineering in place, an integrated product team (IPT) is created that includes people with diverse company backgrounds. The timing of inputs from other functions and suppliers takes place simultaneously with the creation of the performance characteristics of a design. With this approach, a smaller number of changes will take place, most of which will occur early, before the design is finalized. The computer systems interface so that all functions have immediate access to design changes and other information.

A newer practice used at companies like Boeing is to include suppliers and customers on the IPTs (Kolluri 1998).

Product Development Phases

Concept Development Phase

The concept development phase includes defining the market requirements, initial customer contact, and deciding on the project configuration. Requirements that designers and airline companies must study closely are speed, range, payload, furnishings, reliability, cost, maintainability, and commonality with other designs.

Configuration Refinement Phase

The configuration refinement phase starts off with preliminary design concepts and continues through a review of technology, cost, and schedule issues. Product development engineers conduct technical feasibility studies to refine the design configuration. A cross-functional IPT is formed to complete the configuration definition. The IPT consists of members from many disciplines, including design engineering, tooling, manufacturing engineering, quality assurance, customer engineering, finance, and program management.

Design Validation Phase

The design validation phase includes the validation of the product design and the manufacturing processes that go with it. Initial samples of a product are fabricated, assembled, and tested to ensure that customer requirements have been met. During initial production, defects in parts are tracked, the root-cause analyses are performed, and corrective actions are put into place. A new aircraft is first tested using a scaled-down model in a wind tunnel to determine the airframe configuration. Structural tests are performed using a real airplane; avionics are tested while the aircraft is in final assembly.

New Aircraft Development Strategy

Design Guidelines and Goals

By having IPTs, Boeing has been able to convert an idea for a product into something that results in the creation of the guidelines that can be generalized and used to manufacture almost any type of product. The design goals of an IPT should incorporate the following guidelines:

- high customer satisfaction,
- minimum product cost,
- short time to market, and
- high quality and reliability.

Figure 9-1 outlines the adoption of improved technology to produce an aircraft engine. The figure shows an example of an improved manufacturing process. A simple and economical grinding operation was developed for the turbine airfoil. The manufacturing cells have the characteristics of one-piece flow and eliminate non-value-added operations.

Get it Right the First Time

A product's manufacturability is a measure of how easily it can be manufactured:

- with engineering drawings;
- on schedule;
- with a high level of quality; and
- at low enough cost so the manufacturer can make a profit.

Techniques at Boeing used to make sure the 777 is able to be produced are the following:

- simplification—reduces the number and types of parts and part features;
- standardization—having standard parts, tolerances, and part families;
- component selection—selecting preferred sizes, weights, materials, and a near net shape;
- repairability—having a product that requires as little maintenance as possible;

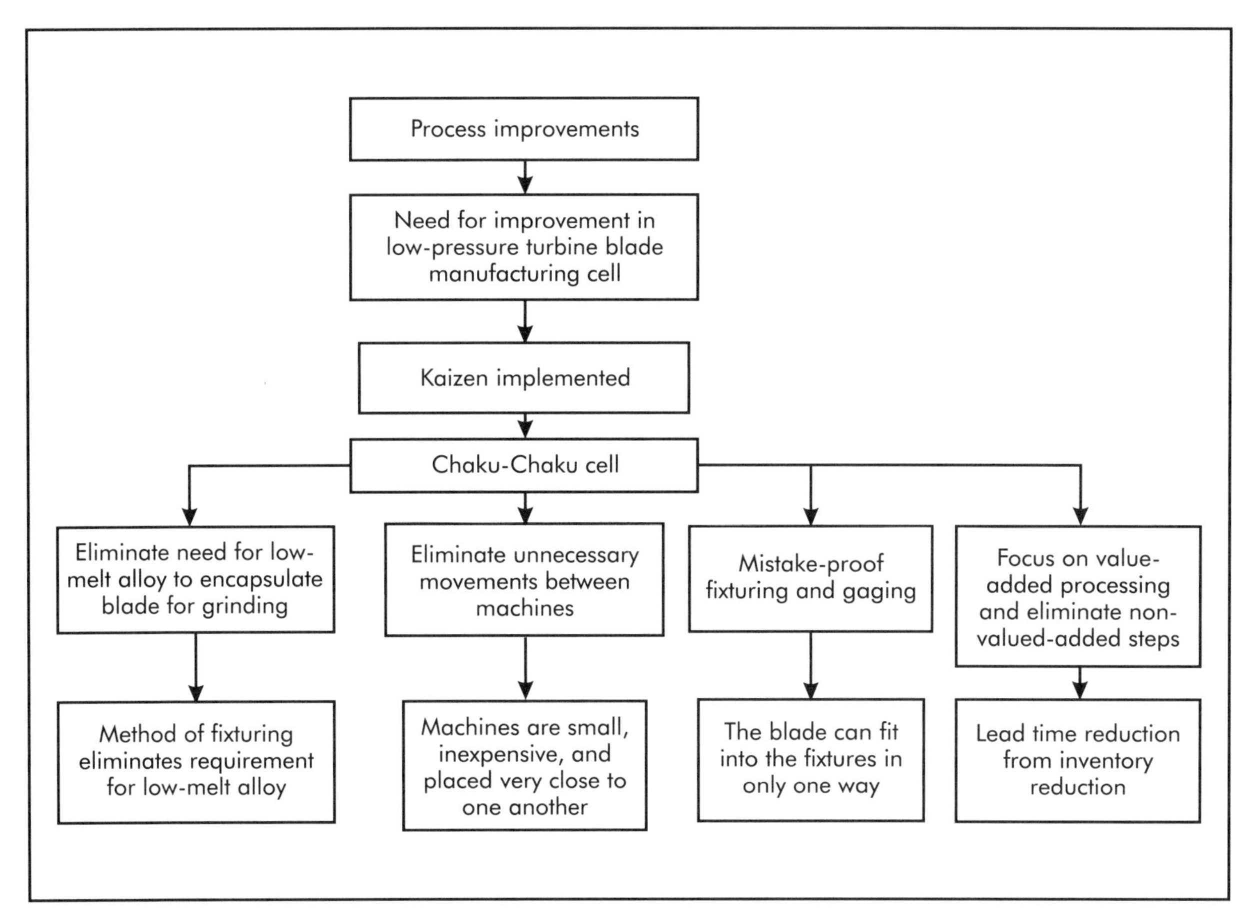

Figure 9-1. Strategy for development of an aircraft engine.

- the use of available high-quality, low-cost, low-risk manufacturing methods and processes; and
- the use of qualified and controlled vendors.

Integrated Product Team

An integrated product team (IPT) must meet the following criteria:

- 10 or fewer members on a team;
- members must volunteer to serve on the team;
- members should serve on a team from the beginning stages of design;
- members are assigned full time to a team;
- members report solely to the team leader;
- the team has a member from each functional organization (marketing, finance, design, purchasing, manufacturing, and repair); and
- members are located relatively close to each other.

Development Techniques Learned

The seven guidelines discussed in the following sections are easily adapted to other industries and an excellent blueprint for a solid product development process (see Figure 9-2).

1. Understanding the Market

Boeing assembled representatives from leading airlines around the world to first identify the market for the new 777 aircraft. It let the airlines tell them exactly what they wanted. This should be true in any industry; it is important to let customers define their market, and what type of product fulfills their requirements.

2. Integrate Product Teams

Personnel from various functional organizations and the customer work together throughout the design, development, manufacturing, and testing phases of a project. Boeing's use of design and build teams resulted in the company optimizing every aspect of the aircraft to meet customer requirements. At the same time, the aerospace company was able to make sure the product was producible, maintainable, and within budget.

3. Computer-aided Design

In recent years, the main mode of representation of designs has shifted from drawings created by computer, to 3D computer-aided design (CAD) models. These CAD models exhibit design as a representation of 3D features. The use of CAD methodology allowed Boeing and its suppliers to design entirely on computer, saving on manufacturing and production costs. Designers were able to make sure all parts fit together properly before anything was made; the design changes were made on a computer before production began. The benefits included the ability to visualize 3D forms of design, and an ability to automatically compute properties such as mass and volume. Historically, the aerospace industry has used a full-scale prototype of a plane to detail geometric interference, using structural elements and components. Using CAD technology, once a digital mock-up of the aircraft is made with CAD, conflicts between the aircraft's structure, brackets, tubing, ductwork, and electrical wiring runs can be identified and corrected before the parts are actually manufactured.

4. Continuous Improvement

Over the years, the concept of quality has changed in the aerospace industry. Instead of inspecting quality at the end of an entire project or process, it is built-in from the very start through the team approach. Today, both the customer and life-cycle cost drive quality improvement. Continuous improvement philosophy focuses on the product's infrastructure. It emphasizes that the only

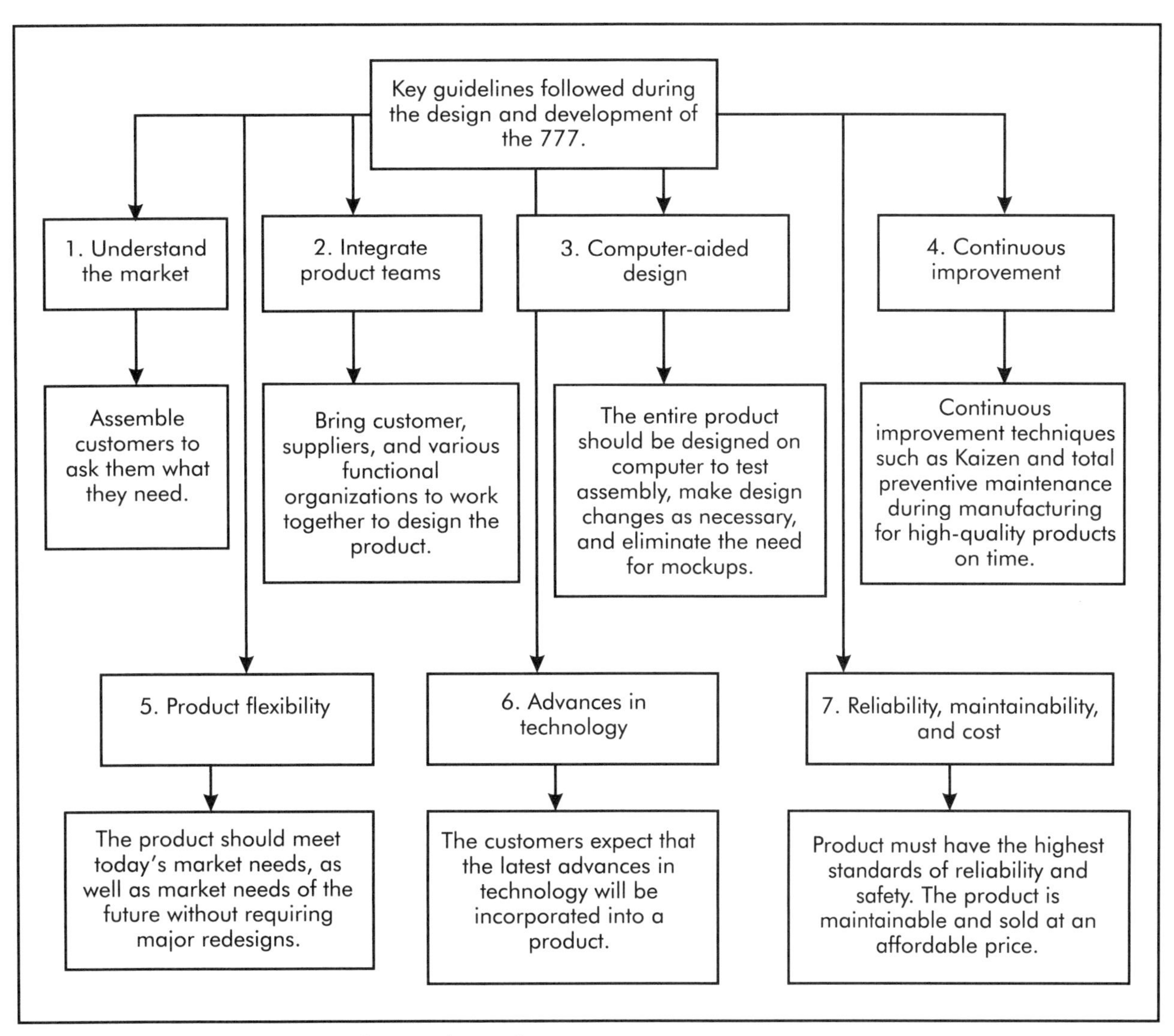

Figure 9-2. Seven development techniques learned.

way to increase value in a process is to continuously seek out and identify waste, and then eliminate it. Each time waste is removed, the process value increases. The regular program of looking for waste and removing it is the foundation of continuous improvement. Through continuous improvements, the process becomes more competitive with lower cost, higher quality, and shorter delivery times.

5. Product Flexibility

Boeing designed the 777 with an incredible amount of flexibility so that it will meet not only today's requirements, but also the future requirements of airlines as market needs change. The 777 can be stretched or shortened, and range and engine power can be increased or decreased to fill niche markets. The idea that a product should be designed flexibly so it can grow or shrink to meet niche and emerging markets with minimal manufacturer investment is an important concept for all industries.

6. Advances in Technology

The airframe manufacturer incorporated the latest advances in avionics, propulsion, wing design, flight entertainment systems, passenger comfort features, etc., into the design while the engine manufacturer provided fuel-efficient, reliable engines for the 777. It was important for Boeing to provide airlines with the latest technologies available. This is true for any industry; the customer expects to get the best reliable and affordable technologies.

7. Reliability, Maintainability, and Cost

No matter how technically advanced a product is, a customer demands a high level of reliability, simple maintenance, and product within a certain cost. Boeing had to design redundant systems into the 777 so that, if one failed, another could perform the job.

IMPROVEMENT USING ASSEMBLY ANALYSIS (CASE STUDY 2)

This case examines an approach to evaluate the installation of a product by highlighting the strengths and weaknesses of an assembly process. It discusses the assembly analysis technique as a tool for installation of a new product. The merits are explored with help from industrial case studies. The first example considered is the field installation of a product. The second example is packaging and handling.

Introduction

Customer satisfaction is a primary factor in successful product development. The modern product development process uses basic customer expectations for input and concurrent engineering as a design approach. The goal is to develop the best product for function, manufacturing, assembling, reliability, and servicing. At the same time, there is increased pressure to get products of higher quality to the market in a shorter time. The product performance to price ratio is scrutinized more carefully. The traditional approach to product design resulted in an insufficient definition of the product, an inadequate cost analysis, and an inability to make design changes. That approach has been replaced by modern product design techniques that compete effectively in the global market.

The new approach initiates the design of a product and its associated processes. Studies show that time spent early in the design stage, when prototyping takes place, is more than compensated for by a savings of time later in the process. Some characteristics of this approach are:

- a better definition of the product, without late changes;
- process knowledge and how to effect product development; and
- precise and accurate cost estimates.

Design is a complex process. It requires a wide variety of knowledge that a single person does not possess. For that reason, it is imperative to form multifunctional design teams for product success. Such design teams have to create a product that addresses the requirements of robustness, design for manufacture and assembly, reliability, and the environment. Using concurrent engineering principles as a guide, the designed product is likely to meet the four basic requirements of quality, cost, just-in-time to market, and customer satisfaction. The product cost is greatly affected by decisions made in the design stage. Some popular design and analysis techniques are the assembly analysis method (Boothroyd and Dewhurst); the assembly evaluation method (Hitachi); the Lucas Method; and the axiomatic method.

In recent years, as design for assembly, disassembly and recycling has gained more recognition, a number of organizations have devised their own sets of suitable guidelines for design practice. Research in life cycle engineering, product design, modeling, and integration has received significant attention. Most previous adaptations of design for manufacturing and assembly were limited to product redesign. These procedures are aimed at minimizing the product assembly time and the number of components. The current study is an attempt to use an assembly analysis technique for the field installation of a product, thereby making the installation more efficient.

There have been some industrial efforts to extend design for manufacturing and assembly programs to more engineering operations, such as design for installation in the elevator industry. The elevator and escalator industries have unique requirements; the final assembly installations must take place at the building sites. Work conditions include weather exposure, confined spaces, and ergonomic hazards, besides large subassemblies requiring lifting devices. The installation is fundamentally an assembly process.

Although the major goal of design for installation techniques is to improve design-related processes such as handling and insertion, the techniques are also effective for design-independent process improvements. The elevator industry has shown that the design for installation procedure is not just design for assembly in a construction environment. The design for installation strategy shown in Figure 9-3 can provide information on design-independent improvements such as improved packaging to eliminate on-site reworking, and better documentation of electrical power transmission layout. It also provides information on installation cost drivers and their variation from location to location (Orelup et al. 1997).

In this case study, a comparative analysis of an existing and a modified electrical conduit (also known as a raceway) is undertaken. Initially, customer feedback using quality-function deployment is used to identify product features. The new design investigated has a special feature that isolates the communication section from the power transmission section. Other highly ranked features of the product are its low installation costs, and flexibility in addition and manipulation of the device on the communications side.

Method

The assembly analysis method is an analytical technique designed to evaluate the potential for automation of an existing product or new product designs after engineering drawings or prototypes have been developed. It is an empirical method. Selecting an assembly method followed by assembly analysis and design improvement is basic to the process. It distinguishes between manual and automatic assemblies. Design improvement is focused on part-number reduction and shortened associated process times. In assembly analysis methodology, a classification and coding system determines the handling, insertion, and fastening times of parts to evaluate the efficiency of the assembly. This methodology provides a graphical interface for the user to input physical characteristics of the assembly items and display analysis results—such as design for assembly efficiency, assembly time, and cost—in the form of a report.

The designer has to evaluate the geometry of each component in the product or its subassemblies and then show the difficulty of part handling and insertion. The results would include an estimated assembly cost and a direction for a redesign to improve the product. The main goal of the technique is to minimize the cost of product within the constraints imposed by design features. The best

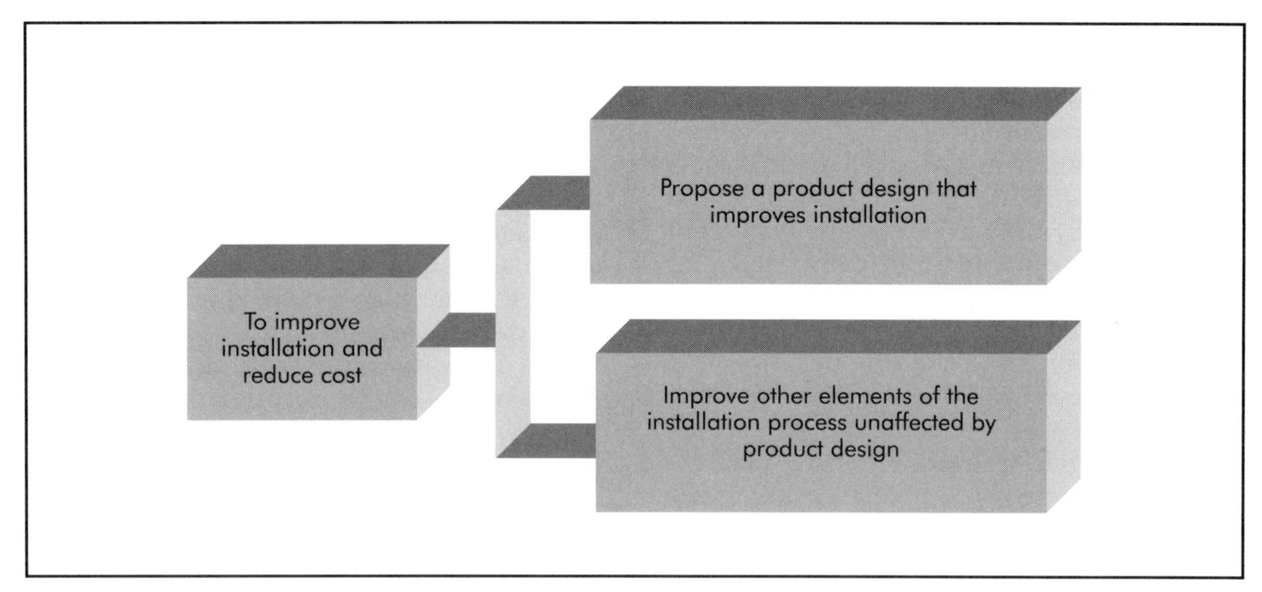

Figure 9-3. Design for installation strategy.

way to achieve this goal is to reduce the number of components to be assembled and then to ensure that the remaining components are easy to install or assemble. In the early stage of design, the designer must evaluate the assembly cost, which means that he or she should be familiar with the assembly processes. The designer should have a logical explanation for requiring parts that result in a longer assembly time, and should be aware that a combination of two or more parts into one will eliminate an assembly operation. Design efficiency is judged on the basis of the strength of an assembly. The whole procedure, shown in Figure 9-4, consists of three basic steps:

1. assembly method selection,
2. assembly analysis, and
3. design improvement.

Due to differences in using human operators versus automated assembly lines for assembly, there are also significant differences between manual and automatic assemblies. The cost is related to both the product's design and assembly. A product's minimum cost can be achieved when the appropriate assembly method has been selected.

Design efficiency is determined by using the appropriate formula. Manual assembly design efficiency is obtained by using the following equation:

$$E_M = 3 \times \frac{N_M}{T_M} \tag{9-1}$$

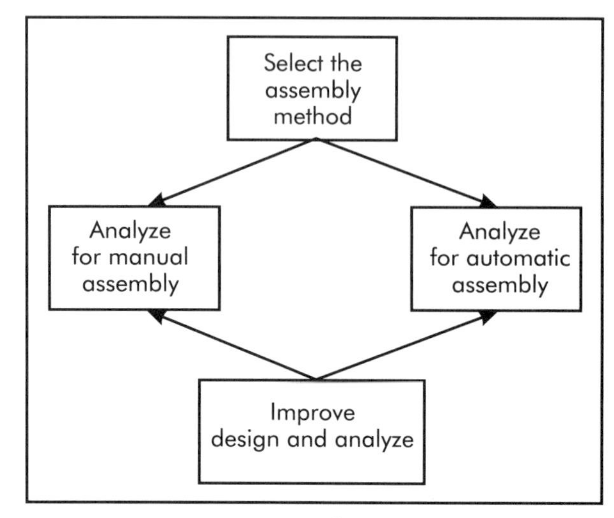

Figure 9-4. Stages in design for assembly.

where:

E_M = manual design efficiency
N_M = theoretical minimum number of parts
T_M = total assembly time

The automatic assembly design efficiency can be determined by using the formula:

$$E_F = 0.09 \times \frac{N_M}{C_A} \quad (9\text{-}2)$$

where:

E_F = automatic design efficiency
N_M = theoretical minimum number of parts
C_A = total assembly time cost

A redesign has to produce a better product with a higher design efficiency. The most effective way of improving design efficiency is through part-number reduction. In the case of manual assembly, operation time reduction is another way of improving the design efficiency. In automatic assembly, feeding and orienting efficiency need to be improved.

Method Applied to Product Installation (Case Study 2A)

In the first example, a comparative analysis of the installation aspects of an existing and modified electrical conduit (raceway) is undertaken. The two models investigated have differing features. The modified design has a special feature that would isolate communication and power lines, as shown in Figure 9-5. In addition, this study explores the possibility of using assembly analysis methodology as a standard tool in the product development department of a company. Design for assembly analyses on the following installations were investigated in the order shown below (Campana and Kondo 1998):

- Example 1A: existing product—initial electrical installation;
- Example 1B: existing product—initial telecommunication installation;
- Example 1C: existing product—secondary telecommunication installation;

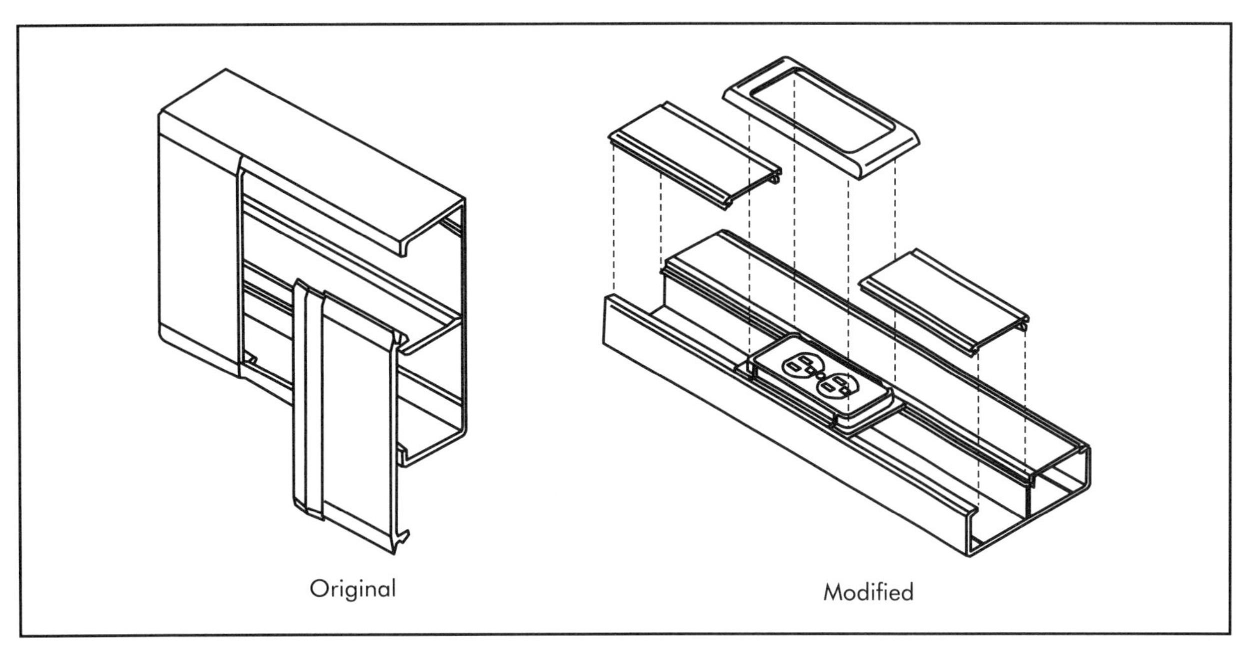

Figure 9-5. Original raceway and modified reaceway with separated communication and power lines.

- Example 2A: modified product—initial electrical installation;
- Example 2B: modified product—initial telecommunication installation; and
- Example 2C: modified product—secondary telecommunication installation.

A typical assembly installation process for a raceway is shown in Table 9-1. In these cases, the experimental setup consisted of an electrical conduit section and placement of internal and external elbows, followed by electrical and telecommunication device installation at midpoint in the raceway. For every product, the electrical contractor completes the electrical installation on the power side. The telecommunications contractor performs communication installations on two occasions, after power installation has been completed. Design for assembly analysis is performed on each of the three scenarios for the two products. The assembly and disassembly operations are defined as customized operations in the analysis database. This includes operations for tool acquisition and other activities. Results include design efficiency, total assembly time, and total assembly cost as displayed in Figures 9-6 through 9-8.

The design efficiency index of the modified product for initial electrical installation is 13% higher than for the existing product. The total assembly time for the modified product system has a 9% higher assembly time than for an initial electrical installation, but a relatively lower assembly time for a telecommunication installation. The total operation cost of the modified product shows exactly the same trend as the total assembly time. It has a 9% higher operation cost than for an initial electrical installation, but a relatively lower operation cost than for a telecommunication installation. The modified product for an initial electrical installation has a 15% lower assembly weight than the existing raceway system for an initial electrical installation. The modified product for an initial electrical installation has a 14% higher total part cost than an existing product system.

The case study shows that the most time-consuming operations for each installation are operations such as screwing the receptacle, wire insertion, cover placement, screwing the bracket, cutting the covers, and connecting the wires followed by stripping the insulation material after measuring the appropriate length. The major activities of a telecommunication installation are operations such as untwisting wires, stripping the electromagnetic shield, placing wires on terminals, inserting communication cables, and disassembling the second cover. The same types of operations are involved during secondary telecommunication installation. Most of the labor-intensive operations are procedures such as cutting wires, untwisting wires, punching wires, placing wires on terminals, connecting wires, stripping insulation materials, stripping electromagnetic (EMI) shield materials, terminal caps placement, and cable insertion. This study gives a clear guideline on the importance of improving cover placement/removal operations, as well as measuring and cutting operations. The case study also identifies the importance of certain time-consuming operations. It highlights operation constraints created by specific industry practices.

Method Applied to Packaging

In the previous study, an important component of an assembly analysis was the identification of time and the assembly process. The aim of this study is to show the normalized times for a sequence of operations involved in packaging. The assembly analysis method for a packaging problem can provide insight into the normalized times for each sequence of operations involved. A list of product-packaging operations is shown in Table 9-2. Their respective time normalized study results are displayed in Figure 9-9.

Table 9-1. Assembly sequences for installation of raceway device in the field

Name	Description	Number of Items	Theoretical Minimum Number of Items	Assembly Time, Seconds
Example 1A: Single Co.	Main assembly	—	—	—
Compartment base	Add and thread	1	1	14.4
Clip	Add and snap fit	2	0	7.2
Wire	Add	1	1	3.2
Wire insertion 1	Operation	3	—	30.0
Wire cutting 1	Operation	3	—	8.0
Bracket	Add and snap fit	1	0	3.6
Subassembly; device	Add	1	—	4.8
Wire	Add	1	1	3.2
Wire cutting 2	Operation	3	—	8.0
Stripping insulation	Operation	6	—	14.0
Device	Add	1	0	3.3
Wire insertion 2	Operation	3	—	6.0
Screw	Add and thread	3	0	32.0
Apply adhesive tape	Operation	1	—	5.0
Wire cutting 3	Operation	3	—	8.0
Stripping insulation	Operation	6	—	14.0
Connecting 9 wires	Operation	3	—	17.0
Device placement	Operation	1	—	7.0
Screw	Add and thread	2	0	22.9
Measure with tape 1	Operation	2	—	12.0
Cutting cover 1	Operation	2	—	15.0
Cover placement 1	Add and snap fit	2	1	24.8
Trim ring	Add and snap fit	1	0	3.6
Face plate	Add	1	0	3.3
Face screw	Add and thread	1	0	12.2
Blank face plate	Add	1	0	3.3
Internal elbow	Add and snap fit	1	0	3.6

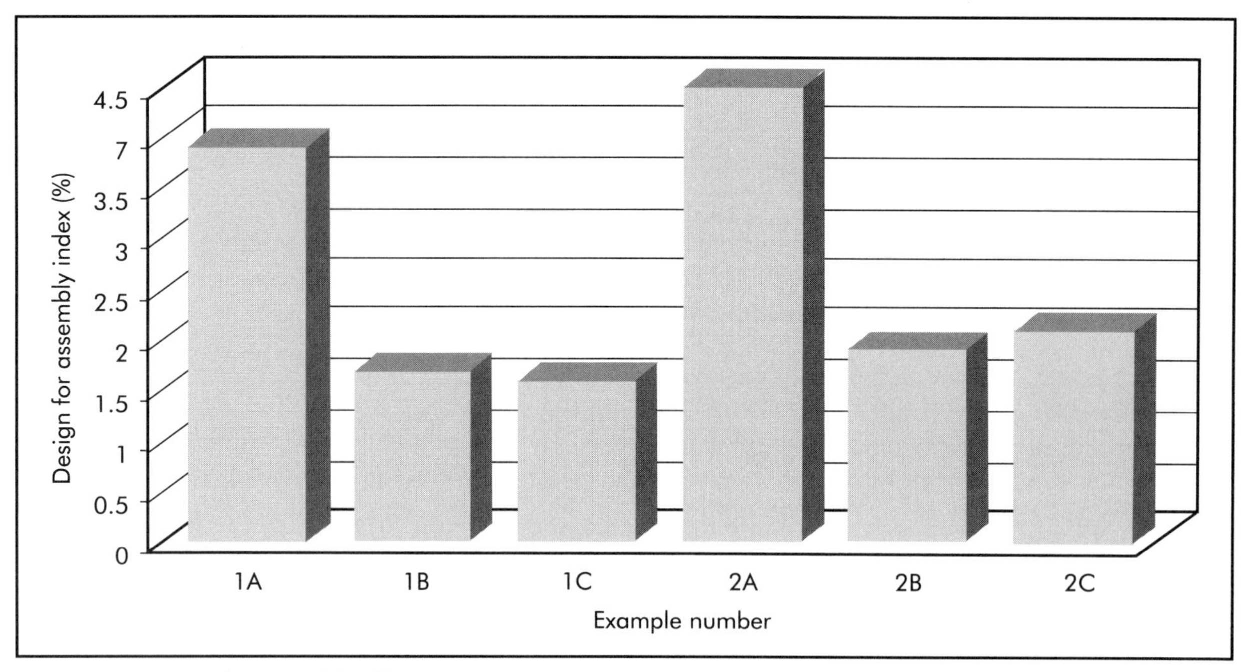

Figure 9-6. Design for assembly efficiency.

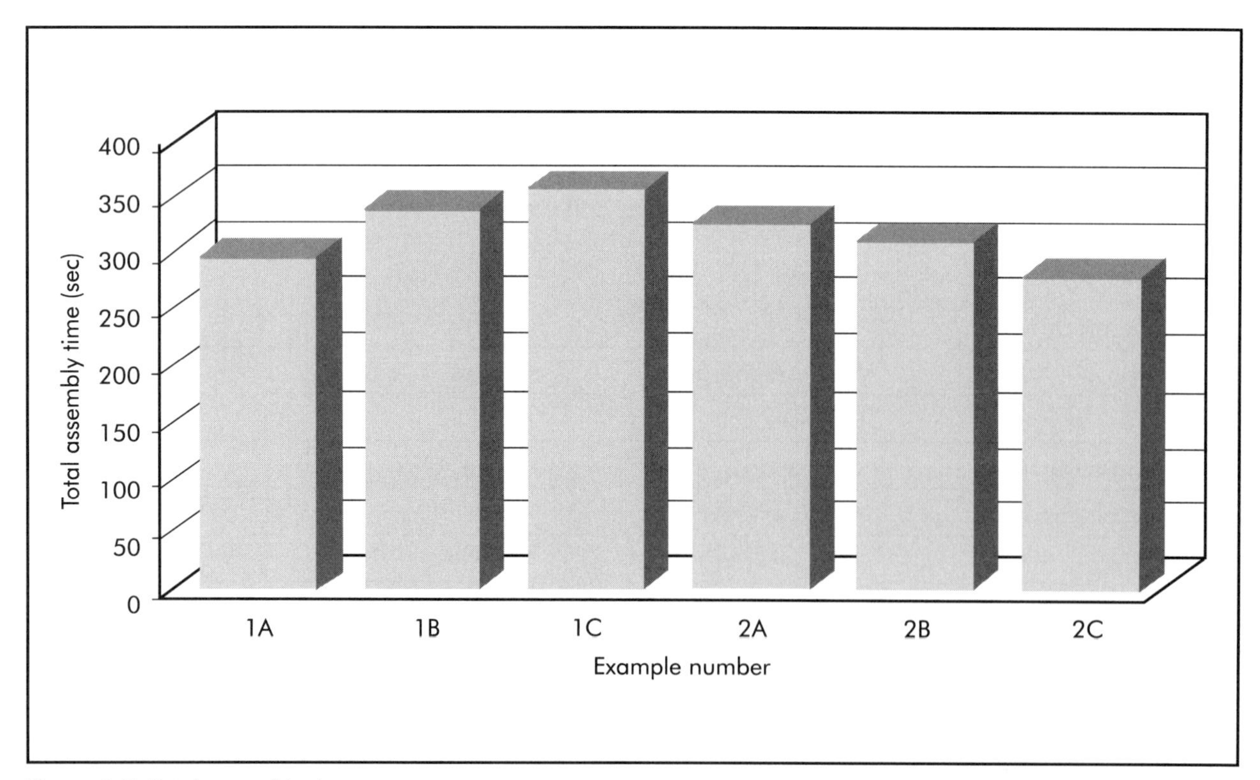

Figure 9-7. Total assembly time.

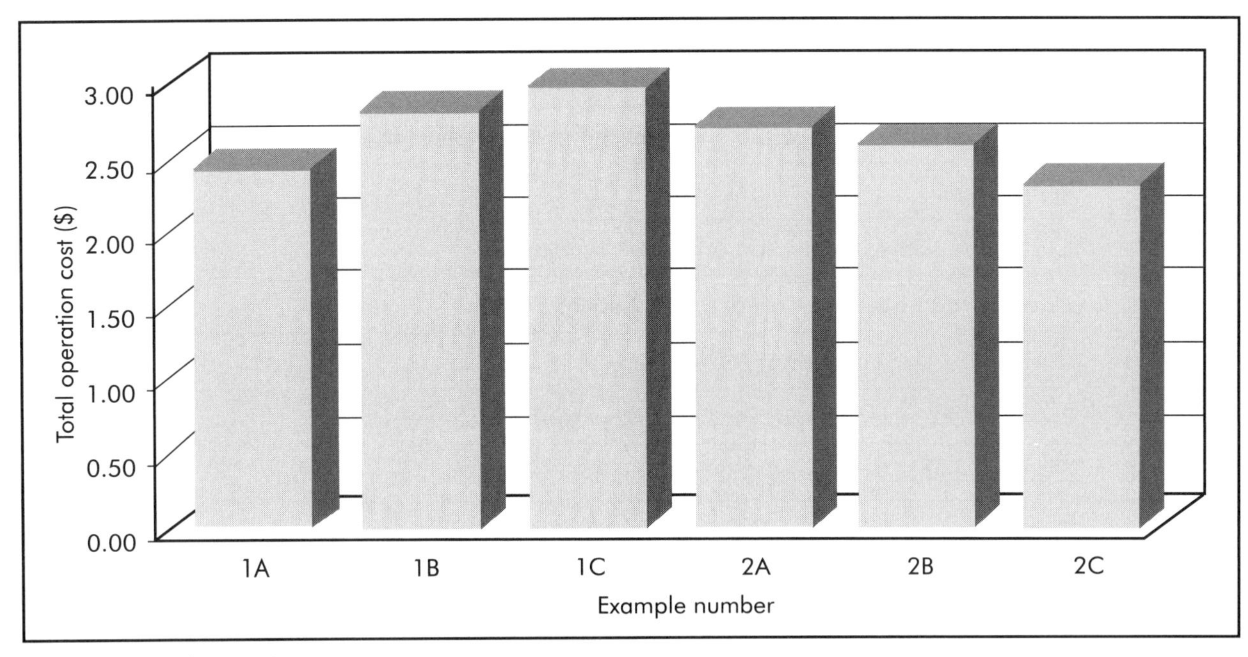

Figure 9-8. Total operation cost.

An assembly analysis method identifies the cycle time, the orders per shift, the operator performance, shipping times, and the optimum number of shifts. The study shows the minimum distance between the operator and order cart to reduce the average time of an operation. It can provide information and predetermine the size and type of a carton before the order form reaches the operator at the handling table. The analysis provides information on the average operation time that can lead to a decision that speeds the overall cycle time for packaging one single order. It can also lead to further decision-making on the types, sizes, and availability of packages leading to a higher level of productivity.

APPLICATION OF DESIGN METHODOLOGY (CASE STUDY 3)

Introduction

The designer proceeds from the recognition of customer needs or desires to be satisfied to satisfaction of them. This process involves a significant amount of reporting and feedback, as well as consideration of the value-related issues of the problem solution. Identifying the problem correctly at this stage saves time and money, and makes it easier to reach a satisfactory solution.

The following questions function as guides in defining the problem:

- From where does this action spring? What is the frame of reference?
- What is the action for?
- Who will it benefit?
- By what means can it be best carried out?
- In what conditions of time and place will it be occurring?
- Who is the independent agent to carry it out?

Playpen Design

In this case study, the problem is that of designing a safer playpen. A typical problem definition would be based on the following

Table 9-2. Elemental breakdown of product-packaging operations

Operation	Definition of Activity
1	Obtain order: pick up the product from the cart and place it on the handling table.
2	Verify order.
3	Cut off the addressee section on the order form.
4	Place the addressee section on an adhesive.
5	Choose the shipping carton/envelope.
6A, B, C	Erect the carton/envelope: FedEx carton—small, medium, large.
6D, E	Erect the carton/envelope: USPS Priority Mail envelope, large carton.
6F	Erect the carton/envelope: bag #5 envelope.
6G, H, I	Erect the carton/envelope: corrugate carton—small, medium, large.
7	Pick up accessories.
8A, B, C	Dispose both product and accessories into the carton/envelope: FedEx carton—small, medium, large.
8D, E	Dispose both product and accessories into the carton/envelope: USPS Priority Mail envelope, large carton.
8F	Dispose both product and accessories into the carton/envelope: bag #5 envelope.
8G, H, I	Dispose both product and accessories into the carton/envelope: corrugate carton—small, medium, large.
9	Remove the strip on the carton/envelope to expose adhesive and close the carton/envelope.
10	Label the carton/envelope with the addressee section after applying adhesive.
11	Dispose the carton to the conveyor or the mailing bag if envelope.

observations: The folding-side playpen presently on the market has been associated with the deaths of several infants up to the age of 12 months. The deficiency in the present playpens is that if the mesh side is either down or up, but not locked, an infant may slip over the edge of the playpen flooring and then become trapped and suffocate in the slash mesh. There are a number of caretaker accidents reported as well. The most common occurrences have resulted in back injuries to the caretaker when he or she bends to lift the infant. Based on this criteria and the data obtained from the Consumer Product Safety Commission (CPSC), a definition of the problem is made. It is proposed that the design of a safer playpen will successfully address the above observations.

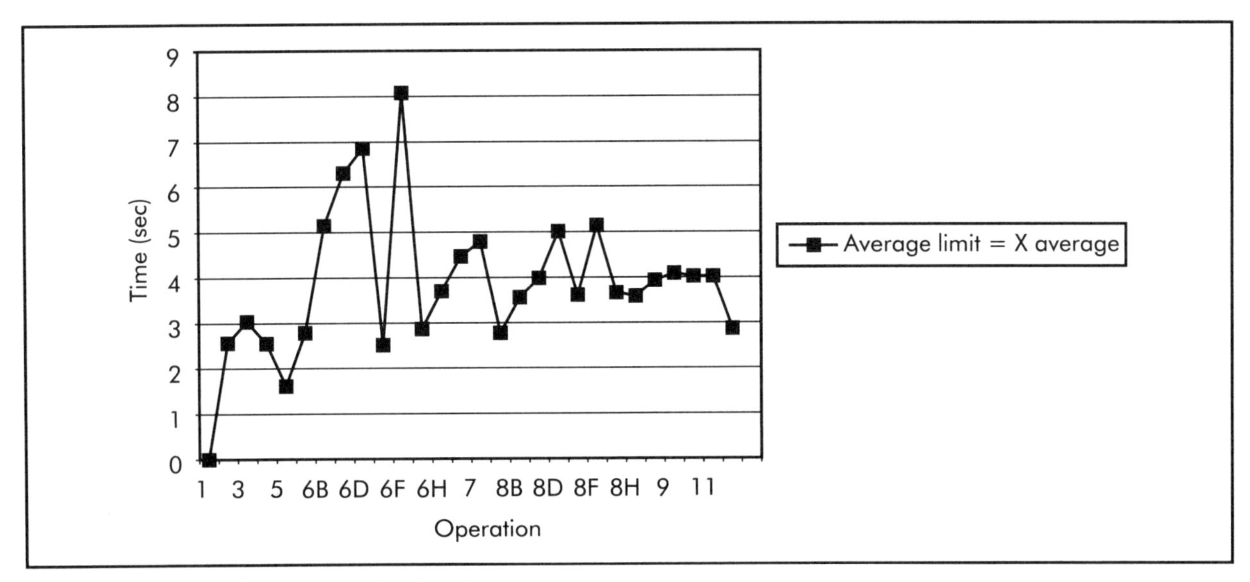

Figure 9-9. Results of time-normalized study.

Establishing Design Criteria

To establish the design criteria, the constraints, assumptions, and facts relevant to the problem are considered:

1. *Constraints* are factors that affect the outcome of a project. They cannot be changed. Constraints are listed after the goal statement. The constraints might show reflections of values and budget.
2. *Assumptions* are facts or statements assumed to be true. The first step in this stage is to clarify assumptions in regard to the problem. Once clarified, assumptions often can be modified to simplify a problem and make it solvable.
3. *Facts* are listed to help clarify what is known, what is not known, and what a designer might want to find out prior to proceeding with a project.

In the playpen design, the design criteria would be:

1. Constraints:
   - limits on the dimensions of the playpen;
   - cost and time factors in redesigning and manufacturing the playpen;
   - ensuring the transportability of the playpen; and
   - meeting minimum government standards on strength and safety.
2. Assumptions:
   - a minimum understanding of product use is expected by the consumer; and
   - the existence of a demand for safer playpens that ensures mass production and profit.
3. Fact:
   - reports from government safety commission show record of playpen safety.

Sources of Information Needed

There are three broad categories of information that engineers have to comprehend:

1. Conceptual—this category covers fundamental principles and laws such as those governing the conversion of mass, energy, momentum, etc. Being principles,

they do not change, unless a fundamental and radical philosophical change in the view of the world's scientific community takes place. The professional engineer must be proficient in these principles.

2. Factual—this represents the kind of information to be found in handbooks, such as properties of substances. It is the sort of information that keeps growing and changing as new substances are invented and new products are developed.
3. Methodological—this represents a link between the above two categories. It is the knowledge of the methods and ways that conceptual information can be applied to generate factual or more conceptual information. It is a skill, a set of attitudes and procedures, which can only be acquired experientially.

The information needed for a safer playpen design would be:

- information on the biomechanics of human anatomy, such as the principles that govern human body movements;
- information obtained from different surveys found in journals or publications from CPSC; and
- information on the use of existing playpens and their deficiencies.

Generating Options and Solutions

The goal of generating options and solutions is to accumulate many potentially useful solutions to the problem. Designers need to be innovative and should maintain an open, receptive mind to new ideas. An appreciation for the unusual or extraordinary is also important. This ability comes from a great deal of practice as designers continue to develop this skill throughout their careers. Solutions have to be feasible and well posed with respect to the problem.

For example, possible solutions for designing a safer playpen would be:

- improving the latch mechanism, which would lock even if the caretaker were to be negligent;
- the design of a detachable, rather than foldable, playpen, which would require assembly;
- the design of a swinging-door playpen; and/or
- an inflatable playpen design.

Evaluation of options. Once a designer has generated solutions, he or she must evaluate the various alternatives and make a decision. Selection of a particular solution is based on understanding the relationship between it and the ecological, social, political, and cultural world in which it takes place. Values play a strong role in this step.

The method to evaluate proposed solutions is to answer the following questions for each option:

- Does the solution satisfy the basic objective of the project?
- Is the solution theoretically feasible?
- Is the solution practical?
- Is the cost within the means?
- Is the proposed solution safe to operate?
- Is it the optimum solution?
- Does it satisfy the constraints?
- Does the solution satisfy all the human, social, and ecological factors involved?
- Is the solution aesthetically acceptable?
- Is the solution legal?
- Can the project be completed in the time allotted?

Feasibility study of the preferred option(s). At this stage, a feasibility study has to be prepared. The study is a short proposal outlining what the designer believes is necessary to analyze. The proposal should indicate:

- what solution(s) is being investigated;
- the reasons for these selections;

- coordination with other groups working on the project;
- what constraints, assumptions, theories, principles, variables, and parameters are being used;
- what the goal is of the analysis;
- the elements of the problem; and
- how transformed information resulting from the analysis will be used by the group and disseminated to other groups (publication, patent, and secrecy need to be thought of at this stage).

Ranking of options. In this step, the designer needs to identify the first-, second-, and higher-order impacts of the solution on society and human beings. Then, based on an understanding of the value issues associated with these impacts, he or she should try to rank alternatives based on the benefits and costs or risks for each.

A useful method for ranking each solution is the creation of a decision table with the following ranking scheme:

- 1: yes/acceptable/good;
- 0: fair; or
- –1: no/unacceptable/poor.

A designer can also devise his or her own ranking system.

Selection of preferred options. After alternative solutions have been evaluated and ranked, the designer chooses the alternative that maximizes the benefits and minimizes the costs and risks. This is quite complicated, since any quantification is based on the values that vary among individuals and groups. A solution need not be the only choice; additional judgment and intuition could lead to a more acceptable solution.

Detailed analysis of preferred option(s). This step involves separating the possible solution into meaningful elements. The chosen solution for the playpen design is broken down into subcomponents:

- size, dimensions, and framework of the solution;
- materials of construction;
- the biomechanical analysis of the caretaker;
- a safety analysis;
- a hazard analysis;
- a cost analysis; and
- the marketability.

HEADER ATTACHMENT IN A HEAT EXCHANGER (CASE STUDY 4)

A major step in the product realization process is the evaluation of alternate concepts from the viewpoint of design requirements. In this example, a design problem with a number of possible solutions is examined to find the best possible solution. This case is about the joining method of a header in a heat exchanger. A heat exchanger is a device in which two or more fluids exchange thermal energy through a heat-transfer surface. Heat exchangers are found in applications such as space heating, air-conditioning, power production, waste-heat recovery, and chemical processing. The header is an integral part of the heat exchanger. This case study analyzes and determines the best method of attaching low-pressure headers to the core, taking into consideration the customer requirements, and ultimately resulting in reduced product cost.

The heat exchanger typically has four headers welded to the brazed core. Hot and cold fluids enter and are exhausted through the headers. The core is made of thin fins, brazed to channels for support. Two end plates hold the assembly together. Core bands are welded at four places to the end plates. The core band provides structural support to the assembly. Headers are welded to the core bands. Due to the frequent need for inspection and repair of the fins, headers have to be removed and replaced (see Figure 9-10).

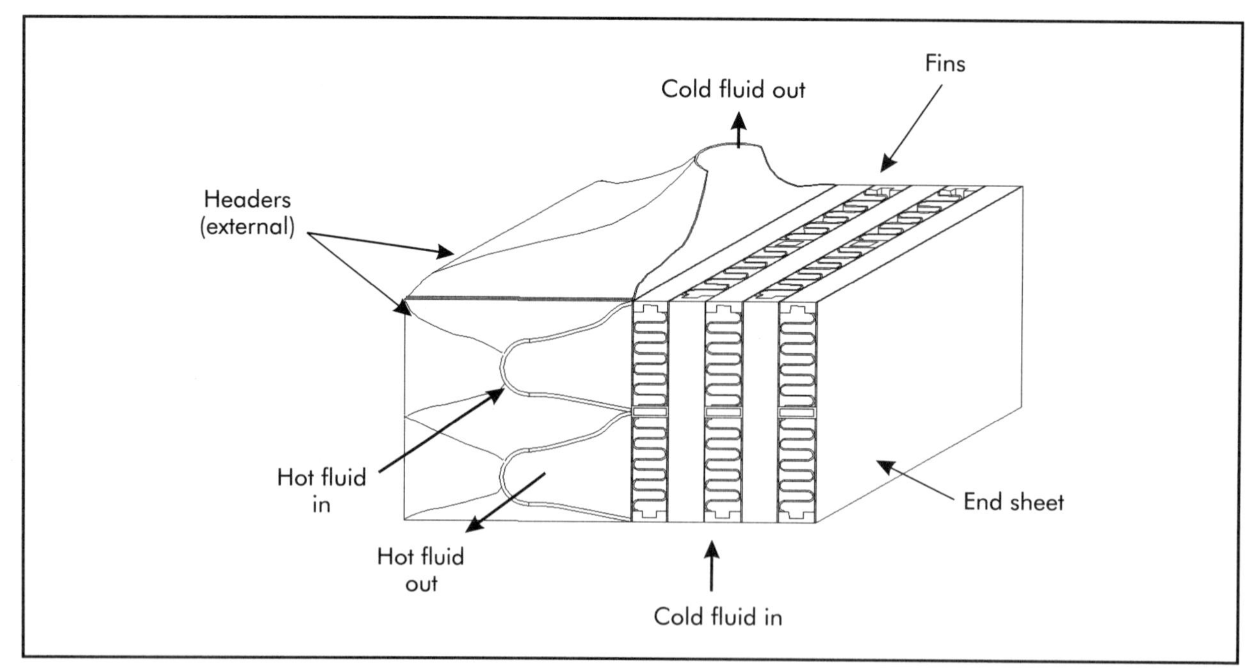

Figure 9-10. Schematic of the heat exchanger.

Most headers are welded to the body. The disadvantages of removing welded headers and re-attaching them are many. The removal of a welded header can cause core damage, which is difficult to repair. Additional finishing work is needed after attachment.

Feedback from the customer indicates that the present design needs improvement and there is a need to replace it with an alternate. The new design should take the simpler approach of attaching headers that would not only reduce the overall cost of rework and repair, but also increase the productivity of the unit.

Requirements for the header connection are:

- easy assembly,
- easy removal,
- easy repair of fins,
- ability to withstand operating conditions,
- no change of material,
- proper fit in the core,
- avoidance of core damage,
- ability to meet leakage requirements,
- reduction of assembly setup for the header joint,
- cost effectiveness,
- durability,
- no increase in weight, and
- ease of manufacture.

Based upon identified customer needs, a house of quality matrix provides some idea about the relationship between different design requirements (see Figure 9-11).

Identifying the Best Alternative

The key factors to be considered in the new design, as assessed from customer requirements, are:

- The header should be easy to assemble.
- The assembly setup time should be reduced.

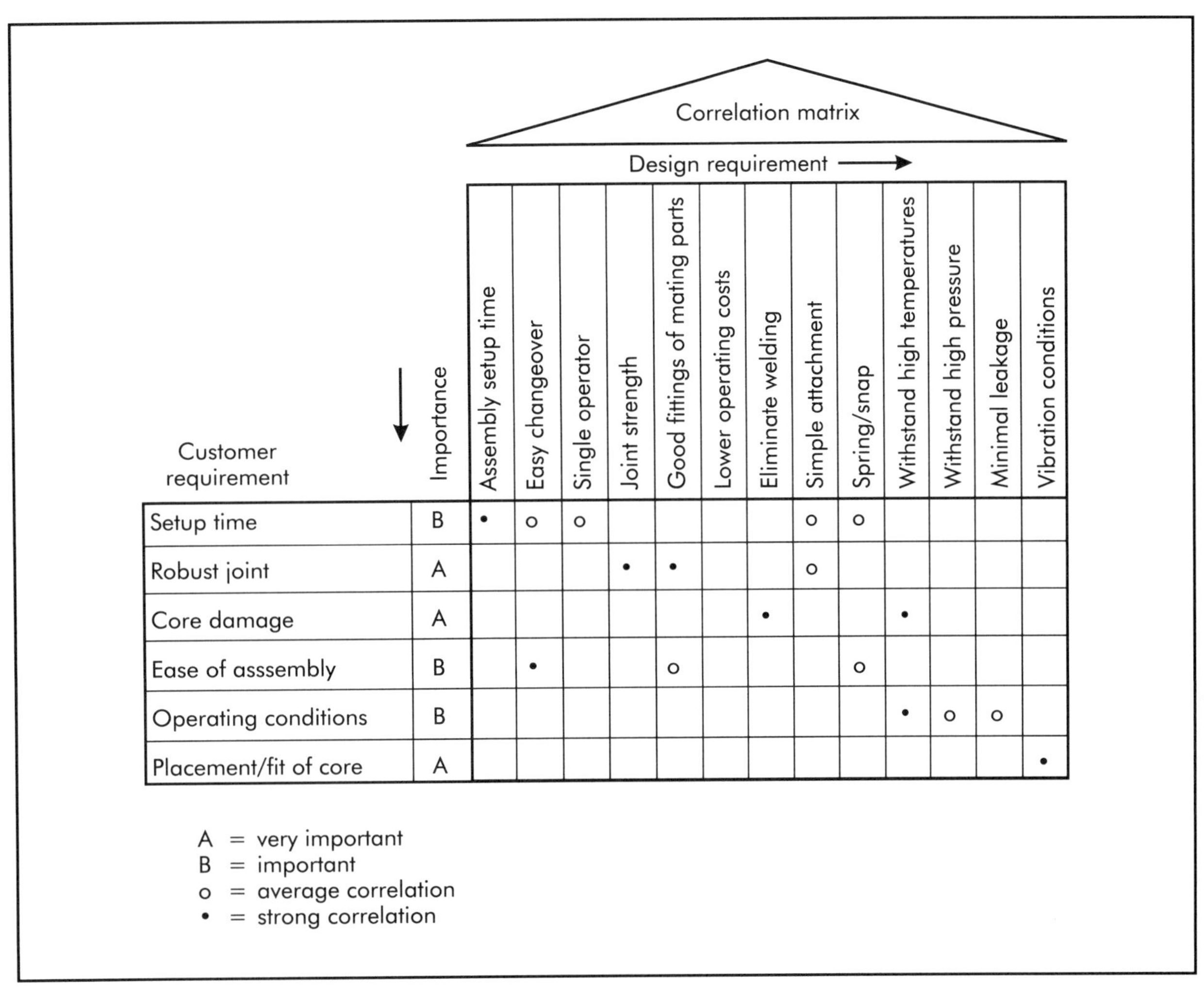

Customer requirement	Importance	Assembly setup time	Easy changeover	Single operator	Joint strength	Good fittings of mating parts	Lower operating costs	Eliminate welding	Simple attachment	Spring/snap	Withstand high temperatures	Withstand high pressure	Minimal leakage	Vibration conditions
Setup time	B	•	o	o					o	o				
Robust joint	A				•	•			o					
Core damage	A							•			•			
Ease of asssembly	B		•			o				o				
Operating conditions	B										•	o	o	
Placement/fit of core	A													•

Figure 9-11. House of quality matrix for the header attachment in a heat exchanger.

- The header-joining method should not damage the core.
- The header-joint design should meet leak criteria.
- The header joint should be free from welding.
- The header should be easily removable.
- The header should withstand operating conditions (pressure/temperature).

After the generation of concepts, one or more designs are selected for further product development. More than one concept is sometimes chosen in cases where manufacturing the prototype and testing it are not expensive. Table 9-3 shows ten possible alternatives that can be used to attach the header to the heat exchanger.

Concepts are selected based on the following criteria:

- interchangeability,
- ease of heat-exchanger repair,
- cost effectiveness,
- process capability,
- durability,
- weight, and
- ease of manufacturing.

Table 9-3. Design solutions/alternatives

1. Snap-fit header to the core and end plates.
2. Rivet header with solid-type rivets.
3. Clinch header to the core using flanges.
4. Snap-fit header into machine groove.
5. Rivet two sides and snap-fit two sides.
6. Snap-on header with finger seal.
7. Bolt header to the core.
8. Rivet or clinch with U-channel.
9. Bond header to the core with epoxy bond.
10. Join header with screws.

The concepts are evaluated using the Pugh table shown in Table 9-4. From the table, the concepts are given a rating on the basis of design criteria. A relative score of better than (+), same as (0), or worse than (–) is placed in each cell of the matrix; (+) is considered as 1, (–) is considered either as –1 or –2, depending on the situation.

From the Pugh table (Table 9-4), the design option with the highest score is further evaluated based on the following factors:

- the thickness,
- the weight,
- the material,
- the joint sealing, and
- the joint strength.

The objective of this case study was to replace the welding joint of a heat exchanger header with an alternate design that was simple, easy to install, and exhibited superior performance characteristics. Of the possible solutions, the snap-on design with finger seal proved to be the most promising design—joining the header with the core of the heat exchanger.

Figure 9-12 provides a view of the snap-on connection.

REFERENCES

Campana, Claudio and Kondo, Jun. 1998. "Engineering Applications Center Project No. 9707." Hartford, CT: College of Engineering, University of Hartford. August.

Kolluri, K. 1998. "Product Development Techniques Used in Boeing 777." Hartford, CT: University of Hartford. Masters Thesis.

Norris, Guy and Wagner, Mark. 1996. *Boeing 777*. Osceola, WI: Motorbooks International.

Table 9-4. Concept evaluation (Pugh)

Design Criteria	1	2	3	4	5	6	7	8	9	10
Performance	–2	–1	–2	–1	–1	0	0	–1	0	–1
Durability	–2	0	–2	–1	–1	0	–1	–1	0	0
Assembly	0	0	–2	–1	–2	+	0	–1	+	0
Manufacturability	–1	0	–1	–2	–1	+	0	–2	0	–1
Complexity	–1	–1	–1	–2	–1	0	0	–1	0	–1
Weight	0	–2	0	0	–1	+	–2	–1	–1	0
Repairability	–1	0	–2	–1	–1	+	0	–1	+	–1
Cost	–1	–1	–1	–2	–2	+	0	–1	–1	–1
Total	– 8	– 5	–11	–10	–10	5	– 3	– 9	0	– 5

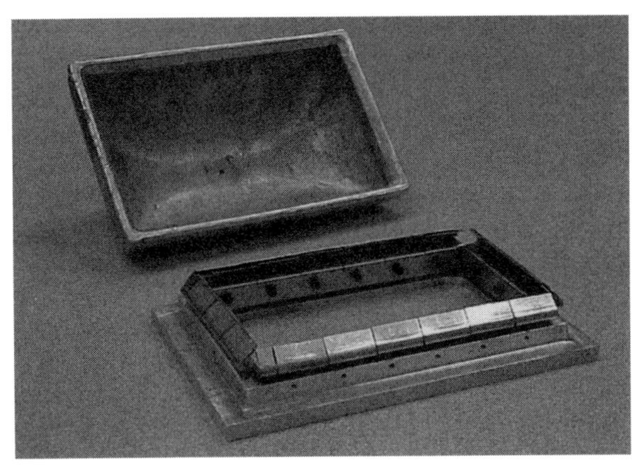

Figure 9-12. Snap-on connection on a heat exchanger.

——. 1997. *Boeing 747—Design and Development Since 1969*. Osceola, WI: Motorbooks International.

Orelup, Mark, et al. 1997. "Worldwide Deployment of Design for Assembly (DFA)." Proceedings of the 1997 International Forum on DFMA, June 9-10, Newport, RI. Wakefield RI: Boothroyd Dewhurst, Inc.

Sabbagh, Karl. 1996. *Twenty-first Century Jet, The Making and Marketing of the Boeing 777*. New York: Scribner.

Questions

1. With the help of sketches, show the basic concepts that can be used for the development of a product. Draw the functional diagram of the whole product. For a typical product known to you, illustrate how you would use the process of morphology to produce alternate designs.
2. Explain how you would select a team to undertake the product design for a multi-component product. What strategies would you use to produce an effective design as speedily as possible?
3. What role does research play in product design? Indicate four courses of research and the desired information that could be expected from each source.
4. How are prototypes used in the product-design process? Describe, with sketches, two rapid prototyping methods, identifying the benefits of each.
5. Using web-based research, identify five corporations and their product lines. In what way do these products support the corporate strategy?
6. During the concept selection stage, how would you use a group of 15 designers to participate in the concept-development work? Can any computer-based tools assist in this process?
7. Select a product with less than 15 parts; estimate the total cost. What types of products are more suitable for direct marketing immediately after the design phase, rather than after the conventional customer test with prototypes of products?
8. What are the characteristics of a product that has a high design-efficiency index? Is it possible to design a product that has a 100% design-efficiency index?
9. The following are ideas for efficient design for manufacturing:
   - facilitate part handling,
   - standardize parts,
   - minimize part count,
   - eliminate adjustments in assembly, and
   - use self-locating screws.

   After examining these ideas, identify situations in each case that may conflict with the other ideas.
10. Provide short answers to the following questions:
    a. What are the major characteristics of each of the following three assembly analysis methods: Boothroyd–Dewhurst method, Lucas method, and Hitachi method?

b. Explain how the design for disassembly methodology can facilitate recycling and promote environmental consciousness. Explain both ideas with the aid of a fishbone diagram.

c. Briefly describe interrelationships, if any, among a product's architecture and its material, manufacturing-process selection, and life-cycle maintenance and reliability. Give examples of products where only one or two of the factors are dominant.

d. List five considerations that have to be taken into account when designing a machined part to ensure efficient manufacturing and easy assembly.

e. What are the considerations that must be taken into account when designing a section for extrusion (a metal-forming operation)?

f. What three questions form the criteria for eliminating a part from the assembly or for combining it?

g. Why is standardization important when you assemble a product?

h. What suggestion should be offered if the designer cannot make the part exactly symmetrical?

i. Can screws be considered an essential part of a product?

j. What design attributes of LEGO® blocks make the product popular?

11. Selecting a product from the list that follows, provide a proposal, outlining a step-by-step product-design procedure. The answers should contain: customer requirements through the house of quality, problem definition by the techniques learned, concept generation, and a function diagram. (The product chosen could be one of the following: solar-powered car/boat, air conditioner, electric stapler, typical lawn equipment, or sports bicycle.)

12. Show by means of a figure, the elements of a Quality Function Deployment (QFD) chart. Explain how a cascade of the QFD charts may be utilized to cover the total design and development process.

13. Form a conceptual design team to come up with a number of concepts for a potato peeler.

    a. Show a simple QFD-matrix relationship identifying three customer requirements and three design features.

    b. Show how to apply the four-step methodology to the problem.

14. Design a manual-assembly line for a new cellular phone produced by a new company. This product has an annual demand of 100,000 units. The line will operate 50 weeks per year, 5 shifts per week, and 7.5 hours per shift. Work units will be attached to a continuously moving conveyor. Work content time will be 42.0 minutes. Assume the line efficiency is 0.97, the balancing efficiency is 0.92, and the repositioning time is 6 seconds. Find:

    a. the hourly production rate to meet the demand; and

    b. the number of workers required.

15. Suppose a manufacturing firm is capable of producing three different products in its factory. Fabricating each product requires several manufacturing operations. Table A-1 shows the time required for each operation, per unit of each product manufactured. Also shown is the unit profit and the maximum available time per day for each operation. Set up the relevant constraint equations.

16. A manufacturing firm is capable of producing two different products in its facility, each requiring three production operations. The company can sell all that is produced and is considering increasing the production capacity. However,

Table A-1.

Operation	Product 1 Time/Unit (minutes)	Product 2 Time/Unit (minutes)	Product 3 Time/Unit (minutes)	Operation Capacity (minutes)
Machining	1	2	1	430
Welding	3	0	2	460
Casting	1	4	0	420
Profit/Unit ($)	3	2	5	

there are constraints on the operation. These constraints, along with other data, are shown in Table A-2. Determine the product mix that maximizes profit.

17. A cup is drawn in 3.5 in. (90 mm) diameter, 1.0 in. (25 mm) deep and 0.06 in. (1.5 mm) thick material using a deep-drawing manufacturing process. Estimate:
    a. the maximum punch force.
    b. the blank diameter (or ultimate tensile stress of the material equal to 47,325 psi [325 N/mm^2]).
18. A 0.5 gal (2 L) jug is manufactured in large numbers as a consumer product. Sketch design alternatives for:
    a. plastic.
    b. metal. Detail four differences between the designs due to manufacturing requirements.
19. Show the axiomatic representation of a product indicating coupled, uncoupled, and decoupled designs. Provide an example. (For example, if considering the design of an air-conditioner, the functional requirements could be providing clean air, keeping the room clean, low-maintenance cost, etc.)
20. What is meant by the "slope" of a learning curve, and what determines the degree of slope for a given learning curve?
21. A company producing defense missiles spent 125,000 hours to produce the first unit. Units two and three were produced with an 86% learning factor. Assuming the same learning factor and at the rate of $40 per hour, what would the cost be for the fourth unit?
22. Sketch the format of a Failure Mode and Effects Analysis (FMEA) chart showing the various column headings. State how the chart is employed for improving product design.
23. Briefly explain what is implied by robust design for a product. Construct an orthogonal array for the robust design

Table A-2.

Product	Time Required Per Unit			Cost ($)	Selling Price
	Operation 1	Operation 2	Operation 3		
5000 series	1.2	2.3	4.5		
6000 series	2.3	6.8	1.9	80	95
Hours available	24	32	24	110	130

experiment on a newly designed surface roughness instrument. Consider the information in Table A-3.

Construct a suitable orthogonal array and substitute appropriate control parameter values.

Table A-3.

Control Parameters		
Laser distance	10	15
Ambient light	Present	Not Present
Laser power	500 MW	1000 MW
Gray scale setting	190	200

24. An automatic camera can have a number of variables associated with it. For example, film speed, flash, and focus are involved. Each of these independent variables will affect a dependent variable, such as the picture quality. The levels of these variables are given in Table A-4.

Table A-4.

Film Speed, A	400 ASA	100ASA
Flash, B	On	Off
Focus, C	In	Out

Construct a suitable orthogonal array and substitute the appropriate control variable values. Which method of tolerance will generate larger tolerances, 100% interchangeability or statistical interchangeability? Why?

25. Identify the correct choice in the following questions.
    a. Value Stream Mapping[SM] is a procedure best described as:
        (1) flexible manufacturing systems and cells mapping.
        (2) material and information flow mapping.
        (3) group technology mapping.
        (4) high-variety and low-volume production mapping.
        (5) medium-volume, medium-variety production mapping.
    b. Flexible manufacturing systems and cells are generally applied in which of the following areas:
        (1) high-variety and low-volume production.
        (2) low variety.
        (3) low volume.
        (4) mass production.
        (5) medium-volume, medium-variety production.
    c. FMEA identifies the following:
        (1) reverse-fault-tree analysis.
        (2) Value Stream Map.
        (3) risk priority number.
        (4) cause and effect.
        (5) robust design.
    d. Production-flow analysis is a method of identifying part families that uses data from which of the following sources:
        (1) bill of materials.
        (2) engineering drawings.
        (3) master schedules.
        (4) production schedules.
        (5) route sheets.
26. Choose a product to evaluate and identify the possible hazards associated with its usage. With the use of sketches, redesign the product and show how these hazards can be eliminated.
27. Specifications of a machined diameter of a locating pin have been given as 3.5 × 0.02 in. (89 × 0.5 mm). A capability test shows a process-standard deviation of 0.004. The capability index is which of the following?
    a. 0.875
    b. 0.6
    c. 1

d. 1.5
e. 1.67

28. How can the use of self-aligning and self-locating features facilitate automatic assembly? In what way does the method of fastening and joining the component of a product affect the feasibility of recycling?
29. Provide brief answers to the following.
    a. A service counter is being designed. How far should employees be expected to reach across the counter?
    b. What is the risk associated with using a screwdriver throughout the workday?
30. Decompose the problem of designing the two-position cordless screwdriver shown in Figure A-1, specified with 4.8 Volts, 50 in./lb (5.7 N/m) torque.

    Suggestions: A designer can take this further by representing a functional diagram and a functional decomposition diagram. The cordless screwdriver can be actuated by the variable-speed-power switch and stored-DC-battery current causing the motor and the shaft to turn. The direction of rotation is reversible through the use of a polarity-reversing switch. The rotating motor shaft employs a spur gear to drive a series of planetary gear trains that deliver power at an increased torque to the output shaft. The output shaft torque is selectively limited to a preset value determined by the operator input to the output shaft clutch. The output shaft delivers rotational energy to various tools through the use of an adjustable chuck.
31. For the cart wheel assembly shown in Figure A-2, perform design for assembly (DFA) analysis. The part details are shown in Table A-5. Recommend an alternate design.

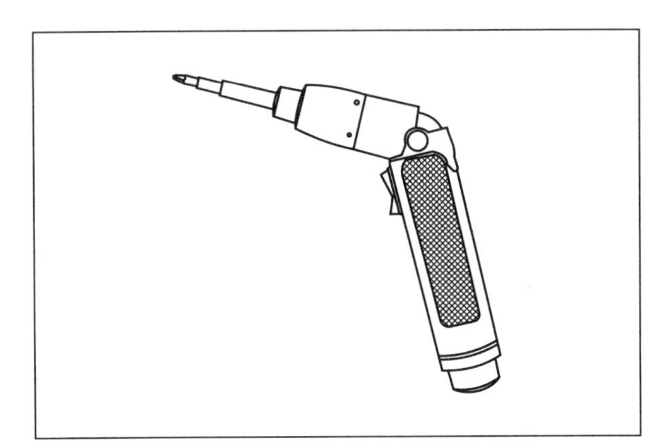

Figure A-1. Two-position cordless screwdriver.

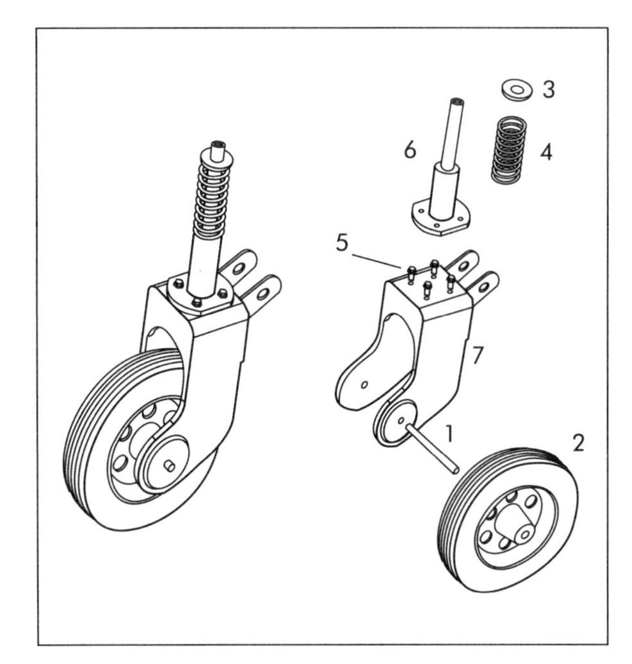

Figure A-2. Cart wheel assembly.

Table A-5.

Component Number	Name	Dimensions in. (mm)	Quantity
1	Axle	6.7 (170) length, 0.6 (15) diameter	1
2	Tire	9.8 (250) diameter, 2.4 (60) thick	1
3	Front spring plate	2.4 (60) diameter, 0.2 (5) thick	1
4	Front spring	5.3 (135) length, 2.0 (50) diameter	1
5	Hex bolt	1.0 (25) length, 0.7 (18) diameter	4
6	Front pillar	11.4 (290) length, 4.7 (120) diameter	1
7	Wheel bracket	12.2 × 8.7 × 5.5 (310 × 220 × 140)	1

Bibliography

Allenby, B.R. 1991. "Design for Environment: A Tool Whose Time Has Come." *SSA Journal*, September, pp. 5-10.

Beitz, W. 1993. "Designing for Ease of Recycling—General Approach and Industrial Applications." The Ninth International Conference on Engineering Design (ICED), August, pp. 325-332.

Bemowski, K. 1991. "The Benchmarking Bandwagon." *Quality Progress*, January.

Campana, C. and Kondo, J. 1998. "Engineering Applications Center Project No. 9707." Hartford, CT: College of Engineering, University of Hartford, August.

Corbert, J., Dooner, M., Meleka, J., and Pym, C. 1990. *Design for Assembly*. Hull, UK: Lucas Engineering & Systems Ltd., University of Hull.

——. 1991. *Design for Manufacture*. London, UK: Addison-Wesley Publishing Co.

Dhillon, B.S. 1996. *Engineering Design—A Modern Approach*. Irvine, CA: Times Mirror Higher Education Group, Inc.

Dieter, George. 2000. *Engineering Design*, 3rd Edition. New York: McGraw-Hill Co.

El Wakil, Sherif. 1998. *Processes and Design for Manufacturing*, 2nd Edition. Boston, MA: PWS Publishing Company, ITP.

Groover, M.P. 1996. *Fundamentals of Modern Manufacturing*. Englewood Cliffs, NJ: Prentice Hall.

Hall, Robert. 1987. *Attaining Manufacturing Excellence: Just-in-Time, Total Quality, Total People Involvement*. Homewood, IL: Dow Jones-Irwin.

Hartley, J. R. 1992. *Concurrent Engineering*. Portland, OR: Productivity Press.

Hrinyak, M., Bras, Bret and Hoffman, W. 1996. "Enhancing Design for Disassembly: A Benchmark of DFD Software Tools." ASME Design Engineering Technical Conference and Computers in Engineering Conference, Irvine, CA. New York: American Society of Mechanical Engineers.

Hundal, Mahendra. 1997. *Systematic Mechanical Designing—A Cost and Management Perspective*. New York: ASME Press.

Hyman, Barry. 1998. *Fundamentals of Engineering Design*. Upper Saddle River, NJ: Prentice Hall.

Imai, Masaaki. 1986. *Kaizen: The Key to Japan's Competitive Success*. New York: McGraw-Hill Co.

Janik, John. 1997. *A Review and Analysis in Line Balancing Methods*. Hartford, CT: University of Hartford.

Jones, Daniel T. 1999. "Macro Value Stream Mapping: Seeing the Whole." http:/ www.lean.org/community/presentations/summit99/Session4B.ppt

Lee, Kunwoo. 1999. *Principles of CAD/CAM/CAE Systems*. Boston, MA: Addison Wesley.

Madhav S. Padke. 1989. *Quality Engineering: Using Robust Design*. AT&T Bell Laboratories. Englewood Cliffs, NJ: Prentice Hall.

Magrab, Edward. 1997. *Integrated Product and Process Design and Development*. Boca Raton, FL: CRC Press.

Norris, Guy and Wagner, Mark. 1997. *Boeing 747—Design and Development Since 1969*. Osceola, WI: Motorbooks International.

Norris, Guy and Wagner, Mark. 1996. *Boeing 777*. Osceola, WI: Motorbooks International.

Orelup, M. et al. 1997. "Worldwide Deployment of Design for Assembly (DFA)." Proceedings of the International Forum on DFMA, June 9-10. Newport RI: Boothroyd-Dewhurst, Inc.

Pryor, L. S. 1989. "Benchmarking: A Self-improvement Strategy." *Journal of Business Strategy*, Nov./Dec., pp. 28-32.

"Reasoning System." 1992. *Computer-aided Design*, Vol. 24, No. 2, pp. 67-78.

Roozenburg, N.F.M and Eekels, J. 1995. *Product Design: Fundamentals and Methods*. Chichester, UK: John Wiley & Sons, Inc.

Rother, Mike and Shook, John. 1998. *Learning to See: Value Stream Mapping to Add Value and Eliminate Muda*. Version 1.1. Brookline, MA: Lean Enterprise Institute.

Sabbagh, Karl. 1996. *Twenty-first Century Jet, The Making and Marketing of the Boeing 777*. New York: Scribner.

Schonberger, Richard. 1982. *Japanese Manufacturing Techniques: Nine Hidden Lessons in Simplicity*. New York: The Free Press.

Schonberger, Richard. 1986. *World-class Manufacturing: The Lessons of Simplicity Applied*. New York: The Free Press.

Shetty, D., et al. 1990. *New Experiments on Non-Contact Inspection of Ground Turbine Blades*. New York: American Society of Mechanical Engineers.

Shetty, D., Neault, H. 1993. United States Patent. "Method and Apparatus for Surface Roughness Measurement Using Laser Diffraction Pattern." Patent Number: 5,189,490, Feb. 23.

Shina S. G. 1991. *Concurrent Engineering and Design for Manufacture*. New York: Van Nostrand Reinhold.

Shina, S. G. 1994. *Successful Implementation of Concurrent Engineering Products and Processes*. New York: Van Nostrand Reinhold.

Singh, Nanua. 1996. *Systems Approach to Computer-integrated Design and Manufacturing*. New York: John Wiley and Sons, Inc.

Stein, M.I. 1968. *Creativity in Handbook of Personality Theory and Research*. Chicago: Rand McNally.

Suh, N. P. 1990. *The Principles of Design*. Oxford, UK: Oxford University Press.

Suzaki, Kiyoshi. 1987. *The New Manufacturing Challenge: Techniques for Continuous Improvement*. New York: The Free Press.

Tompkins, James A. and White, John. 1984. *Facilities Planning*. New York: John Wiley & Sons, Inc., pp. 49.

Tuttle, B. Lee. 1990. "Creative Thinking: The Touchstone of DFMA." Newport, RI: International Forum on DFMA, May 14-15.

Ulrich, K. T. and Eppinger, D. 1991. *Product Design and Development*. New York: McGraw-Hill.

Usher, John, Roy, U., and Parsaei, H., eds. 1998. *Integrated Product and Process Development: Methods, Tools, and Technologies*. New York: John Wiley & Sons, Inc.

Voland, George. 1999. *Engineering by Design*. Reading, MA: Addison Wesley.

Womack, James, Jones, Daniel T., and Roos, Daniel. 1990. *The Machine that Changed the World: The Story of Lean Production*. New York: Harper Perennial.

Womack, James P. and Jones, Daniel T. 1996. *Lean Thinking: Banish Waste and Create Wealth in Your Corporation*. New York: Simon & Schuster.

Yoo, S.M., Dornfeld, D.A., and Lemaster, R.L. 1990. "Analysis and Modeling of Laser Measurement System Performance for Wood Surface." *Journal of Engineering for Industry*. Vol. 112, February, pp. 69-76.

Zeid, Ibrahim. 1991. *CAD/CAM Theory and Practice*. New York: McGraw-Hill, Inc.

Index

N

O

P

Q

R

S

T

U

V

W